Anonymous

The review of insanity and nervous diseases:

A quarterly compendium of the current literature of neurology and psychiatry - Vol.

1

Anonymous

The review of insanity and nervous diseases:
A quarterly compendium of the current literature of neurology and psychiatry - Vol. 1

ISBN/EAN: 9783337713515

Printed in Europe, USA, Canada, Australia, Japan

Cover: Foto ©berggeist007 / pixelio.de

More available books at **www.hansebooks.com**

No. 1.   AUGUST, 1890.   Vol. 1.

THE REVIEW

OF

INSANITY＊NERVOUS DISEASE

A QUARTERLY COMPENDIUM OF THE CURRENT LITERATURE OF
NEUROLOGY AND PSYCHIATRY.

EDITED BY

JAMES H. McBRIDE, M. D., MILWAUKEE, WIS.

ASSOCIATE EDITORS :

LANDON CARTER GRAY, M. D.   C. K. MILLS, M. D.
     NEW YORK CITY.   PHILADELPHIA.

C. EUGENE RIGGS, M. D.   W. A. JONES, M. D.
     ST. PAUL, MINN.   MINNEAPOLIS, MINN.

     H. M. BANNISTER, M. D., KANKAKEE, ILL:

$2.00 PER YEAR.   SINGLE NUMBERS, 50c.

SWAIN & TATE,
BOOK AND JOB PRINTERS,
MILWAUKEE.

INDEX.

REVIEW

OF

INSANITY AND NERVOUS DISEASE.

ORIGINAL ARTICLE.

AMYOTROPHIC LATERAL SCLEROSIS.

By LANDON CARTER GRAY, M. D.

(Professor of Nervous and Mental Disease in the New York Polyclinic.)

Amyotrophic lateral sclerosis is a barbarously compound and complicated name whose meaning can only be unraveled by a combined knowledge of Greek and neurology. The word amyotrophia is derived from *a priv.*, *a muscle*, and *nutrition*, the meaning being a muscle lacking nutrition; so that amyotrophic lateral sclerosis means *atrophic muscles plus a lateral sclerosis.* What a lateral sclerosis is, involves a somewhat detailed explanation, half anatomical, half clinical.

It is a thoroughly established fact in nervous anatomy that the motor and the sensory nerve strands go from the cortex of the brain to their respective motor and sensory structures in the periphery. And the so-called "centres," which have been mapped out in the last twenty years, are nothing more than terminal stations in the cortex of these different motor and sensory nerve strands. The motor nerve strands have been more accurately studied than the sensory ones, because it is much easier to detect an impairment of motion which is

2

objective than an impairment of sensation which is very
largely subjective. The motor nerves have been found
to centre in the so-called motor convolutions behind and
in front of the fissure of Rolando. From these motor
convolutions the motor nerve strands proceed to the
base of the brain where they converge into what is known
as the internal capsule, occupying its anterior portion.
Below this internal capsule they pass to the crura cerebri, thence through the pons, and down to the upper
portion of the spinal cord, where the majority pass over
to the other side—decussate, as it is technically called,
and pass down the cord in what is known as the Lateral
Pyramidal Column. From this Lateral Pyramidal Column they pass into the anterior horns of gray matter,
having direct connection with the great ganglion cells
which stud these horns in groups, the connection being
probably made through what is known as the axis cylinder process, which runs directly into the nucleolus of the
cell. From these anterior horns again proceed the
motor nerves, which pass out through the anterior roots
and terminate in the muscular fibril in what is known as
the terminal plate, which is a granular protoplasmic
mass resting upon the muscular fibril. It will be seen
from this brief description that we can trace the motor
nerve from its departure in the great pyramidal cells of
the motor convolutions until it buries itself in the end-
plate of the muscle. It should, however, be stated that
the rule that most of the motor fibres pass over to the
opposite side of the cord and go down through the Lateral Pyramidal Column is not invariable. It has been
shown by Flechsig that in almost all cases a small number of these motor fibres pass down on the same side of
the cord in the small columns on each side of the anterior median fissure known as the columns of Turck, and

that in some exceptional cases it may happen that almost all the fibres will pass down upon the same side instead of decussating. It is this latter fact which explains Brown–Sequard's cases of lesion on the same side of the brain with the paralysis; indeed I have known of two cases in which an operation was done upon the brain assuming that the lesion would be found on the side opposite the paralysis, and in which the autopsy disclosed a lesion upon the same side; moreover, I have three cords in my possession in which the decussation is very scanty and not visible to the naked eye. It has been assumed for many years that these Lateral Pyramidal Columns can be primarily diseased. In 1874 Dr. Erb, of Heidelberg, called attention to a certain group of symptoms which he attributed to disease of this Lateral Pyramidal Column, and consisting of exaggerated motor reflexes, contractures, and a so-called spastic or spasmodic gait. These motor reflexes consisted mainly of two types, one the so-called tendon reflex or knee-jerk, and another called the foot clonus. The former consisted simply of an exaggerated reflex action in the tendon of the quadriceps extensor muscle, so that when the latter playing over the flat surface of the tibia, just below the patella, was tapped upon whilst the leg was held dangling, a jerk forward of the foot and leg resulted. The foot clonus consisted of a similarly exaggerated reflex in the tendons of the foot and was evoked in one or both of two ways, viz., first by extending the leg and then flexing the foot quickly and forcibly upon the leg and holding the foot so flexed, when a rhythmical tremor of the foot would ensue; or, secondly, by making the patient tap the toe upon the floor, when a similar rhythmical tremor would result. Whenever the foot clonus is found it may be assumed to be abnormal, but the

knee-jerk or tendon reflex is found in most healthy individuals, and it is only when it is exaggerated that it is to
be considered abnormal, whilst the question of what is
abnormal and what is normal is often a very nice one. It
must also be made plain that the word *contracture* is a
technical name indicating something quite distinct from
what is meant by the word *contraction*. A contracture
is a muscular condition in which the muscles are rigid,
but the rigidity is yielding and wax-like, and gives usually the same sensation of resistance as that which is
offered by wax, so that the name of the waxy flexibility
or *flexibilitas cerea* of the older authors was an excellent
one. This contracture is almost invariably of nervous
origin, and differs very distinctly from the muscular contraction or rigidity which may be found around an
anchylosed joint or a fractured or inflamed limb, inasmuch as the resistance of the latter cannot be overcome
without pain or great exertion. When any patient presents these two symptoms of exaggerated reflexes and
contractures in both lower limbs he is very apt to also
present what is known as the spastic, or spasmodic gait,
which consists of a peculiarly stiff walk, with a springy
action as if strong bed-springs were under the feet, so
that the individual walks in a way that was happily designated by the nick name of " Jiggery John," irrever-
ently given to one of my unfortunate patients. This
spastic gait is the result of the exaggerated reflexes and
the contracture. Of course, as all the muscles have
tendons, it follows that all the muscles must have tendon reflexes, but it is not every muscle in which the tendon reflex can be evoked as happily as in the quadriceps
extensor, whose tendon plays over a broad, flat, bony
surface, and which has suspended below it the flail-like
leg and foot to act as an index of its reflex excitability,

and as a matter of fact the tendon reflex of most muscles cannot be evoked unless they are considerably exaggerated. We are therefore practically reduced largely
to a consideration of the knee-jerk as almost the only
tendon reflex available for clinical purposes. A slight
degree of exaggeration of the tendon reflex cannot therefore be discovered in other muscles, although contractures are readily detected even in a slight degree. These
contractures, exaggerated reflexes, and spastic gait were
the symptoms in the cases that were described by Erb,
and which were believed by him to be due to a lesion of
the Lateral Pyramidal Columns of the cord, a so-called
primary lateral sclerosis. But in the sixteen years that
have elapsed since the reading of Erb's paper, not a single solitary autopsy has supported the claim that there
is such a disease as a primary sclerosis of the Lateral
Pyramidal Columns. The nearest approach to proof of
this was furnished by a case of Dr. Dreschfeld's, but the
defect in this was, that no examination was made above
the spinal cord. On the contrary, the many autopsies
that have been made have shown that lesions anywhere
in the motor tract, in its course between the motor convolutions and the anterior horns of gray matter, are
capable of setting up disease of this motor tract in the
Lateral Pyramidal Column, and thus giving rise to the
above triad of symptoms—exaggerated tendon reflexes,
contractures, and spastic gait. The name of the
lesions is legion that may thus cause descending degeneration of the motor tract—tumors of the brain or cord,
intracranial hemorrhage, hydrocephalus, various forms
of encephalitis, especially that which has the peculiar
tendency to the formation of cavities (porencephalitis),
many different forms of myelitis, Potts' disease, etc.

So much for the so-called lateral sclerosis, which is,

as we have seen, simply a degeneration in the majority
of instances of the lateral pyramidal column, although,
if the decussation of the motor strands was as imperfect
as we have sometimes seen that it is, the degeneration
might be mainly found in the column of Türck.   We
have also seen that these motor strands pass into the
anterior horns of gray matter, making direct connection with the great ganglion cells, from which
ganglion cells pass out the motor nerves along
the anterior roots to the terminal end plates,
lying upon the muscular fibrils. Suppose that this
descending degeneration of the motor tract should
extend into these great ganglion cells that stud the
anterior horns of gray matter, what symptoms should
we expect then?   If there is anything in nervous
pathology that is beyond the shadow of a doubt, it is
the fact that atrophy of the ganglion cells of the anterior horns of gray matter produces muscular atrophy.
This is precisely what takes place in the so-called cases
of amyotrophic lateral sclerosis.   To the triad of symptoms upon which we have been dwelling—exaggerated
reflexes, contractures, spastic gait—add muscular atrophy, and you have in the main the clinical picture of a
case of amyotrophic lateral sclerosis.

The typical cases usually present three somewhat
fairly defined stages : namely, first a paresis, second of
atrophy, third of contracture and exaggerated reflexes.
But there are many exceptions to this rule. The onset
is generally in the upper extremities, sometimes in one
of the fingers. The extension to the lower extremities
and also to the medulla oblongata is usually made within
six to twelve months. Sometimes, however, the onset
may be in the medulla oblongata or in the lower
extremities. The paralysis is purely motor. The mus-

cular atrophy is invariably of the whole body of the muscle and not fibre by fibre as in progressive muscular atrophy. The tendon reflexes are exaggerated and ankle clonus is often obtainable; but these symptoms disappear in the later stages. Commingled contracture and atrophy give rise to characteristic deformities, such as the claw-hand, etc., and there is generally a fibrillary tremor of the so-called intentional type, i. e. a tremor caused by voluntary movements. The progress of the disease is gradual. The affected muscles usually present some phases of the electrical reaction of degeneration. Death ensues from implication of the vital nuclei in the medulla oblongata, from general debility, or from pulmonary complications. The prognosis is very unfavorable, if not hopeless, both as regards life or improvement.

The lesions in amyotrophic lateral sclerosis always consist of a degeneration in the lateral and anterior pyramidal columns, and of destruction of the ganglion cells in the anterior cornua of the spinal cord, and the analogous motor nuclei of the medulla oblongata. The implication of the lateral columns is always more marked than that of the anterior pyamidal columns, and the latter are sometimes scarcely affected at all. The lesions of the ganglion cells are the same as are seen in poliomyelitis anterior, (essential infantile paralysis), and consist of morphological alterations in their structure. Some of the cells disappear entirely, some lose their processes, some become smaller in size, some are highly pigmented. It has been believed by many authors that the similar alterations in these anterior cornua are due to precedent arterial alterations, but there has been no proof advanced of this view. The lesions in the pyramidal columns consist of impairment or destruction of the nerve tubules and proliferation of the connective tissue.

Inasmuch as the lesions are always bilateral in the pyramidal columns, in the anterior horn, or the nuclei of the medulla oblongata, there has always been a suspicion in the neurological mind that the affection was in the nature of a secondary degeneration of the pyramidal columns extending into the anterior horns. In some cases there has been found to be a marked degeneration of the fibres running from the motor convolutions, whilst in other cases the degeneration has been traced up into the pons, the crura cerebri, and the internal capsule, whilst in still other cases the degeneration has not been traced above the decussation. In none of these cases has there been any grave organic disease of the upper part of the pyramidal tract. Whether, therefore, the cases reported as being without lesion above the decussation were not so minutely examined as to detect the slight alteration which others have found, is a question that very naturally arises. At the present time we cannot, for this reason, distinctly affirm or deny that the lesion in the pyramidal columns is a secondary lesion, although the probabilities are strongly in favor of its being so. In many cases of myelitis of the anterior horn, (essential infantile paralysis), the Lateral Pyramidal Column has been found secondarily affected; in many cases of lesion in the cerebrum the anterior horns have been found diseased, although the pyramidal tracts were intact. In other cases of lesion in the cerebrum muscular atrophy has supervened, without intermediate lesion of the pyramidal tract or the anterior horn, or even in some cases without lesion of the peripheral nerves—from all which facts it is easily seen that there is a close connection pathologically between the pyramidal tract, the anterior horns, the motor nerves and the muscles.

1890.] BY LANDON CARTER GRAY, M. D.   II

The disease is to be differentiated from :
Progressive muscular atrophy;
Myelitis of the anterior horn;
Ordinary myelitis;
Postero-lateral sclerosis.

Progressive muscular atrophy begins with a fibrillary
atrophy of the muscles, the paralysis is in proportion to
this fibrillary atrophy, and there is no contracture or
exaggerated reflexes whatsoever, whilst in amyotrophic
lateral sclerosis the paralysis is usually a primary symptom, and the atrophy succeeds and is in the body of the
muscle, or muscles, and is not fibrillary, and there super-
vene contracture and exaggerated reflexes.

Myelitis of the anterior horn in the child should never
be confounded with amyotrophic lateral sclerosis, for
the former is always sudden in onset, is monoplegic in
its distribution, and affects only a certain muscular
group, or groups, within that one limb, whilst the paralysis is a flaccid one, and there are no contractures or
exaggerated reflexes. This same myelitis of the anterior horn in the adult usually begins with motor paralysis of one or more limbs of the whole body, reaching its
acme in a few days, muscular atrophy succeeding
within a fortnight, and the paralysis remains a flaccid
one, contractures and exaggerated reflexes very seldom
supervening.

An ordinary myelitis in its earlier stage might be
confounded with amyotrophic lateral sclerosis, but the
progress of the former affection should render a mistake
impossible, as ordinary myelitis has pronounced sensory,
rectal, and vesical symptoms, which are never present in
amyotrophic lateral sclerosis.

Postero-lateral sclerosis, that is, that form of combined
myelitis in which the posterior and lateral columns are

simultaneously affected, would have superadded to the lateral sclerosis the symptoms of implication of the posterior column, *i. e.* certain stabbing and lightning-like pains, severe, sudden, seldom confined to one locality long, together with impairment of one or more of the sensations of tact, muscular sense, pain or temperature, ataxia, and some vesical and rectal weakness.

6 East Forty-ninth Street, New York.

NEUROLOGICAL.

—————

Translations are furnished as follows:

ITALIAN.

H. M. BANNISTER, M. D., Asst. Physician Eastern Illinois Hospital for Insane.

GERMAN.

G. J. KAUMHEIMER, M. D., Milwaukee.

RUSSIAN.

T. KACZOROUSKI-PORAY, Chicago.

SCANDINAVIAN AND FRENCH.

C. FRITHIOF LARSON, M. D., Chicago.

SPANISH.

ALFRED RAPHAEL, Chicago.

ANATOMY AND PHYSIOLOGY.

ON THE CONDITIONS OF TEMPERATURE IN NERVES DUR-ING ACTIVITY AND DURING THE PROCESS OF DYING.

Dr. Rolleston writes on this subject in *The Journal of Physiology* for March. The results of his observations may be summarized as follows: The energy of nerve metabolism is less than that of muscle, and a much smaller part of it is wasted as heat. During the passage of the nervous impulse there is no evidence of any heat being evolved from the nerve trunk. In dying, the nerves evolve heat. There is some evidence to show that nerves die at different rates.

ON THE NATURE OF THE TENDON REFLEX.

Mr. Sternberg has found the tendon reflex to be composed of an osseous and a muscular element.

A blow upon a bone, especially upon an articular surface and in the direction of its longitudinal axis, causes a contraction of all attached muscles.

A tense muscle will contract if it receives a blow, especially in the direction of its fibres.

The tendon plays only a mechanical part.

There exist cerebral centres which inhibit these reflexes.

There are two varieties of muscular contractures which may be differentiated by the tendon reflex.

In certain contractures due to local degenerations in the brain, in certain spinal affections and articular contractures, the reflexes are increased.

In those due to cerebral hemorrhages, tumors or abcesses, or to uræmia or meningitis, the reflexes are frequently diminished, never increased.

Pure facial reflexes do not exist. (Wiener Mediz. Presse, No. 17, 1890).

NOTES ON THE COURSE OF THE SENSORY TRACTS IN THE CENTRAL NERVOUS SYSTEM.

The true origin of the spinal sensory fibres is in the ganglia on the posterior roots.

The greater part of the sensory fibres enter the posterior columns, while the remainder enter the posterior cornua. Of the latter, some certainly enter the columns of Clarke and from there pass to the lateral cerebellar column. Others have been traced to the anterior horns, while those entering the cord most laterally end in the substantia gelatinosa Rolandi. The fibres separate and enter the cord at different levels. The posterior columns are the route by which the greater part of the sensory fibres reach the medulla.

After entering nuclei there, numerous fibres cross to the opposite side (fibræ arciformes internæ) and form a part of the structure denominated by Flechsig as the intermediate olivary layer, and by others, the lemniscus.

The fibres which enter the posterior cornua cross to the opposite side in the anterior gray commissure and ascending in the antero lateral columns, join those which crossed above.—(DR. L. EDINGER, *Deutsche Med. Wochen*, No. 20, 1890.)

PHYSIOLOGY OF THE LEMNISCUS.

CASE OF GLIOMA OF A POSTERIOR HORN OF THE CORD.

Patient was a girl of eighteen who six months previous to admission noticed her left arm was insensible to changes of temperature. Tendon reflexes in left leg increased. Analgesia absolute in left arm. Partial on left side of neck, trunk and thigh. Elsewhere sensation normal. Temperature sense abolished where analgesia existed, also sensations of location and pressure; sensation of touch normal, no motor difficulty. On autopsy small recent clot in right temporal lobe was found, and another in left frontal. Lumbar enlargement normal, but just above this the left posterior horn was replaced by a transparent, gelatinous tissue which contained no fissures or cavities and showed structure of a glioma with no intact nerve fibres. The new formation, while mainly confined to the posterior horn, involved in the cervical enlargement the posterior gray commissure and surrounded the central canal. At the level of the fifth cervical nerve it sent a linear prolongation into the columns of Gall on both sides. Slight diffuse atrophy of the left lateral column was also found at this level. At the level of the second cervical nerve the sclerosis had also invaded the posterior column. At the medulla a sharply defined atrophy of fibres throughout the extent of the intermediate olivary layer was found on the right side. Right pyramid showed diminution of fibres. Atrophy of lemniscus on right side in pons and peduncles. All other parts healthy. The author believes this case to prove, first, that the columns of Gall cannot be regarded as a direct continuation of the posterior roots, as they were intact to the cervical region, while on the left side the posterior roots were destroyed throughout the dorsal cord. Second, the median part of the lemniscus contains the principal fibres for the conduction of cutaneous sensation on the opposite side of the body.—(DR. G. ROSSOLIMO, *Archiv. f. Psychiatry*, Band 21, Heft 3.)

AUDITORY SPHERE AND AURICULAR MOVEMENTS.

(*Preliminary Communication.*)

Dr. B. Baginsky reports (1.) Irritation of Monk's auditory sphere, especially of the lower angle of the temporal lobe, is followed by lifting of the eyelids and movements of the opposite ear.

2. The reactions obtained by irritation of the auditory sphere differ
materially from those described by Hitzig, Ferrier and others as resulting
from the irritation of the ear centre.—(*Neurolog. Centralblatt*, No 15, 1890.)

The following is a summary (*Lancet*, June 7, *et seq.*) of the Croonian Lectures on Cerebral Localization by Dr. Ferrier. In the first lecture he refers
to the results of experimentation on fishes, frogs and birds. He reports two
recent experiments of Schader which seem to show that removal of the
hemispheres of a frog does not deprive the animal of either spontaneity, or
of special instincts, or of the ability to feed itself, as has been stated by
others. If this is so it would appear that the principal point of distinction
between the brainless and normal frog cannot be upheld. In relation to
the result of experiments on birds he mentions Monk's recent investigations
which show that complete extirpation of the hemispheres produced complete and absolute blindness. He, however, quotes Schader as stating that
in his experiments complete extirpation of the cerebral hemispheres in
birds did not produce blindness. Ferrier seems to incline to the view of
Schader and says that we must therefore "class birds with fishes and frogs,
which without doubt retain their sense of sight and guide their movements
accordingly, notwithstanding the complete removal of their cerebral hemispheres." Beginning with mammals, Dr. Ferrier reviews the chief
experiments that have been made up to date. He gives a brief summary of the phenomena of electrical irritation of the different regions
of the cerebral cortex and says that this method of experimentation
indicates some form of functional differentiation. This method will not
show the existing boundaries of different centres because contiguous
regions with different functions are apt to be discharged together. The
individual movements of a limb, which are dependent upon a certain
centre, may be produced when a contiguous centre is stimulated. "This
may be interpreted either on the supposition that the particular movements (say of the thumb) are represented throughout the whole of
the arm area, or that it is only a case of the diffusion of the stimulus
from one part to another." It is difficult to decide which of these views is
the correct one, and it may be that neither represents the whole truth. In
regard to the probable location of the visual centres he says, "I contend
that the only hypothesis which seems to harmonize with all the facts is,
that the angular gyri are more particularly the centres of clear vision, each
simply for the eye of the opposite side. Whether the other portions of the
retina, upper, lower, outer, and inner, are especially represented in corresponding regions of the occipital lobe according to the hypothesis of Monk
and Schafer, cannot as yet be said to be established." In relation to the
location of the auditory sense, Ferrier concludes that the facts of human
pathology support the view that the sense of hearing is localized in the
temporal lobe. In relation to the decussation of sensory fibres in the cord

he is not inclined to accept Brown-Séquard's contention that the paths of muscular sense do not cross with other sensory tracts, but ascend in the cord on the same side.

Ferrier is inclined to the view that the forms of tactile and common sensation are localized in the falciform lobe, and it is possible that common sensibility, may to some extent, be bilaterally represented, so that the destruction of one lobe may be compensated for by the other. He considers, however, that this question requires further investigation. The olfactory sense is probably located in the hippo-campel lobe. He gives the usual localization for the motor centres and combats Dr. Bastion's view that the so-called motor centres are in reality sensory centres that excite the true motor centres in the spinal cord through the pyramidal tracts. He concludes that the motor centres of the cortex are not the centres of tactile or general sensibility, nor are they the centres of the muscular sense. In regard to the function of the frontal centres there is still some obscurity.

THERMAL MECHANISMS AND MOTOR LEVELS.

Dr. Hale White (*Brit. Med. Journal*, April 26) writes on the parallelism between the three thermic mechanisms and Dr. Jackson's three levels. The heat mechanism consists of three parts: (1) The heat loss, or thermolytic mechanism; this is the lowest developed, for most animals lower in the scale than birds have no other heat mechanism, their temperature rising and falling with the external medium. This mechanism is probably located in the region of the pons and medulla. (2) The thermo-genetic or heat producing mechanism; this is probably located in the corpus striatum. (3) The thermotaxic, or heat regulating mechanism, which is probably located in the cortical region.

In the same journal of May 3, Dr. Donald McAllister states that he was the first to state the analogy between the thermo and the motor mechanism, a fact which Dr. White did not mention. McAllister's theory was propounded in the Gulstonian Lectures of 1887 on the nature of fever.

Prof. N. Mislavski (*Kowalewskij's Archives of Psychi. and Neurol.* Nov. 3, 1890,) writes on innervation of the stomach. Experimentation and pathological investigation long since demonstrated that the stomach is innervated by the vagus. Irritation of this nerve produces strong and rhythmical contractions of the plyoric end. On the cardiac end the effect is different. Very strong electrical or mechanical irritation of the cardiac end produces contraction of that portion, while a weaker irritation produces dilatation. The action of the sympathetic upon the stomach is the reverse of that of the pneumogastric. Irritation of the sympathetic produces not only arrest of the rhythmical contractions of the pylorus, but it arrests the movements of the stomach as a whole. Similar results are obtained from irritation of the gray matter of the spinal cord, of the medulla, and of the anterior por-

tion of the optic lobes, where the centres for the inhibition of the pylorus
are located.   It is also found that irritation of the "first convolution" induces strong contraction of the pylorus.   Irritation of peripheral nerves
tends to arrest the normal movements of the stomach.

Dr. Gowers *(Lancet,* May 3, *et seq.)* writes on the *Functions of the Nervous System.*   His first article is on "How does the cerebellum co-ordinate ?"   He states that we are unable to conceive how it is possible for the
cerebellum to do its work by downward action or in any other way but
by upward action directed to the cerebral cortex.   He mentions the fact
that the only track of the spinal cord that has been proved to pass through
the cerebellum degenerates upwards and that therefore it conducts upwards.
Whatever the middle lobe does to co-ordinate movements it must do by
acting upon the cerebral cells.   We naturally think of the nerve cells as
acting only under stimulation, but it is doubtful if they are ever inactive.
This applies both to motor and sensory cells.

Dr. E. D. Fisher (*N. Y. Med. Jour.* March 29, 1890) at a meeting of the
"New York Academy of Medicine," read an article on the *Functions of the
Cerebellum as Indicated by Recent Research.*   He said:
"The general functions of the cerebellum in animals and in man are the
same, although they may be somewhat modified by the special characteristics of the mammals under consideration.   The experiments which the
speaker recorded had been carried on in the Loomis Laboratory with every
precaution against sepsis, following the same methods observed in cerebral
surgery in man.   The operations were eight in number—seven on dogs and
one on a monkey, the latter animal being exhibited.   In summarizing the
results so far obtained, one fact had stood out very clearly—namely, that it
required a considerable lesion of the cerebellum to produce any symptoms,
and that cortex lesions could not be localized as in the cerebrum.   Loss of
equilibrium and inco-ordination were present in every case, more or less
markedly, but no sensory disturbance.   Recovery always occurred after a
certain time, the cerebrum seeming to take on the function of the cerebellum, although this was never absolute, as even in the dog, which lived some
six months after the operation, some inco-ordination and loss of equilibrium
were present.   The psychical functions, so far as could be studied, were
not affected, and the sexual desires were neither lessened nor increased "

Experiments on the *Blood Supply of the Brain* made by Roy and Sher-
rington are published in the April number of the *Journal of Physiology.*
They found that stimulation of any sensory nerve of the body with the
induced current produced expansion of the brain and they state that this is
due mainly, if not entirely, to pressure or elastic distension of the cerebral
vessels as a result of the rise of blood pressure in the systemic arteries.

Closure of both the carotid arteries caused enormous contraction of the brain. Closure of one cartoid artery produced no appreciable effect on the volume of the brain nor did closure of one jugular vein   Closure of both external jugulars caused expansion of the brain.   It is therefore important to remember that pressure in the veins influences the volume of the brain.   Lowering of the arterial pressure from loss of blood or other causes was accompanied by diminution of the volume of the brain, and this must be looked upon as resulting from passive contraction of the cerebral vessels.   During asphyxia there is active expansion of the cerebral vessels.   In the animals while curarized stimulation of the medulla with induced currents caused cerebral congestion, induced currents passed through the spinal cord had the same effect as stimulation of the medulla. This congestion of the brain was not active but a result of increased tension of the systemic arteries.   If the animals struggled there ensued cerebral congestion, this being proportioned to the muscular effort.   Chloral produced contraction of the brain which was gradual and continued some time.   Chloroform produced contraction, and ether expansion of the brain. The experiments with opium and morphine gave results that were not always uniform, but they state that on the whole the most constant effect of the intravenous injection of opium was contraction of the brain.   Intravenous injection of solution of bromide of potassium caused primary and temporary contraction, followed by expansion of the brain.   Strychnine produced enormous expansion of the brain.   The intravenous injection of diluted acids, sulphuric, nitric or lactic, produced immediate and enormous expansion of the brain.   The effect of alkalies was to produce prompt and decided anæmia.   Liquor ammonia produced cerebral congestion · and quinine produced slight congestion.   Alcohol and atropia produced congestion.   One of the most evident facts observed by them is that the ''blood supply of the brain varies with the pressure of the systemic arteries." From this we learn that the central circulation is constantly influenced by the pressure in the systemic arteries.   They conclude that there are no special vaso-motor nerves from the sympathetic supplying the brain directly, and further that the chemical products of cerebral metabolism contained in the lymph, which bathes the walls of the arterioles of the brain, can cause variations of the calibre of the central vessels, and that in this reaction the brain possesses an intrinsic mechanism by which its vascular supply can be varied locally, in correspondence with local variation of functional activity. There are then two mechanisms by which the cerebral supply is regulated: first, one by which the blood supply is varied through a *local* mechanism, and secondly, the *vasor-motor* by which the blood supply varies with systemic arterial pressure.   Changes in the venous pressure also affect the cerebral vessels, so there are really three mechanisms for central vascular control.   The practical conclusions are of course numerous.   The fact that

3

in all cases where the systemic blood pressure is high we have congestion of the brain, explains why we have cerebral hemorrhage in diseases where the blood pressure is raised as in Bright's disease. Venesection in cerebral hemorrhage, while it might stop the bleeding, might kill the patient by interfering with a protective mechanism. It is shown that compression of a carotid artery does not diminish the blood supply of the brain. The depressing effect of alkalies is probably due to the anæmia of the brain which they produce.

Dr. F. W. Mott (*Brit. Med. Jour.*, May 17) publishes a report on bilateral associated movements and on the functional relation of the corpus callosum to the motor cortex. Following is a brief summary of the report. (1) Stimulation of centres connected with associated movements of the head and eyes, and with adduction of the vocal cords, produces the same effect after as before section of the corpus callosum. (2) Stimulation of the corpus callosum by regular induction currents produces localized bialateral movements in all parts of the body, the muscles which respond to the stimulus depending upon the position of the electrodes along the commissure. (3) Stimulation of the intact corpus callosum after removal of the motor area of one side produces movements of the corresponding muscles on that side of the body only, that is on the side of the body the cortical motor area of which is not injured. (4) Direct stimulation of the fibres of the corpus callosum in their transverse area, produces localized movements on the side opposite to that hemisphere with which the stimulated side is still connected. (5) Epilepsy is produced either by stimulation of the intact corpus callosum or by direct stimulation of the fibres in section with a strong current.

In the same journal for June 21 Dr. Mott reports researches on associated eye movements produced by unilateral and bilateral cortical faradization of the monkey's brain. (1) The area of the cortex of the frontal lobe, which when stimulated gives rise to associated eye movements can be divided into three zones : (a) a middle zone immediately below the horizontal part of the precentral sulcus, faradization of which is followed by simple lateral deviation to the opposite side. (b) An upper zone immediately above this which may extend to and include part of the marginal gyrus. This gives on faradization downward inclination usually combined with lateral deviation. (c) A motor zone immediately below the middle one and sometimes extending nearly down to the margin of the hemisphere. This gives upward inclination usually combined with lateral deviation. (2) That simultaneous bilateral faradization of the frontal cortex at identical points (that is, those points which when singly excited give rise to conjugate deviation of the eyes to the opposite side) invariably brings the eyes into position of looking straight forward. (3) Bilateral faradization of identical points in the occiptital visual area produces results similar to those obtained in the frontal area.   (4) It was found that a weak stimulation of the frontal area sufficed to overcome a strong one of the occipital visual area.

___

ON THE INEQUALITY OF THE PUPILS IN HEALTHY INDIVIDUALS.

(*Archiv de Neurol.*, No, 50.)

Moebius finds the pupil smaller among the aged.   In soldiers M. Ivanoff finds the two pupils rarely equal; only in twenty cases out of one hundred and fifty observed.   In two-thirds of these cases the left pupil and the left half of the face are more developed than the right.   The diameter of the pupil is in direct relationship with the development of the face of the same side and the members of the side opposite.

___

In the first report of the Loomis Laboratory for 1890 is an interesting article on the *Centre for Vision*, by Gilman Thompson and Sanger Brown. The following are some of their conclusions   Lesions of the posterior part of the occipital regions of cats and dogs invariably produced total blindness of the opposite eye with no impairment of vision on the operated side. Decussation of the optic nerves at the chiasma in these animals is therefore complete.   The blindness is permanent.   Removal of the entire occipital lobe posterior to the angular gyrus in monkeys produces permanent homonymous hemianopsia.   In monkeys, there is therefore incomplete decussation of the optic nerve.   Hemianæsthesia in cats and dogs results from extensive lesions in the posterior part of the occipital lobes.   Hemianæsthesia in the monkey results from the removal of the angular gyrus and from extensive removal of the occipital lobe.   In both cases it is transient.

___

ON THE PHYSIOLOGY AND PATHOLOGY OF SLEEP.

Prof. Mauthner considers sleep to be due to an interruption of conduction between the cortex and the periphery.   The seat of this interruption is in the central gray matter and its cause is not anæmia or hyperæmia, but an accumulation of waste products, as in muscle fatigue.—(PROF. MAUTHNER, *Wiener Medizin Presse*, No. 22, 1890.)

___

AN INSTRUMENT FOR THE DETERMINATION OF SENSIBILITY TO PASSIVE MOTION.

Dr. Goldscheider, assistant at Leyden's clinic, has invented an instrument for this purpose.   It consists of a padded board from which depends a frame bearing a graduated sector.   A pendulum swinging in front of the latter denotes the arc of movement.   It is impossible to judge of finer disturbances of sensation of passive motion without this instrument and a knowledge of the normal sensibility.   In the shoulder a movement of 0.3° is felt, in the elbow of 0.5° to 0.8°; in the wrist of 0.3° to 0.6°; in the hip 0.5° to 1°; knee the same, and in the ankle 1.3° to 1.6°.—(*Berlin Klin.*, *Wochenschrift*, No. 14, 1890.)

ON RAPID HARDENING OF THE SPINAL CORD BY THE ELECTRIC CURRENT.

1. Pieces of cord, immersed in a bichromate solution and exposed for four or five days to the action of the positive pole, became as hard as if treated two or three months in the ordinary way.

2. Pieces of the same cord in the same solution, at the negative pole, became swollen and of a soft consistence.

3. Control pieces barely changed.—(DR. L. MINOR, *Neurolog. Central-blatt*, No. 10, 1890.)

PATHOLOGY AND SYMPTOMATOLOGY.

Blocq (*Annales Medico-Psychologiques*, July and Aug.) was awarded the Moreau prize for an essay on *Contractures.*   He divides the subject into three parts.   The first part is a general symptomatic study of contracture; the second part treats of spasmodic contractures considered in general as a morbid entity; the third part is devoted to the study of morbid rigidities. Those that have not yet been classed he calls pseudo-contractures.   He divides contractures into three chief categories from a clinical point of view. In the first he considers whether there exist spasmodic phenomena; second, as to whether the nervous system is involved; the third is devoted to the alterations that occur in the muscular structure.   The special clinical characters of spasmodic contracture are: a constant sensation of tension in the parts, its distribution to certain associated muscles, its tendency to become general, the exaggeration of the tendon reflexes, its disappearance during chloroform narcosis.   Spasmodic contracture does not depend upon any anatomical lesion of muscles but upon various morbid conditions of the nervous system producing irritation of the anterior horns of the cord.   The mechanism of this irritation may be direct (as toxic, or absence of inhibition) or indirect (peripheral nerves, centripital, pyramidal columns).   In the development of spasmodic contracture there is in certain cases fibrous retraction of the tendons, probably due to rupture and cicatrization of tendonous fibres which result in malformations, and which alone justify operation.   Pseudo-contractures are observed in the course of most traumatisms and inflammations of muscles.   In certain circulatory troubles of Parkinson's disease and in primary amyotrophies, contractures are characterized clinically by special rigidity to the touch, by their irregular distribution, the absence of any tendency to become general, the non-exaggeration of the re-flexes, and their non-disappearance during chloroform narcosis. Ischæmic pseudo-contracture follows prolonged deprivation of the part of arterial blood, and is characterized by rigidity and coldness of the limb and is due to a morbid change in the muscular fibres. The pseudo-contracture of paralysis–agitans is an element in the disease and may exist without the tremor.

CONTRIBUTION TO THE STUDY OF CONJUGATE DEVIATION OF THE EYES.

Conjugate deviation of the eyes has been observed after unilateral apoplectic attacks, in which case the eyes are directed to the sound side; and with unilateral spasms, when they usually are turned toward the con-vulsed side.

The author has observed a case with right sided clonic convulsions, which soon became tonic, in which the eyes were turned to the left.

On autopsy, a hemorrhage of the size of a pigeon's egg was found in the left frontal lobe. This had destroyed the nucleus caudatus and broken into the ventricle.—(Dr. J. NEUMAN, *Berlin Klin., Wochenschr. Trocheusch*, No. 18, 1890.)

ON APHASIA WITH INTACT VISUAL PERCEPTION.

The author reports two cases, one of which was caused by a tumor of the left occipital lobe and corpus callosum, the other case being still living. There was inability to recognize objects visually, although named articles could be selected with ease. The patients could also describe articles which they could not name. For the theoretical explanation of this phenomenon the author assumes a loss of continuity between the optic centre and that pre-siding over speech, while that from the centres of other sensations was normal.—(Dr. MOELI, *Berlin Klin., Wochenschr.*, No. 17, 1890.)

In the *Lancet* for May 31, was published a synopsis of a lecture on apha-sic and amnesic defects of speech, by Dr. Bastion. The four forms of ver-bal memory are these: first, the auditory memories of words which are probably registered in the posterior part of the upper temporal convolution; second, the visual memories of words registered in some part of the occipi-tal lobe; third, the memories of articulatory sensations registered in Brocas' convolution; fourth, memories of things connected with writing movements registered in the hinder part of the second frontal convolution. Concern-ing localization Dr. Bastion said he believed that no motor centres exist in the cerebral cortex, and that what we term cortical motor centres are cen-tres in which the sensory results of different movements are registered, and

that a re-excitation constitutes a necessary antecedent, the stimulus for
the reproduction of the movements to which they are related. The excitation of these parts of the cortex constitutes a terminal part of a voluntary
act so far as it has to do with the cerebral cortex. The chief
word centres are seldom called into play singly. They are usually roused
through commissural fibres almost simultaneously though really in rapid
succession. We therefore think in words whether the result be silent without speech or writing. In examining a patient we shonld observe first,
that if the patient understands what is said to him and can repeat words
that he has just heard pronounced, it shows that the auditory centres and
the centres of the memories of articulatory sensations with the commissures
joining them, as well as the out-going channels and the corresponding
motor centres, are free from serious damage: second, if the patient can
understand written speech and can write a word which has been repeated
to him, it will show that the visual centres and the centres for the memories of sensations connected with writing movements and with their commissural fibres are free from serious damage. Third, if a patient can read
aloud or name objects at sight it shows that the visuo-auditory commissure
must be intact. Fourth, if the patient can write from dictation it shows
that the auditory visual commissure must be undamaged.

———

A synopsis of the Lumleian Lectures on *Convulsive Seizures*, by Dr. Jackson, is published in successive numbers of the *British Medical Journal* beginning with March 29.

Convulsions and other paroxysms are owing to sudden, excessive nervous
discharges. Convulsions differ in kind according as centres discharged differ in rank, or as the centres differ in evolutionary levels. There are three
levels of the central nervous system. The first consists roughly of cord,
medulla and pons. This level represents the simplest movements of all
parts of the body. This level is cerebro-cerebellar, and is at once the lowest level of the cerebral and of the cerrebellar system. The middle, or second level, is composed of centres of the Rolandic region and possibly of the
ganglia of the corpus striatum. It represents complex movements of all
parts of the body. The highest, or third level of the cerebral system is
made up of the centres of the pre-frontal lobes. It represents the most
complex movements of all parts of the body. These highest centres are the
acme of evolution. They have the same kind of constitution as lower centres and are sensori-motor as certainly as the lumbar enlargement is.
Psychical states are not functions of any centre, but are simply concomitant with functioning of the most complex sensori-motor nervous arrange-
ments. There are three kinds of fits corresponding to the three levels. He
classifies the three kinds of fits as (1) ponto-bulbar, or lowest level fits; (2)
epileptiform, or middle level fits; (3) epileptic, or highest level fits, the latter probably due to excessive discharge from pre-frontal lobes.

Dr. Jackson reiterates his theory that both sides of the body are represented in each cerebral hemisphere, which is but an expansion of Broadbent's hypothesis of the double representation of the bilaterally acting muscles. Thus in certain cases of hemiplegia one series of movements is lost and another series of movements of the same muscles is retained. Contracture following hemiplegia is due to loss of the complex movements with retention of the simpler movements, which are shown in the chronic muscular spasm.

There are three varieties of epilepti*form* seizures: (1) Those starting in the hand, usually in the thumb or finger; (2) Those starting on one side of the face, usually near the mouth or in the tongue, or in both parts; (3) Those starting in the foot, usually in the great toe. The starting point is usually the same in each patient, but not always. The patient whose fits sometimes begin in the hand may occasionally have them begin in the face. Epilepti*form* fits usually begin in those parts which have the most voluntary uses. It has been shown that those parts having many small and greatly changing movements are represented by many small cells, and those parts having few and little changing movements are represented by a few large cells. The size of cells is important in regard to their nutrition: both in health and in disease small cells will be nourished more quickly than large ones when both are bathed in the same nutrient fluid. The small cells will become highly unstable sooner than large ones during nutrition. It is a fact that most epilepti*form* seizures begin in parts having "small movements," in parts represented by areas of the cortex having most small cells. Post epilepti*form* paralysis includes all paralyses the result of the seizures, and is due to exhaustion of the nerve cells which have been discharged during the fit. There is probably a relation between the severity of the fit and the subsequent temporary paralysis. It is important to remember that we may lose some *movements* of certain muscles without disability of those muscles. For instance, in aphasia from a lesion of Broca's convolution there is paralysis of certain combinations of muscular movements, though the muscles remain normal for certain other combinations of movements. It is to be remembered that what is psychic aphasia is physically paralysis however it may appear.

ON THE RELATIONS OF JACKSONIAN AND COMMON EPILEPSY.

There are two points at which the clinical pictures of these diseases may fuse. 1. The convulsions of Jacksonian epilepsy may become generalized. 2. Those of common epilepsy may become restricted.

A consideration of their distinguishing symptoms will show that they have something more in common than the name.

1. Extent of the spasms. While the restriction of the spasms to a single limb, or at least the origin in it, is rare in common, it is the rule in Jacksonian epilepsy, although exceptions occur on both sides.

2. Consciousness is usually lost at the beginning of an attack in common epilepsy, and is preserved in Jacksonian epilepsy, although exceptions are common.

3. The initial cry may be present in Jacksonian epilepsy.

4. Rapidly repeated attacks of Jacksonian epilepsy will indnce a *status epilepticus* identical with that succeeding common epilepsy, and as fatal.

Hughlings-Jackson and others have for a long time held that idiopathic epilepsy was due to the same causes as Jacksonian epilepsy, although the irritation would necessarily be of greater extent and duration.

It is, however, necessary to hold fast to the distinction between the two, although *the most insignificant cortical spasm involves an epileptic element as well as the most trivial petit mal, the momentary disturbances of consciousness and the fleeting isolated prodromal sensations.*—(DR. L. LOWENFELD, *Arch. of Psychiatrie*, Bd. XXI, H. 2.)

VARIATIONS OF SENSIBILITY IN EPILEPTICS. The following are the conclusions of an elaborate paper illustrated with clinical histories by Dr. Cæsar Agastini in the *Rivista Sperimentale di Freniatria*, XVI–I–II, 1890.

The tactile sensibility and power, as well as the sense of weight are less in epileptics than in normal individuals, and this difference is increased after the attacks with phenomena of laterality in relation to the plagiocephaly, and with the prevalence of the convulsive discharges on one or the other side of the body.

Thermic sensibility usually normal, is very little altered after severe epileptic attacks.

The gustatory sensibility is lessened as compared with healthy individuals, and frequently the phenomena of agensia and hemipogensia appear after the attacks.

The olfactory sense is diminished in epileptics under the influence of bromides, and the hyphasmia and anosmia are more marked after the convulsions.

The auditory sensibility is lessened, usually on the side opposite the plagiocephaly and the difference is accentuated after the attack.

Visual acuteness is found usually unimpaired and only slightly involved with the convulsions, with phenomena of narrowing of the visual field and more rarely disturbance of the chromatic sense.

GENERAL ORGANIC SENSIBILITY.—The sensibility to pain is usually normal and only slightly disturbed by the epileptic attack. This may also be said of the muscular sense.

The sense of equilibrium is generally disordered and more noticeably so after the convulsions.

The electrical sensibility is diminished and the hypoalgesia is increased after the attack with phenomena of laterality.

REFLEXES.—The pupils are dilated in a notable proportion (31.25%) of epileptics during the interparoxysmal periods, and their reaction to stimuli is rather sluggish, but after the convulsions, in the majority of cases there is greater dilatation and a more rapid reaction.

The reflexes of the external ear, the olfactory reflexes, those of the palate as well as the abdominal, cremasteric, and planter reflexes are excited generally less quickly in epileptics during interparoxysmal stage, but become more pronounced, especially the planter, after the convulsive phase.

The rotulion reflex is in general quicker than normal and is notably increased after the attack.

The foot and ankle clonus are generally lacking during the interparoxysmal periods as compared with the greater number of cases after the attack.

DISTURBANCES OF MOTILITY.—Disturbances of walking, writing, and electro muscular contractility are observed, and the latter is still more diminished after the attacks.

SOMATIC AND PSYCHIC FEATURES.—There are sufficient characters to place epilepsy among the degenerative neuroses, and to demonstrate its affinities with moral insanity and congenital delinquency.

Dr. J. J. Putnam (*Boston Med. and Surg. Jour.*, April 10) reports three cases of Cerebral Tumor with autopsy.

CASE I.   Male patient, married, thirty-nine years of age.   Tumor of a sarcomatous character involving the posterior half of the right middle frontal convolution.   Symptoms: intense and incessant headache, almost exactly corresponding with the situation of the tumor; one well-marked attack characterized by partial loss of consciousness, and convulsive movements limited to the left shoulder; double optic neuritis, which was very much more intense on the side of the tumor; duration of the illness, six months.

. The autopsy showed a sarcomatous tumor which occupied the entire width of the middle convolution and extended from about the middle of the external convex surface of the frontal lobe back to the pre-frontal sulcus.   It also encroached upon the inferior and superior frontal convolutions subcortically.

CASE II.   Two subcortical sarcomatous tumors, one occupying nearly the whole of the second temporal convolution, the other, the posterior part of the parietal lobe.   The symptoms consisted in alteration of mental character; aphasia, partly sensory and partly motor, with alexia; a general convulsion, probably due to invasion of the membranes of the temporal lobe by the new growth; epilepti*form* seizures, consisting in tremor of the right arm and leg; right hemi-paresis of shifting intensity, terminating in complete paralysis with contracture; long continued coma; Cheyne-Stokes respiration; death, with rapid elevation of the temperature. The duration of the illness, so far as recognized symptoms are concerned, was five months.

Autopsy showed tumor involving entire second left temporal convolution and the posterior portion of the left parietal lobe. An interesting feature of this case was that there was no apparent anæsthesia, but this may be accounted for on the ground that the new growth was not large and was situated at the posterior border of the parietal lobe and had perhaps been of slow growth.

CASE III. Showed attacks of the arrest of the power of speech for short periods without loss of consciousness or of power of expression in writing or of comprehension of simple speech or written signs, preceded by trifling tremor of right hand, permanent slight hesitancy in speech with slight paraphasia. Occasional general convulsions with aura. Frontal and occipital headache, mental failure, finally repeated attacks of slight convulsive action of right hand with tendency to contracture, paresis of extensors. Operation by Dr. Beach. Tumor not found at the point thought to be indicated by the symptoms (just above the motor speech area) but found after death in the supra marginal gyrus at its posterior end. The exact boundaries of the tumor were as follows; its posterior edge was limited by the posterior ascending branch of the fissure of Sylvius. The anterior bond was formed by a small sulcus leaving the fissure of Sylvius a little behind the anterior extremity of the interparietal fissure, which ran down parallel to the fissure of Rolando and was continued by the sulcus into the Sylvian fissure.

TWO CASES OF TUMOR OF THE POSTERIOR FOSSA OF THE SKULL.

The first case derives its interest from the fact that the lesion was sharply localized, and that cases in which hemianopsia was the result of such localized lesion are few in number, only seven having been previously published.

A female, aged 61, was seized on the street with vertigo and epileptiform convulsions, followed by violent hallucinatory mania. Becoming rational in a few days she was discharged, no physical alterations being noticed, eight weeks after she returned, complaining of headache, vertigo, vomiting, and ataxia of the left hand. Mind was clear, although apathetic. Occipital region on right side was painful on percussion. The right side of the body was hyperæsthetic. There was bilateral left homonymous hemianopsia and a tendency to fall backward and to the left on standing or walking, as well as a paresis of the left half of the body, which later on, became complete. Hallucinations of sight occurred on the left side only. Choked disc developed in the right eye three weeks after admission, and in the left one a week before death. The state of her mind varied between apathy and excitability. During the last weeks of life partial deafness and conjugate deviation of the eyes to the right occurred. Death in coma two months after last, four months after first admission.

Autopsy disclosed a glio-sarcoma of the size and shape of a hen's egg, in the right occipital lobe. Posteriorly the tumor occupied the point of the lobe, having destroyed the cortex and become adherent to the dura. Inferiorly, the cortex was also destroyed. Above, laterally and toward the commissure, the tumor did not reach the cortex, while anteriorly, its point projected just beyond the first occipital sulcus. There was no softening around the tumor.

The accurate localization of the tumor did not cause difficulty after the physical symptoms had developed. Lesions of the optic tract were excluded by the absence of implication of other cranial nerves. The pain, choked disc, left hemianopsia and hemiplegia all pointed to the right side. Nothnagel has stated, "in such cases the hallucinations are due to the anatomical changes in the cortex." The loss of co-ordination was due to pressure upon the cerebellum.

CASE II. Female, single, aged 39, who denied infection and traumatism suffered for a year before admission with vertigo, increasing headache and difficulty of vision, which soon increased to total blindness. Six weeks before admission to the clinic, she complained of nausea and vomiting and showed slight stupor. A gradual decrease of strength was noticed, first in the lower, then in the upper extremities. She then had several attacks of unconsciousness, swaying upon standing and walking, tendency to fall backward, nausea. Left angle of mouth lower than right, tongue deviated to left, left palpebral fissure narrower than right. Pupils did not react to light, right larger than left. Motion of eyes impaired, except in a downward direction. Nystagmus in the terminal positions. The ophthalmoscope showed optic atrophy on both sides, sense of smell abolished, hearing in left ear impaired, speech was nasal, deglutition difficult. A decided motor impairment existed on the left side with ataxia of the left arm. Knee jerk was absent on both sides. Later the neck became stiff and the left half of the tongue atrophied. The left cornea became totally anæsthetic the right almost so, sensation being otherwise intact.

Before death, spasmodic movements of the tongue and right angle of the mouth occurred. Death, eight weeks after admission to clinic. Diagnosis, tumor of left half of cerebellum. (Locomotor ataxia?)

Autopsy showed hydrocephalus internus and a tumor somewhat larger than a plum, at the base of the left half of the cerebellum. This had compressed and flattened the left half of the pons and medulla. The nerves stretched across the tumor. Microscopic examination showed decided atrophy of the ascending trifacial root. The left abducens and trifacial showed some of the axis cylinders swollen. The left oculomotor was on the whole intact. The right oculomotor showed many more degenerated fibres than the left. The nature of the tumor is not mentioned. The posterior columns of the cord were degenerated throughout their whole length.

The diagnosis in this case was made on (1) the disturbance of equlibrium; (2) the early and pronounced vertigo; (3) the headache which could not be exactly localized, but was combined with stiffness of the neck.

While nausea was constant and pronounced, vomiting was rare. The symptoms of irritation and paresis of the cranial nerves were due to the pressure effects of the tumor.

While the weakness of the limbs and ataxia of the left arm might have been due to the tumor, the absence of the knee jerk suggested the additional diagnosis of locomotor ataxia. The coincidence was probably accidental. —(Dr. R. Wollenberg, *Arch. f. Psychiatric*, Band XXI, H. 3.)

ON THE PATHOLOGY OF TUMORS OF THE CEREBRUM.

While the cases studied were not selected they were from hospital wards, and all had pronounced symptoms when admitted.

Of 23 cases, the neoplasm was either sarcoma, glioma, or glio-sarcoma in 18; metastatic carcinoma in 3; solitary tubercle in 1, and gumma in 1.

Basal miliary gumma and tumors of the cerebellum, pons and medulla are not included.

The diagnosis of an endocranial neoplasm was definitely made in 20 cases.

Typical choked disc was found in 14 and optic neuritis in 5 cases. Care is required in deducing a diagnosis from the choked disc, as the author has seen 3 cases of uræmic coma in which there was simply a papillitis albumenurica without retinal changes.

Encephalopathia saturnina and poliencephalitis acuta superior alcoholica may also simulate an advanced stage of brain tumor.

The choked disc may appear at any stage of the trouble, often only a short time before death. The time of appearance and extent of the change have no relation to the size or location of the tumor.

Probably next in value to choked disc as a diagnostic sign is the stupidity and somnolence.

Melancholia or mental irritability may occur, although a tendency to make "funny," or would be witty remarks, is more frequent.

Aphasia, both motor and sensory, of various degrees of severity was found in 10 cases of left sided and 2 of right sided tumor. This symptom is of no assistance in accurate localization of the tumor. We can only say that the tumor affects certain parts either by pressure or direct presence. Neither will the motor symptoms assist us in accurate localization. Coarse disturbances of sensation are rare, while the finer ones are difficult to detect on account of the mental stupor and somnolence.

Oculomotor paralysis was found in three cases, due in all to direct pressure upon the nerve.

The headache shows no constant relation to the location of the tumor, and may be wholly on the opposite side.

Tenderness upon percussion is a more reliable and constant sign, being present in almost all cases óver the seat of the tumor.

Bergmann excludes from the possibility of operation, all multiple, and metastatic tumors, syphilomata and tubercle.

He also advises against operating when pressure symptoms are present or when the tumor is supposed to be very large. Before deciding on the advisability of operation, the surgeon must have information as to the size, extent, and location of the tumor, its multiplicity and probable character, its situation, whether in the cortex, or immediately below it, and the possibility of its being surrounded by a zone of degenerated brain tissue.

As has been seen it is only in rare cases that the neurologist is able to furnish satisfactory data. Most of these cases had been seen by Westphal, and on autopsy only one of them was found to be so situated that an operation would have been advisable.

Operations for tumors of the brain have been reported in fifteen cases, with seven deaths and eight recoveries, at least partial. This is exclusive of cysts, of tumors complicating tumors of the skull, and of one exploratory operation.—(DR. H. OPPENHEIM, *Archiv. f. Psychiatrie*, Bd. XXII, H. I.)

CONTRIBUTION TO THE KNOWLEDGE OF THE SYPHILITIC AFFECTIONS OF THE CENTRAL NERVOUS SYSTEM.

The most common form of cerebral syphilis is the diffuse basal gummous meningitis. On autopsy, the base of the cranial cavity appears as if filled with a turbid jelly. The nerves are discolored and more or less swollen. The lumen of the arteries is irregular, their walls thickened. The effusion may have the character of a tumor; in places softening and hemorrhages are frequent.

Histologically we find the new formation to be composed of granulation tissue, rich in cells and vessels and containing fibrous and caseous foci. Circumscribed gumma is rare. The arteries are always involved. The entire vascular wall is thickened by a small-celled infiltration. Infiltrations into the intima contract the lumen of the vessels.

The nerves show microscopically that their degeneration is dependent upon infiltrations which radiate from the epineurium between the fibres. These are dense fasciculi of fibrous tissue containing small cells and numerous vessels. While the nerve trunks enlarge the fibres atrophy, the cerebral substance suffers by the rupture and obliteration of its vessels.

The symptoms of diffuse basal syphilis are rather clear. Headache, vomiting, vertigo, occasionally fainting or spasms, moderate dementia, loss of memory, mental apathy. These mental symptoms do not increase. This is a very valuable point in differentiating this condition from cerebral tumor.

Intercurrent disturbances of consciousness, somnolence, attacks of confusion or mania, polydipsia and polyuria have been frequently observed.

All these symptoms may be of very short duration. The paralytic symptoms are usually found in the optic and motor nerves of the eye. The third nerve is involved most frequently. The fifth nerve is involved oftener than the olfactory. The auditory and facial are involved only when the process extends very far back. The temperature rarely rises, which is a point that will distinguish this disease from tubercular meningitis. The most valuable diagnostic point is, however, the inconstancy of the symptoms. Especially is this noticeable in the optic nerve. The visual field contracts and expands. Pupillary reactions are lost and return. Ptosis and facial paralysis appear and disappear over and over again. If the endarteritic proliferation leads to obliteration of the sylvian artery, we get hemiplegia or aphasia. If the basilar and vertebral arteries are involved, bulbar symptoms will appear.

In syphilitic meningitis of the convexity the meninges are the seat of granulations. Important diagnostic points are the frequent absence of general pressure symptoms and the "progress by jumps." In general it may be said that the clinical picture is not as clear as that of basal syphilis. Syphilis of the cord alone is very rare. We generally find the cerebro-spinal type. Pathologically it is a diffuse gummous meningitis, exceptionally it is circumscribed. A lardaceous or gelatinous exudation occasionally fibrinous, is found between the different meninges and around the cord. It extends over the roots of the nerves. The diagnosis of spinal syphilis depends on the preceding or coincident symptoms of cerebral syphilis and the "progress by jumps." Westphal's symptom may appear aud disappear. Brown-Séquards hemi-paresis is often suggested. Hyperæsthesia, stiffness of the spine and neuralgic pains may accompany the meningeal thickening. The impossibility of associating the symptoms with any localized lesion must be borne in mind. Disturbances of the temperature sense have been observed. The differential diagnosis between cerebral syphilis and multiple sclerosis must be based on the fact that the first developes around, the latter in, the cord, and the presence of nystagmus, scanning speech and tremor in sclerosis. The distinction from tabes may be difficult. Pseudo-tabes syphilitica is a clinical and pathological entity. The differentiation from degeneration of the lateral and posterior columns is only possible late, if at all. The prognosis of cerebro-spinal syphilis is not as favorable as that of the purely cerebral type.

An energetic mercurial treatment offers the best prospects if employed early. The author offers a word of caution in regard to patients who have had syphilis and complain of neurasthenic symptoms. These are often the first symptoms of cerebral syphilis. Such patients should be carefully examined and closely watched, but should not be put on antisyphilitic treatment until the diagnosis is sure.—(DR. H. OPPENHEIM. Read at a meeting of "Hufeland Gesellsch. für Heilkunde." *Neurolog. Centralblatt.* No. 15, 1890.)

ON ACROMEGALY.

Recklinghausen discusses Acromegaly as follows:

1. The most characteristic change is the extraordinary and disproportionate growth of the terminal members, *i. e.* hands and feet, and of certain other parts; nose, ears, lips, malar eminences and chin. Even the tongue, penis and clitoris may be involved.

2. This hypertrophy begins at a period when the general growth of the body is usually complete (after 30th year) while giants usually show their increased size (local or general) even during adolescence.

3. In acromegaly we find a thickening of the involved parts disproportionate to their length.

It has not been definitely demonstrated that there is always an increase in length of the long bones, although it has been observed. In the published accounts of autopsies, no notice seems to have been taken of the condition of the spinal cord or peripheral nerves, so that it is not possible to establish a causative relation between the changes observed in this case and the acromegaly.

In several cases an enlargement of the hypophysis cerebri has been noticed, but it was absent in this case, and in some other published cases.

As between the opinion of Freund (local giant growth) and Maric (distinct nervous disease) Recklinghausen inclines to the latter.—(PROF. F. VON RECKLINGHAUSEN, *Virchow's Archiv.*, Band 119, H. 1.)

ON PUPILLARY VIBRATION IN AFFECTIONS OF THE CENTRAL NERVOUS SYSTEM.

This phenomenon, known to opthalmologists as *hippus* is observed best in diffuse light and parallel visual axes. The vibrations are irregular in extent and duration. They are especially noticeable in multiple sclerosis, in acute meningitis (initial stage) and in those focal diseases of the brain, which begin as apoplexies with motor disturbance in the paralyzed muscles. In the latter it usually occurs on the side opposite the paralysis. It was also found in a number of cases of neurasthenia. It is usually found with increased tendon reflexes.—(PROF. OTTO DAMSCH, *Neurolog. Centralblatt*, No. 9, 1890.)

Investigations by the sanitary authorities of numerous cases of "Nona" reported in Austria show that the disease has no existence. The cases are cases of coma or stupor occurring in the course of pneumonia, typhoid fever, meningitis, or severe infectious diseases.—(*Wierner Medizin Blaetter*, No. 14, 1890.)

———

Dr. Lee reports *(Lancet)* a case of heart disease resulting from emotional cause. The patient was a nurse. One day when sitting in the nursery with the children their father came into the room in a state of mental derangement and seized the youngest child with the apparent intention of throwing it out of the window. The nurse was so frightened that she could not move or speak and immediately after this her heart troubles began. The case resembled one of Grave's disease so far as the cardiac symptons were concerned. She could not walk without distress from shortness of breath. Her pulse was generally 140. If she got up to walk it was 150. Digitalis had no effect, but opium had a slight effect in quieting the heart. The case was considered incurable.

———

Dr. C. L. Dana (Proceedings of "Am. Neurologlical Society") reports a case of chronic softening of the spinal cord with paraplegia. The case seemed to show there was softening of the anterior horns which was due to obliterating arteritis. He stated that the case established upon a firm foundation for the first time that in the gray matter of the cord there might exist progressive softening from obliterating arteritis just as was found in the brain.

———

M. Ballet (*Jour. des Soc. Scientifique*, May 14th) reported the case of a patient whose left hand and arm presented atrophy of muscles, tendons and bones. Patient also had complete hemianæsthesia and had had several attacks of hysterical mono-plegia. The amyotrophy was attributed to hysteria.

———

Dr. Loomis reported (New York Pathological Society, April 5th) results of an autopsy on a case of hydrophobia:

"The autopsy was made twelve hours after death, and with the exception of an engorgement of the cerebral vessels and slight pulmonary congestion and œdema, no pathological condition could be found in any organ or tissue of the body. A microscopical examination of a section of the lower portion of the medulla showed congestion of the capillary vessels, bnt no structural changes. A portion of the spinal cord was removed, and was reduced by Pasteur's method to an emulsion in a porcelain crucible, and injected into the subdural space of a healthy rabbit. The animal remained perfectly well for twenty-one days, and then developed the characteristic paralysis of hydrophobia, commencing in the posterior extremities and rapidly progressing, until death occurred from paralysis of the muscles of respiration. A second inoculated rabbit died in the same manner after remaining perfectly well for twenty-three days. The classical course of the disease in the rabbits proved conclusively that they died from the virus of hydrophobia."

HYPOTHESIS AND FACT REGARDING THE NATURE OF HYSTERIA.

The basis of the hysterical state is an irritable weakness, so that the influence of external and internal stimuli is increased and made easier. The influence of the emotions upon the motor, vaso-motor, sensory and secretory functions, is physiological. In fact certain functions, such as crying, laughing or blushing, are in most persons, purely under the control of the emotions. In hysteria, however, the physiological resistance is so reduced that slight emotions of this sort produce maximum effects.

Reflex excitability is also increased, minimum stimuli causing maximum reflexes.

Hysterical paralysis is either emotional or reflex in its nature. As the centres are easier excited, they are also exhausted easier. The hysterical paralysis is a true paralysis in that there is an interruption of conduction somewhere between the seat of the will and the motor centres, so that the patient is not able to bring the paralyzed part under the power of the will.

In a case of hysterical aphonia, while the patient was unable to talk, she could sing or give a cry of pain. In the first case the emotion of singing was enough to overcome the obstacle to will conduction, in the second the cry was reflex.

Hysterical anæsthesia is due to an inhibition of the perceptive centres themselves, so that ordinary stimuli are not perceived.—(DR. H. OPPEN-HEIM, *Berlin. Klinisch. Wochenschrift.* No. 25, 1890.)

HYSTERICAL APHONIA.

By E. FLETCHER INGALS, M. D. *(Jour. Am. Med. Assoc., Aug.)*

One of the most interesting manifestations of hysteria is paralysis of the adductor muscles of the vocal cords. It is a singular fact that in nerve trunks not larger than a knitting needle, only those fibres are involved in this disease which supply the adductor muscles, while adjoining fibres supplying the abductor muscles are not implicated. This affection usually occurs in young women. It is sometimes observed in children and men, and

4

may occur at any age. It is usually associated with some form of hysteria and sometimes follows violent inflammation of the larynx, and may result from lead or arsenical poisoning. Hysterical aphonia is usually suddenly developed, often the voluntary power of phonation being lost while invol. untary movements, as sneezing and coughing, are normal. Some recover promptly, others are obstinate, remaining for months or years. In treatment, sprays, pigments, and powders are beneficial, especially when there is catarrh. Sometimes tonics, iron, quinine, and strychnine are advisable. Dr. Ingals gives the latter in large doses. He advises the use of the static current.

Ballet and Tissier contribute (*Arch. de Neurol.*, No. 58) an article on *Stuttering in Hysteria*. They conclude: (1) That a transitory speech disturbance resembling ordinary stuttering is sometimes observed in hysteria. (2) So far it has only been observed in men. (3) It usually follows aphasia (hysterical). (4) It persists during several weeks or months and is variable in degree. (5) It is accompanied by difficulty in the tongue movements, such as paresis, deviation, and trembling.

Investigations of Gilles de la Tourette and Cathelineau *(Progres Med.*, March 1st), on the urine of hysterics show that there is a diminution of the fixed residue and of the phosphates. The relation between the earthy and alkaline phosphates, which is usually 1 to 3, becomes 1 to 2 or 1 to 1. The quantity of urine is diminished though micturition is frequent. The weight of the patient is reduced 200 to 300 grammes daily. In epilepsy there is an increase in the solid constituents of the urine and this difference may be useful in diagnosis.

ON LEAD PALSY.

It has long been a debatable question whether the primary changes in this disease occur in the muscles, nerves or the central nervous system. Prof. Eichhorst has lately had the opportunity of examining the nerves and muscles of a case.

The muscles were found absolutely normal, as well as the spinal cord. In the radial nerve numerous aggregations of nuclei were found, which were always situated within the sheath of Schwann. The axis cylinder had either totally disappeared or was represented by detritus.

The vessels of the epi—, peri—, and endo-neurium showed marked thickening of their walls.

Prof. Eichhorst proposes to make further investigations to determine whether the atrophic neuritis or the vessel change is the primary one. (*Virchow's Arch.*, Bd. 120, Heft 2.)

ON NEURITIC PARALYSIS OCCURRING IN THE COURSE OF DIABETES MELITUS.

The occurrence of nervous disturbances in diabetes has been observed for a long time, although Leyden was the first to refer them to a peripheral origin.

The author relates four cases, all of which involved the lower extremity, usually the crural nerve was the one involved.   There were typical symptoms of neuritis, with wasting and the reaction of degeneration.

All improved under an anti-diabetic diet, the improvement of the paralysis and pain, keeping pace with the disappearance of the sugar.   While the process was receding on one side, the other side might become involved.

The prognosis is, on the whole, favorable.

In all cases of neuritis, of obscure origin, the urine should be examined for sugar.—(Dr. Ludwig Bruns, *Berlin. Klin. Wochenschrift*, No. 23, 1890.)

———

DIABETIC PARAPLEGIA.

(Arch. de Neu., No. 57.)

Charcot, in a clinical lecture on this subject says that the nerve lesions in diabetes may be either vascular in their origin or due to the toxic effect of sugar in the system or to the want of liquids.   Diabetic coma is probably the result of auto-intoxication.   Monoplegias due to local cerebral softening are probably vascular in their inception.   Diabetic paraplegia resembles alcoholic paralysis, except that in the latter disease there are troubles of sensibility and none in the former.   Ataxia and diabetes may coexist.

———

Tillemans reported *(Soc. Biology*, 19 Apr.,) observations on the osseous alterations of neuro-pathological origin in two cases of syringomyelia. Spasmodic paralysis of legs slowly developed, associated with the ordinary symptoms of the disease.   He had a felon of right index and loss of pieces of bone.   Pain developed in hands and arms followed by deformities of the hands.   Ten years subsequently there were sloughing of ends of fingers and sub-luxations of wrists and large sloughs on back.   There was abolition of muscular sense in limbs and trunk with loss of temperature sense in lower limbs.   Knee jerk lost.

A second case reported presented analogous symptoms.

———

A CONTRIBUTION TO THE STUDY OF CHRONIC PROGRESSIVE PARALYSIS OF THE OCULAR MUSCLES, AND THE FINER ANATOMY OF THE BRAIN.

Male, aged 62.   Complained of poor sight for two or three years before he was first seen.   Ptosis, first on right, later on left side.   In June, 1886, he showed symptoms of mental derangement.   One month after, he had an attack of unconsciousness with spasms of extremities.

When first seen, ptosis was worse on right side. Bilateral rigid myosis, almost complete oculomotor and trochlear paralysis, ataxic movements of tongue, tremor of upper and inco-ordination of lower extremities, swaying with closed eyes. Sensation seemed normal except in nasal branch of 5th nerve. Ammonia and essential oil of mustard were barely felt. Left optic disc somewhat discolored. Taste almost abolished. Reflexes exaggerated on left side. Senile dementia four months later, another attack of unconsciousness, with left sided spasm and facial paresis and conjugate deviation of eyes and of head to the right. After spasm ceased, eyes and head turned to left. Later on, somnolence, ulcer of cornea, nystagmus. Death five months after admission. Autopsy showed some adhesion of dura to bone. Gelatinous effusion 1–25 inch thick between dura and pia on left side. Effusion extends over left base. Pia of the convexity turbid and thickened. Arachnoid œdematous. Ventricles filled with fluid, ependyma thickened and granular. Centrum ovale of a pale pink color, studded with many fine points of blood. Some senile atrophy of cortex. Cord normal, with exception of ascending root of trigeminus on left side.

Nucleus of hypoglossus degenerated on both sides. Posterior nuclei of vagus and glossopharyngeus showed diminution of their ganglion cells and great atrophy of nerve fibres. Nuclei of facial and abducens showed fewer ganglion cells on left side. Left ascending trifacial root degenerated throughout its whole extent. Posterior roots of trochlear nerve had disappeared; their nuclei were very much atrophied especially on right side. Roots and nuclei of motor oculi highly atrophic on both sides. Westphal's nuclei were intact. Many ganglion cells showed vacuolation.

Capillaries were distended with blood, especially in the gray matter, throughout the entire brain axis. Numerous punctiform hemorrhages in the floor and walls of the third ventricle. Right " bundle from the pes. to tegmentum " was only one third the size of the left. Interstitial neuritis (slight) of left optic nerve.

Wernicke was the first to apply the term poliencephalïtis to the pathological process in this affection. He distinguished poliencephalitis hemor_ rhagica acuta and chronica, and with regard to location. P. superior, involving the roots of the 3rd to 6th nerves, or ophthalmoplegia nuclearis, and P. inferior, involving the 7th to 12th nerves with bulbar symptoms. Six cases of poliomyelitis superior hemorrhagica acuta have been reported. In five, alcoholism is given as a cause, in one, poisoning by sulphuric acid, death ensuing in from 10 to 20 days. Autopsy showed in all, numerous punctiform hemorrhages in the 3rd and 4th ventricles and aqueduct of Sylvius. In Thomsen's case, which lasted 20 days, commencing degeneration of the nuclei was found, being most advanced in the abducens and hypoglossus, less in the motor oculi, least in the trochlear nerve,

Gayet has reported a subacute case following severe fright.

At the autopsy, a general grayish discoloration of the brain axis, much congestion and a large number of miliary hemorrhages were fouud, extending from the roots of the third pair to the point of the calamus scriptorius. Both thalami were involved.

The chronic progressive form was first observed by A. Von Graefe, who referred the symptoms to a cerebral tumor.

Including the present case, 25 cases with 9 autopsies have been published.

In 5 cases, tabetic symptoms preceded or accompanied the paralysis of the eye muscles. In Eisenlohr's case, although it had lasted over two years, the only change found was hyperæmia around the abducens nuclei.

In general, the changes found consisted in atrophy and degeneration of the nuclei and roots, of greater or lesser extent and intensity, although some intact ganglion cells were found in all cases.

In one case, the trouble was due to disseminated sclerosis, throughout the brain, medulla, and cord.

Mauthner, in his work on ocular paralysis, assigns the following causes, besides tumor, for a nuclear paralysis; ependymitis with secondary degeneration of the gray floor of the ventricle, diffuse sclerosis, multiple sclerosis, and that process which causes atrophy of the ganglion cells.

The first two have not been observed clinically, but he deduces their possibility from analogy with tabes with bulbar symptoms.

Wernicke, Strümpell and others believe the degeneration of the ganglion cells to be primary. Benedikt regards it as secondary to an inflammatory process which disappears if life is sufficiently prolonged.

In only 3 cases was hyperæmia of the nuclei and their vicinity present.

The author is inclined to occupy a position intermediate to these authors, believing that inflammation was present at least in some cases.

Hutchinson, who has reported 17 cases, regards syphilis as the chief etiological factor.

In the author's case, although there was almost complete oculomotor and trochlear paralysis, a comparatively large number of intact ganglion cells was found at the spot where the nuclei of these two nerves adjoin. He throws out the suggestion that this may be the nucleus for the ocular branches of the facial nerve. Mendel and others have found a similar collection of cells atrophied after destruction of the ocular branches of the facial nerve in animals.

To Westphal's nuclei (median and lateral nuclei) he assigns the innervation of the internal ocular muscles. The spastic myosis he regards as due to irritation of these nuclei. Mental disturbance was present in 4 cases.

The author is in doubt whether the deviation of the head and eyes, observed before death was of paralytic or of irritative nature.

The author adds some remarks on the composition and course of the "solitary fasciculus." He states that his sections verify Rollers account of its

course.   He believes that in at least three-fourths of its bulk it represents an
ascending root of the glossopharyngeal nerve.   The dorso-median quadrant
however, has different functions.   Its fibres are much finer.   It furnishes
some fibres to the root of the vagus and then runs toward the cerebrum,
furnishing numerous fibres to the sensory root of the trifacial.   Its further
course is not clear.—(Dr. A. Bœttiger, *Archiv. f. Psychiatrie*, Band XXI,
Heft 2).

OCULAR TROUBLES IN MULTIPLE SCLEROSIS.

The investigations of the author were made upon 100 cases of undoubted
multiple sclerosis.   Exhaustive microscopic examination of the optic nerve
in 5 cases lead the author to the following conclusions: the changes observed
in the optic nerve in multiple sclerosis are in many respects peculiar and
different from other degenerative affections of the same part.   The primary
change in multiple sclerosis seems to be a proliferation of the finer reticu-
lum of connective tissue between the individual fibres with increase of
nuclei, which change may secondarily affect the larger fibrous septa and
inner nerve sheath.   The atrophy of the nerve fibres must be considered
secondary.   Degeneration of the myelin sheath is rapid and complete,
while the axis cylinders are often preserved intact.   Vessels are increased
in number and small ones dilated.   Changes in their walls and in perivascular spaces.   The papillæ may be perfectly normal even if decided atrophic
changes exist in the nerve immediately behind them, or they may show partial discoloration.   In the 100 cases there was complete optic nerve atrophy
in 3 cases; twice binocular, once monocular; decided impairment of vision
in all these cases: in 19 cases the temporal sectors were atrophied, inner
ones partly so: in 7 cases both eyes were affected; in 8 cases one eye; in 4
cases atrophy one sided, with fading to the temporal half of the other: in 8
of these cases there was no pronounced disturbance of vision ; partial
atrophic discoloration of the temporal half of the papillæ with normal inner
half existed in 18 cases; 7 times monocular, 11 times binocular; vision
normal in 6 of these cases: optic neuritis 5 times.   Ophthalmoscope showed
no change in 48 cases; 5 of these had visual difficulties and in 1, changes
were found on *post mortem* examination.   The author has found optic atrophy in 18% of all cases of tabes, and in over 8% of all cases of progressive
paralysis examined in various hospitals.   He found contraction of visual
field 24 times; central scotoma with normal periphery absolute 4 times;
relative 9 times.; central scotoma with contraction of periphery twice;
irregular peripheral contraction with relatively intact central vision 5
times in both eyes, 3 times in one.   The degree of central scotoma varied.
In some all colors and white were not seen; in others red and green in various degrees; in others all colors were recognized but were foggy.   The

history of origin of eye troubles was obtained in 22 cases; in 11 its origin was sudden, the course was rapid, trouble attained some severity and soon improved; in the remaining 11 it began gradually in one eye in 4 cases, in both eyes in 7 cases. Improvement of vision occurred in 12 out of 22 cases, and in 2 complete recovery took place. In 4 of these cases the ocular symptoms preceded the disease. The history of the cases shows that the amblyopia of multiple sclerosis may occur in a variety of forms but oftenest resembling retro-bulbar neuritis. There is little or no resemblance between the amblyopia of multiple sclerosis and that of tabes. Central scotoma was not observed once in the 100 cases, while progressive concentric contraction of the visual field was found in 8% of all cases of tabes presenting eye troubles. Hemianopsia was not noticed. Paralysis of the ocular muscles and nystagmous and nystagmic tremor may exist separately or combined. Paralysis of the ocular muscles was never complete and was found in 17 cases in the form of paresis. Pronounced ophthalmoplegia externa was found twice. In all, the pupil reacted normally to light and accommodation, slight contraction upon convergence was barely perceptible in 2 cases, and absent in left eye in the third. In 100 cases of locomotor ataxia from Westphal's clinic ocular paralysis was found in 20%. It was combined with nystagmous in 6 cases. In the 17 cases of sclerosis with ocular paralysis nystagmous was found 13 times. In 100 cases of tabes with eye symptoms, from Schoeler's Eye Clinic, ocular paralysis was found 41 times. Nystagmous was found 12 times, once vertical, in all others horizontal. Nystagmic tremor was found at the end of motions of the eye ball 46 times. In 16 of these there was decided paresis of muscles. Nystagmic tremor is rare in other nervous diseases. It occurs in 12% of cases of tabes as against 46% in sclerosis. True nystagmous was found in only 3 out of 500 cases other than multiple sclerosis examined in nervous clinic. There was no abnormality of pupils in the majority of cases, it being less frequent in multiple sclerosis than in tabes. Various morbid conditions of pupils found in 11 cases.—(Dr. W. Uhthoff, *Archiv. f. Psychiatrie*, Band 21, Heft 1.)

CASE OF MORVAN'S DISEASE WITH AUTOPSY.

By Joffroy and Achard.

(*Archiv. de Med. Experimentale et D'Anatomie Path.*, *July.*)

The authors give an elaborate account of a case that came under their observation, of which the following is a brief resumé. Patient was a woman who died at seventy-three. At thirty had inflammation of tissue of fingers with loss of bones followed by deformities. Sensibility to heat and pain diminished equally in hands and forearms. Tactile sense also diminished. On autopsy there were found "syringomyelic cavities" in cervical

cord, and involving chiefly the posterior horns and slightly the posterior
columns. Blood vessels thickened and in many places obliterated. Anterior roots of spinal nerves degenerated and the nerves of the arms showed
marked degeneracy. The latter was considered secondary to the cord
lesions. In discussing the case the authors say that it is yet difficult to distinguish between syringomyelia and Morvan's disease. They hold that the
accepted view of the preservation of some of the sensory elements and the
loss of others in syringomyelia does not hold good in all cases.

MORVAN'S DISEASE.—Charcot (*Progres. Med.*, Mch. 15), says Morvan, who
first described this disease, called it analgesic paresis with inflammation of
the fingers.

Its chief characteristics are as follows: Sudden access of pain in fingers,
paresis and analgesia of one hand, and deep inflammation of tissue of fingers of the nature of felon. There are blisters on the hand, ulceration and
necrosis of phalanges. Lower extremities seldom affected. Fingers of one
hand are attacked successively and in some cases the other hand is attacked
twenty years afterward. It occurs at any age, is sometimes traumatic and
affects men oftener than women.

Dr. Hughes (*Alien. and Neurologist, July*) writes on *Extra-Neural Nerve
Disease.* His view is, that many forms of nervous disease are due to pathological conditions originally outside of the nerve tissue. There may
be extra-neural as well as intra-neural morbid nervous phenomena.
He quotes the statement of Dr. DeWitt of Cincinnati, that the excitement
in the primary forms of insanity is the cause of the pathological changes
in the cortex and not the result of it. We quote the following: "Thus
we see how much disease of the nervous system may proceed from causes
extra-neural or adneural as distinguished from intra-neural, and how difficult
it is to determine how much or how little of the *post-mortem* appearance
is the cause, as distinguished from the sequence of the morbid action, how
much is indirect and how well we are justified in taking a binocular view
and of employing the direct and indirect designations of intra-neural and
extra-neural nervous disease."

Dr. Landon Carter Gray in his presidential inaugural address before the
"N. Y. Neu. Soc." took for his subject, *Can We Diagnose Hyperæmia or Anæmia
of the Brain or Cord?*" After noting the opinions of numerous authors on
the subject, he says, "After this necessary cursory review we may ask ourselves again, 'Can hyperæmia and anæmia of the brain and cord be diagnosticated?' I, for my part, would answer very positively that the diagnosis is not possible by means of the nerve symptoms alone. In other
words, the symptoms as I have re-narrated them above, are not sufficient of

themselves to warrant a diagnosis of anæmia or hyperæmia of the brain or cord.   If we have evidence of some intra cranial disturbance, such as headache, delirum, vertigo, tinnitus aurium, insomnia, flushing or pallor of the face, and we find we have some concomitant conditions, that would make it reasonable to suppose that there might exist a ccngestion or anæmia of the brain or cord, then we should be warranted in regarding such a diag- nosis as probable."

———

Dr. H. J. Berkely (*Am. Jour. Med. Sci.*, June), publishes a history of two cases of tumor of the corpus callosum.   The first case was an insane man who had exalted delusions, otherwise perfectly healthy.   At autopsy tumor the size of a filbert was found in the posterior part of the corpus callosum; the frontal lobes were exceedingly small.   The atrophy of the frontal lobes was the inexplicable feature of this case.   He refers to a case observed by Dr. Councilman of a tumor which occupied the whole central region of the corpus callosum in which there was distinct degeneracy of the commissural fibres, and which accounts for at least a portion of the atrophy of the fron- tal lobes in the first case.   These cases accord with the view that the cal- losal system binds together identical portions of the hemispheres, for both frontal halves were nearly symmetrically atrophied.   They seem to show that the function of the fibres of the corpus callosum is to connect corres- ponding parts of the hemispheres.

In the *Boston Med. and Surg. Jour.* for April 24, Dr. Dwight has an article on the *Closure of Cranial Sutures as a Sign of Age.*   His examinations include the skulls of 100 paupers.   His conclusions are, that the sutures begin to close much earlier than has been supposed.   It was apparent in several cases under thirty.   The closure of the cranial sutures began on the inside with very few exceptions.   The closure of a suture on the out- side does not always correspond to the closure of the suture on the inside. Sometimes a suture is closed in one part on the inside and in another part on the outside.   The order of closing of the cranial sutures is very uncer- tain, and while for the purpose of determining age the condition of the sutures might be of some use in the hands of an expert they would be abso- lutely worthless to a person not experienced in anatomy.

Humphrey "Roy. Med. Chir. Soc." May 27, directed attention to two changes of an opposite nature occurring in skulls of elderly persons.   One consisted of an increase of thickness and of density, and therefore of an increase of weight, which seemed to be attributed to a shrinking of the brain with consequent lessening of pressure, and increase of congestion of the vessels of the sknll which are supplied from the interior.   This conges- tion leads to bone deposition on the inside.   The other change consisted of atrophy from without, whereas atrophy in other bones proceeds in the main from within.   The skull becomes smaller and lighter and reduced in thick-

ness. This change might be uniform or it might take place in some parts
more than in others. Where it is not uniform the thinning of the skull
produces depressions, and fracture of the skull of old persons sometimes
occurs at this point.

_____

Dr. McEwen (*Brit. Med. Jour.*, Apr. 5) publishes an interesting case of
limited lesion of the spinal cord. The patient was a man who had fallen
fifteen feet and was partially unconscious from the fall. He lay on his
back with both arms abducted to nearly a right angle. Fore arms flexed
on arms, hands lying on chest, wrists flexed, also fingers, with exception of
the index. He had no grasp and could not extend his arms, but when pas-
sively extended could flex the fore arm. He could raise his upper arm by
the shoulder muscles. He could not supinate below arm. Breathing dia-
phragmatic. Intercostal muscles entirely inactive. Power of lower limbs
good. Sensation of the arms was slightly dulled, in fore arm and hand
dullness was increased from outer toward inner side. Cutaneous reflexes
impaired with the exception of planter, deep reflexes absent, sphincters
paralyzed. Pupils in state of stabile myosis, paralysis of light reflexes·.
Slight contraction to accommodation. Fundus normal. Fibrellar twich-
ings all over the body. At times complained of pain in arms. Movement
of left leg at one time impaired. At times temperature subnormal.
Patient improved considerably and was discharged from hospital but had
very imperfect use of arms and hands, the former of which remained
flexed. The lesion was probably one affecting principally the direct columns and did not implicate the lateral crossed columns. It probably involved the lowest two cervical and first two dorsal nerves.

_____

Dr. Walton (*Jour. Ner. and Ment. Dis.*, July) makes a contribution to the
study of traumatic neuro-psychoses. The article is an interesting review of
the literature of the subject and contains a resumè of cases that have been
under the author's observation. He places no value upon rapid pulse in
the diagnosis of these disorders, as he considers it may be due to temporary
excitement of the examination and is known to occur in various functional
nervous disorders independent of traumatism. .The author states that he
has never seen a case in which he felt justified in making a diagnosis of
spinal concussion. We should be extremely careful in attributing locomotor ataxia to traumatic origin. Dr. Walton's views on this subject seem to
coincide in the main with those of Dr. Page, whose work on "Injuries to
the Spine and Spinal Cord" was published in 1883. In the discussion
which followed the reading of this paper before the Neurological Association Dr. C. K. Mills said there were at least three classes of cases resulting
from injury: (1) pure fright; (2) cases where the indications were clear that
fracture or hemmorrhage or other serious lesion had taken place; (3) cases

in which the symptoms presented were both objective and subjective, with a preponderance of the latter. As to the third class, the true explanation seemed to be that there was a real lesion, giving rise to a certain set of symptoms forming a ground work upon which the patients erected for themselves an enormous psychical superstructure. He thought that the existence · of some organic lesion, whether myelitis or the result of hemmorrhagic pressure upon some delicate parts of the nervous system, would explain many of the symptoms peculiar to this class. He has seen patients presenting the same phenomena for months, which it would have been practically impossible that they could have repeatedly enacted upon the same lines. To assume it was an unscientific argument. Fear he did not think could produce the symptoms. Suggestion might do much in certain cases, but in most there were elements which precluded the idea of continuous suggestion.

CONTRIBUTION TO THE STUDY OF TRAUMATIC NEUROSES.

Dr. Jul. Ritter reports 2 cases from Prof. Brieger's clinic which present many points of interest and do not fit into any accepted classfication of this trouble.

The first case was that of a girl who, a fortnight after a trivial injury to 2 fingers of the right hand, complained of pain and weakness in the right arm, which was found to be several centimeters smaller than the other at all points. There was no disturbance of mobility, sensation, or electric reaction.

The second case followed 2 weeks after a glancing blow on the side of the nose. There was found concentric contraction of the visual field on both sides, and almost absolute abolition of smell, taste and hearing on the injured side. ‸ A symptom not noticed in any reported case was the implication of the facial and hypoglossal nerves. Sensation was normal.—*(Berlin. Klin. Wochenschrift*, No. 16, 1890.)

OPHTHALMIC MIGRAINE.—Dr. J. C. DeCosta (*Jour. Ner. and Ment. Dis.* Apr.) reports a case of this disease with remarks. The patient was a woman aged thirty-two. The summary of her symptoms is as follows: attacks preceded for some hours by dull headache, nausea, languor, sometimes numbness, paræsthesia of right arm, and muscular weakness. Occasionally transient aphasia, transient hemianopsia, violent pain in ophthalmic division of fifth nerves, scintillating scotoma, and vomiting.

According to Liveling epilepsy and migraine are nearly related. The attack is due to accumulation of nerve force and to unstable nerve elements,

the accummulated force violently exploding in a storm of pain. According
to this view an explosion in the motor sphere means epilepsy. In the
psychic, epileptic mania. In the sensory, neuralgia.

———

Dr. M. Allen Starr (*N. Y. Med. Jour.*, April 19), read an article before
the "New York Academy of Medicine," on *Peripheral·Irritation in Relation
to Reflex Neuroses*. He said that in the various reflex neuroses, in which
peripheral irritation was the supposed cause of the disease, he was of the
opinion that if there was any actual lesion producing the trouble, nature
would indicate it by calling attention to the seat of the trouble by discom-
fort or pain. He thought that when the source of irritation was not evident,
or only found by extraordinary effort, the chances were that the neurosis
was not produced by any direct peripheral irritation, but was due to
impaired control and defective nutrition of higher centres. He was confi-
dent that the large majority of patients, who presented symptoms of reflex
neuroses, had been exposed to influences which undermined the strength
and nutrition of the nervous system, and that their symptoms were the
result of defective central control. These were to be regarded as suffering
from functional or organic diseases of the most highly developed portion of
the nervous system, and were not merely victims of peripheral irritation.

———

FUNCTIONAL NERVOUS DISEASES IN THEIR RELATION TO GASTRO-INTESTINAL DERANGEMENTS.

By W. H. Thompson, M. D. (*Jour. Nerv. and Ment. Dis., Apr.*)

Dr. Thompson holds that we should not hope to find that what we call
functional nervous diseases are actual faults of structure but are manifestations of disordered sources of nerve energy. He calls attention to the fact
that no structural organic nervous disease is ever truly intermittent, and
therefore no truly intermittent disease can have an organic basis in the
nervous system. In such a case we must look elsewhere for the causation
of the morbid phenomena. In supposed functional nervous disease where
manifestations are continuous, the presumption is strong that it has an organic basis. He seems to hold that functional nervous disorders are due to
auto-infection, or self poisoning, from absorption into the blood either of
septic materials from the alimentary canal or from some other part of the
body. We know that decompositions are constantly occurring in our digestive laboratory which produce definite and virulent poisons, and that serious
illness, and sometimes death, result from their absorption. He quotes Dr.
Hunter as saying in regard to pernicious anæmia that the seat of disintegration is chiefly in the portal circulation, and is due to poisonous agents absorbed from the intestinal tract. Our most important remedies for functional nervous affections perhaps owe their action largely to their power as
disinfectants.

CONTRIBUTION TO THE STUDY OF PERIODIC RESPIRATION AND CHEYNE-STOKES PHENOMENON.—The different varieties of periodic respiration can be observed in an animal in which there has been sub-bulbar section of the cord. We may consider that in an uninjured animal when the phenomenon of Cheyne-Stokes is produced the functional trouble is not limited to a circumscribed part of the bulb but spreads to the whole nervous respiratory mechanism. Second, the Cheyne-Stokes phenomenon has for its cause diminished excitability of these centres; in a pathological case it may be due to troubles of nutrition, or to a dynamic inflammation and an incomplete inhibition of these centres.—By M. E. WERTHEIMER, (*Brown-Sequard's Archives Physiologie*, Jan., 1890.)

The following is a synopsis of the Gulstonian lectures by G. Newton Pitt, M. D., on *Cerebral Lesions (Lancet*, April 5, *et seq.).* His lectures are based on records of fatal cases occurring at Guy's hospital. He first deals with cases of ear disease proving fatal from complications set up in the cranial cavity. The most common complications are cerebral abscess and thrombosis of the sinuses. Out of 57 cases considered, death occurred before thirty years of age in all except 9. There were 18 cases of cerebral abscess and in only 2 of the cases had the otorrhœa existed less than a year. Three of the abscesses were in the cerebellum, 1 in the pons, 2 in the centrum ovale and the remaining 12 in the tempor-spenoidal lobe. The temperature is rarely high in uncomplicated cerebral abscess. In six of the cases it was not above normal. Thrombosis of the lateral sinuses occurred 22 times. When this complication arises followed by acute pulmonary trouble the internal jugular should be ligated in the neck, the lateral sinus opened and the clot scraped out. Abscess is more frequent than meningitis or thrombosis as a complication of ear disease. Agonizing headache usually indicates abscess; rigors, thrombosis of the sinus, and pyrexia, either thrombosis or meningitis. More than half the cases of abscess due to causes other than ear disease resulted from pyæmia. Abscesses due to cranial lesions are usually cortical. He reports examination of 79 cases of cerebral embolism. In 68 cases it was due to heart disease. 75% of the emboli were found in the middle cerebral vessels, the two sides being affected with equal frequency. In 11 cases there were emboli in the arteries of both sides. In regard to cerebral aneurism it was found only 19 times in 9,000 autopsies. In 75% of the cases there was heart disease. The conclusion is justified that septic embolus is a starting point in the changes which result in cerebral aneurism. The clot inflames and disappears, and the vessel dilates from the inflammatory changes in its coats. Cerebral hemmorhage in young people generally results from aneurism.

Dr. Preston (*Alien. and Neu.*, July) writes on Landry's paryalsis. He publishes the history of a case that came under his observation and summarizes the symptoms as follows: " 1. Sudden onset of the paralysis, and its progressive tendency, beginning with the lower extremities and involving nearly the whole body. 2. Perfectly normal electric reactions. 3. Greatly exaggerated deep reflexes. 4. Great diminution of general sensibility. 5. Much impaired muscular sense. 6. Probable involvement of medulla. 7. Absence of muscular atrophy. 8. Recovery, which was probably perfect, as the man left the hospital and presumably returned to his work."

Very little is known of the etiology of the affection. It seems to have no constant pathological lesion. The mind is nearly always clear. A grave prognosis should always be given.

———

AN UNUSUAL CASE OF ATROPHY OF THE SKIN is the title of a paper (*Alien. and Neu.*, July) by Dr. Ohrmann-Dumesnil. The patient was a child who had some years previously sustained a severe burn on the anterior aspect of the right wrist. The right arm presented atrophic spots situated on the anterior surface of the arm and fore arm and over the brachial and radial nerves. These spots were five in number, distinctly depressed, and were paler than the normal skin surrounding them. The skin of these spots was much thinner than in the healthy parts. There was marked atrophy of the muscles of the arm and hand. The doctor attributes the atrophy of the skin to the injury to the radial nerve produced by the burn.

———

Pilliet (*Tribune Med.*, May 15), writes on the *Histological Lesions of Chronic Encephalitis of Infancy*. The lesions of this disease are divided into three groups: 1. Cellular lesions of the gray substance. 2. Lesions of the neuroglia. 3. Lesions of the nerve fibres. In many cases there is adherence of the meninges to the brain, thinning of convolutions with wasting of proper nerve elements. In others there is observed the so-called *disseminated atrophic sclerosis*, which consists in a connective tissue new formation, and which occupies and replaces the superficial layers of the cortex. In lobar sclerosis there may be found simple replacement of normal with the sclerosed tissue. There seems to be a concurrent sclerosis of the vessels of the brain tissue proper. The pyramidal cells are greatly atrophied and in some places disappear.

———

Dr. W. Squire (*Brit. Med. Jour.*), discusses paralyses of childhood, in which he relates briefly the histories of several cases. He closes his article with the following remarks:

"Thus two forms of spinal paralysis occur in infancy, the one, more accidental and infinitely rarer than the other, requires rest, sedatives, and delay in the use of mechanical means; the other demands mechanical aid from the earliest period to guard against associated deformities that would otherwise follow, and in this respect differs from diphtheritic paralysis, which may be expected to amend if immediate dangers be guarded against; in both, local stimulation is required, and the employment of tonics, including the use of strychnine, is advantageous.   Excepting the rarer instances where diphtheria leads to lesion of the nerve centre, diphtheritic paralysis is associated with a local degeneration of the small fibres only of nerve and muscle, leaving no permanent disability or deformity of the bodily frame and limbs; but in the anterior poliomyelitis of childhood an entire muscle or group of muscles is cut off from its centre of innervation, and degenerative changes that are irremediable follow, inducing collateral effects, often of the most serious consequence, many of which could be remedied or averted by timely care and treatment."

Dr. Mantle (*Brit. Med. Jour.*), discusses the cause of laryngismus in children.   He says there is no doubt that rachitis predisposes to the attacks as also does deformed chest wall, which interferes with the proper aeration of the blood as pointed out by Dr. Jackson.   It may also be caused by enlarged bronchial glands pressing upon the vagi.   The doctor calls special attention to another cause of laryngismus, which he says is not infrequent, namely, elongation of the uvula, which produces the spasm by irritation. He relates a case of laryngismus that developed into general convulsions, where removal of the uvula resulted in complete cure.

Sachs and Peterson (*Jour. Nervous and Mental Disease*, May) have a paper on the *Cerebral Palsies of Early Life*, based on the study of one hundred and forty cases.   Diplegias and paraplegias are more likely to be of congenital origin.   Hemiplegia to be acquired within the first three or four years of life.   Epilepsy occurred in forty-four per cent. of their cases. The initial lesions were either hemorrhage, thrombosis, or embolism.   In regard to the polio-encephalitis of Strümpell, they did not believe it was well founded.   Meningial and cortical hemorrhages are more frequent in children; other cerebral hemorrhages more frequent in adults.

PATHOLOGY OF THE BRAIN IN IDIOCY.

Dr. H. Köster contributes to the *Upsala Lakar Forhand, Nos. 6 and 7*, 1890, an elaborate article on this subject.   The chief value of the article is in the collection of the literature of the subject, the literary references alone occupying several pages.   He discusses the size of the idiot brain as compared with the normal brain, the relation of the hemispheres normally and pathologically, abnormalities of lobes, defects of internal structure, and defects of corpus callosum.

He finds that whereas the brains of idiots are smaller than the average of normal brains, yet they are sometimes larger. In regard to the corpus callosum, defects of this structure are observed in idiocy, but it has also been observed that people possessing considerable intelligence have had this part defective.

Though giving a resumé of the literature of the subject, Prof. Köster's conclusions are chiefly based on his own investigations. He finds that a pretty constant pathological condition of the brain of idiots is an abnormal growth of neuroglia, replacing healthy structure. Dilatation of perivascular and pericellular lymph spaces are also found, but he is not certain that this is an anti-mortem change. There is also found marked atrophy of the brain cells with pigmentation, and vacuolation of the same. The most constant and marked pathological conditions are irregular position of pyramidal cells and thickening of the vessel walls. Prof. Köster considers that our knowledge of the pathology of idiocy is very incomplete. His article is illustrated and refers to the works of 151 authors.

Dr. Ashby *(Brit. Med. Jour.)* writes on *Paralysis Occuring During the First Two Years of Life.* He classes them as follows. First, intra-uterine lesions (meningo-encephalitis); second, meningial hemmorrhage; third, syphilitic arteritis and softening; fourth, acute cerebral paralysis (encephalitis and embolism, etc.); fifth, acute spinal paralysis (anterior polio-myelitis); sixth, peripheral paralysis. He calls attention to the fact that asphyxia is a frequent cause of meningeal hemmorrhage during early life, and is also a not infrequent cause of such hemmorrhages in protracted labor. The hemmorrhage is usually bilateral, most commonly involving the parietal region. In some cases an extended meningial hemmorrhage occurs without any definite symptoms, this being probably due to the uudeveloped condition of the cortical centres. When the child begins to walk there is then found some difficulty. Acute cerebral paralysis sometimes occurs during an attack of acute illness, the immediate pathological condition being either meningeal hemmorrhage or embolism of the middle cerebral artery. In some cases there is also thrombosis of the arteries. In other cases after an attack of socalled brain fever the child is left permanently deaf. Strümpell has suggested that this condition is perhaps due to inflammation of the gray matter of the cortex.

Macnamara *(British Medical Journal)* reports several cases of malarial neuritis and neuro-retinitis. The first case was a man who during an attack of ague suddenly became almost totally blind. On examination the optic disks were found almost completely obscured by effusion which extended into the retina and the retinal veins were tortuous and congested. Treatment with quinine relieved the trouble. The second case was a man who had suffered from frequent attacks of intermittent fever for five years and who was found to have neuro-retinitis of both eyes. Full doses of arsenic combined with strychnine were ordered and he was sent to a bracing climate where he soon recovered. The third was the case of a lad ten years of age who during an attack of intermittent fever became completely blind. The optic papillæ of both eyes were obscured by effusion, retinal veins were distended and serious effusions extended into the retina. Treatment with arsenic and strychnine soon relieved the ague and he entirely recovered his sight. The fourth was the case of a man, who, during an attack of intermittent fever, became almost entirely blind. The muscles supplied by the left ulnar nerve were also paralyzed. Under appropriate treatment he entirely recovered. All these persons were residents of India at the time of attack.

Mr. Lawford read a paper on *Optic Nerve Atrophy in Smokers* before "The Ophthalmological Society of the United Kingdom," May 1. The paper was based upon the history of several cases of atrophy in which the symptoms in the earlier stages resembled those of toxic amblyopia. The patients were men and smokers. Discontinuance of the use of tobacco led to no improvement, sight becoming progressively worse. There was gradual failure of vision with central negative scotomata for form and color. Mr. Adam Frost said he looked upon these as cases of tobacco neuritis in which atrophy supervened. He was of the opinion that if vision failed beyond a certain point in tobacco amblyopia recovery did not take place.

Mr. Bowlby recently read a paper before the "Roy. Med. and Chir. Soc." on the *Condition of the Reflexes in Cases of Injury to the Spinal Cord.* In 11 cases of spinal injury the limbs were flaccid and tendon reflexes were abolished and there was no rigidity. In most of the cases the spinal reflexes were also lost, but this was not the case in all, and in some they soon returned. In cases of spinal cord injury, where division is not complete, the reflexes are increased. In all cases of complete crushing of the spinal cord the reflexes were lost. As spinal cords so injured are incapable of repair, surgical interference will only be of use when the deep reflexes are not abolished. Dr. Bastian claimed that these facts supported his assertion that the deep reflexes are abolished when there is complete transverse destruction of the cord.

Dr. H. C. Bastion read a paper before the "Roy. Med. and Chir. Soc." on *Transverse Lesions of the Spinal Cord with Special Reference to the Reflexes.* He claimed that in total transverse lesions of the lower cervical or upper dorsal region that the deep reflexes were totally abolished. He insisted that the loss of sensibility must be absolute as well as the loss of voluntary power in order to bring about the abolition of reflexes. The

5

subject is of great interest in view of the importance of obtaining a correct
symptomatology for total transverse lesion of the cord, and has important
bearing on some other points in the diagnosis of diseases of the cord, and
because it tends to throw light upon the origin of rigidities and exaggerated
reflexes, and seems also to cast a side light upon one of the functions of
the cerebellum. He claimed that the abolition of reflexes in these cases
is due to the simultaneous cutting off from the lumbar region of the cord
the influence not only of the cerebrum but also of the cerebellum.

ON PARAMYOCLONUS MULTIPLEX.

Patient 29 years old; was born asphyxiated. The trouble was noticed
when he began to walk and talk. There were clonic spasms of a number
of symmetrical muscles of the face and extremities, which ceased during
voluntary efforts and sleep, and were aggravated by mental excitement·
Sensation and the motor efficiency, as well as the electric irritability were
intact. The knee jerk was exaggerated.

This case is the 29th reported.

Only one case (Friedreich's original case) has been submitted to autopsy,
with a negative result.

Friedreich assumes an abnormal excitability of the motor ganglion cells
in anterior horns. If the face is implicated the nuclei in the medulla share
this excitability.—DR. ERICH PEIPER, *Deutsche Medizin Wochenschrift*, No.
19, 1890.

Dr. George Herschel exhibited to the London Medical Society, January
27, cases of Thomsen's disease in two brothers, aged eighteen and twenty-
eight respectively, and he said that another brother had also been affected.
Dr. Herschel said that Erb's statements respecting the electrical reactions in
this disease were not correct.

Cook and Sweeten (*Brit. Med. Jour.*) report a case of Thomsen's disease
in a lad of nineteen. The father of this patient and the father's second
cousin, and the patient's sister have all had the disease. The boy seems to
have had the affection from childhood. The eyes were unaffected. The
muscles of the mouth were involved and the mouth could not be properly
opened. In the case of his father the eyes were affected.

Dr. A. B. Ball writes ("*Trans. of Assn. of Amer. Physicians*," Vol. 4)
on *Thrombosis of the Cerebral Sinuses and Veins*. In regard to the influence of the arrest of the blood current in producing thrombosis, it is to
be regarded as a secondary element. The fundamental conditions are to be
found in vessel lesions and in changes in the composition of the blood.
Lesions of the inner walls of vessels are probably essential to the throm-

botic process and it is quite possible that the initial step in every case is a radical alteration of the endothelium. It is also probable that morbid states of the blood may have a share in the production of thrombosis. Cerebral thrombosis may be divided into two classes; the first is secondary to the thrombotic process, starting outside of the intra-cranial circulation. They usually have their starting point in caries of the petrous portion of the temporal bone, or in caries of any part of the cranium, or purulent inflammation, etc. The second group comprises what has been called marantic thrombosis, or non-inflammatory form. This form of thrombosis usually occurs by preference in the superior longitudinal sinus or in the cortical veins which empty into it. It frequently occurs in infants, also in the later stages of phthisis after profuse hemorrhages, and in chronic cachexia of any kind.

Dr. S. Weir Mitchell (*Trans. Assn. of Am. Physicians, Vol. 4*), reports a number of cases of subjective false sensations of cold. This peculiar sensation in one case followed injury; in other cases the causation was obscure, but the doctor believed it was of neurotic origin. In one case the sensation was confined to one side of the body, in other cases to the buttocks, and in others to the feet and various parts of the body. In some cases the feeling of cold was accompanied by an actual rise of temperature of the part. One rather remarkable case he reports, was that of a lawyer fifty-seven years of age, who had had scarlet fever at seventeen, with renal complications. Albumen and casts have been found in his urine since that date, although he has enjoyed excellent health. During the last few years he has been annoyed by subjective sensations of cold, which are increased by mental application. To relieve these sensations he wears three suits of the heaviest kind of woolen underwear, three pair of the heaviest woolen socks, felt boots of the heaviest material over his ordinary boots and shoes, and a flannel bandage around his body. At night he wears two of the above mentioned suits, a flannel bandage, woolen socks, and sleeps under five woolen blankets on a feather mattress, with a hair one under it. He always keeps the night temperature of his room 80, and after a hard day at court from 90 to 95. The sensation of cold is positively painful, though his surface temperature is normal.

Dr. Beverly Robinson (*N. Y. Med. Rec.*, April 19,) publishes an article on the *Relations of Peripheral Irritation to Disease as Manifested in the Throat and Nose.* He states that during the past few years specialists in laryngology have given too much importance to irritations of the throat and nasal passages in the causation of disease. In regard to hay fever and hay asthma he says, ''Can those diseases be cured frequently by the removal of nasal obstructions?'' and answers certainly and unhesitatingly, ''No.'' He says peripheral irritation in the nose is without much doubt a predisposing cause of asthma in a certain proportion of cases. In a very large percentage it should be regarded as a direct and efficient cause of this disease. Asthma may of course be occasioned by various peripheral irritations, much as a disordered stomach or a decayed tooth, and likewise an irritated or obstructed nose may produce it. He says he has seen many cases in which the removal of mucous polypi and large turbinated bones had seemed to result in the cure of asthma for a time, but it returned later.

Bocq and Marinesco (*Arch. de Neurol.* No. 57) write on the pathological anatomy of Friedreich's disease. They conclude that the disease is due to a morbid heredity exhibited in degeneracy of certain regions of the cord, and due to a vascular disease primarily. The posterior columns and pyramidal tracts being later in development, and therefore more differentiated, more readily suffer from a defect of nutrition which the vascular disease produces.

L. MINOR, OF MOSCOW, IN *Archives de Neurology*, No. 50, ON THE CAUSES OF LOCOMOTOR ATAXIA.—L. Minor reports 8 cases of locomotor ataxia in which there was a history of syphilis in 7 and its probable existence in the other one. In estimating the frequency of nervous diseases among Russians and Jews he found that while Jews have a larger amount of nervous disease than Russians, there is not among them so much locomotor ataxia and general paralysis. In 383 cases of nervous disease, 260 were Jews and 123 were Russians, which showed the relative frequency. Among Russians syphilis occurs four times more frequently than among Jews. Ataxia and progressive paralysis are also five times as frequent. Prof. Kojewnikoff furnished Dr. Minor with the result of his investigations extending over a period of three years, and including 2,403 patients, of whom 347 were Jews. There were 67 cases of ataxia, and of these at least 60% were syphilitic and not one Jew among them. He had 31 cases of progressive paralysis in which previous history was reliable. In these there was a syphilitic history in at least 45%, and it is at least very probable in 60%. Among the cases of progressive paralysis there were 3 Jews. It is seen that in ataxia and progressive paralysis syphilis is met with in 60% of the cases. It is also seen that while among 2,056 Russians there was found a large number of cases of ataxia and progressive paralysis, among the 347 Jews there was no ataxia and only 3 cases of progressive paralysis. In statistics furnished by Dr. Korsakoff, which included 2,610 patients, there were 89 Jews. Among these cases there were 66 of ataxia, 4 of whom were Jews. 70% had had syphilis, among the 4 Jews 3 had syphilis. Sixty-nine of these cases had progressive paralysis, 1 being a Jew. 72%

were syphilitic. Combining the statistics furnished by these physicians
we have the following:

4,700 Russians of whom 2.90% had ataxia.
696 Jews of whom .8% had ataxia.
4,700 Russians of whom 2.6% had general paralysis.
696 Jews of whom .8% had progressive paralysis. .

Dr. Vermel (*Progres Med.*, Feb. 22) writing on the relation of paresis
and ataxia to syphilis, objects to the use of mercury in the two former diseases claiming that it tends to aggravate the pathological condition. Mendel by placing dogs on a rapidly rotating table with their heads towards the
circumference produced artificial general paresis. These dogs had the
dementia, paralysis of extremities, convulsion of facial muscles with paralysis, alteration of voice, and at autopsy there were found characteristic
alterations of meninges and cortex. By reversing the position of the dogs
anæmia of the brain was produced and no symptoms of paresis.

In certain dogs that were under treatment with injections of mercury it
was found that the symptoms of paresis were much more easily developed
than in dogs not under treatment.

SYMPTOMATOLOGY OF TABES, WITH SPECIAL REFERENCE
TO EAR, PHARYNX, AND LARYNX.

The author has not found in literature any reference to the exact and
exhaustive examination of the ear in this disease. Most authors state that
partial or total deafness and labarynthine vertigo may occur. He made
minute examinations of the ear in 40 consecutive cases of tabes with the
following results: both ears normal in 7, positive bilateral affection of inner
ear in 15, suspected affection of both in 4, suspected affection of 1 in 2,
beginning affection of internal ear on both sides in 1, beginning affection of
internal ear on both sides suspected in 2, beginning affection of internal
ear on 1 side in 1, affection of internal and middle ear in 5, pure disease
of middle ear in 4.* The majority of the patients did not complain of
subjective noises or of deafness. He is not able to determine whether the
symptoms are due to a real affection of the internal ear or to a neuritis of
the auditory nerve. In 8 out of 11 cases there was hyperæsthesia to the
electrical current. Pharynx and larynx examined in 36 cases and in no
case were they normal. Anæsthesia or hyperæsthesia of the velum was
found 30 times, sensibility of the pharynx reduced in 14 cases, of the larynx
in 9, epiglottis deeply depressed in 5, diminished pharyngeal reflexes in 1
case, in 4 cases increased reflexes with hyperæsthesia, paresis of adductors
of vocal cords found 10 times, absolutely immovable 4 times, decreased
abduction of vocal cords found in 8 cases, ataxia of tongue found in 9 cases.
Pupillary symptoms observed in 92 cases as follows: absolute immobility

---

*The above table makes 41 though but 40 are mentioned in the text.

found 41 times, Argyle-Robertson symptom found 45 times, inequality found 24 times, myosis found 39 times, mydriasis found 6 times, ptosis found 6 times, paralysis of ocular muscles 12 times, optic nerve atrophy 9 times, paradoxical pupil reaction once, 1 case had exophthalmic goitre which had developed 10 years before the ataxia.

THERAPEUTICS.—Nitrate of silver has given best results in the author's hands. Iodide of potassium,-strychnine and ergotin are of little value. Antipyrine, antefibrin, methacetin and salycilate of soda stop pains in early stages, mitigate them in later. Galvanism of great benefit, cold water treatment of value in beginning, harmful at later stage.—DR. A. MARINA, (*Archiv. f. Psychiatrie*, Band 21, Heft 1.)

Huet (*Annales Medico-Psychologiques*, Jul. and Aug.), in contesting for the Moreau prize, took for his subject "Chronic Chorea." He proposes to ascertain if chronic chorea due to heredity, may be distinguished from chronic chorea without heredity, and if it constitutes a distinct morbid form. He comes to the conclusion that it is only a variety of chronic chorea. He concludes that there exists, principally in adults and old people, and occurring sometimes, but seldom, in adolescence, a chronic form of chorea. It is slowly progressive, and is usually accompanied with loss of mental vigor. A nervous heredity is a strong etiological factor. The motor troubles are much the same as in Sydenham's chorea, though sometimes the movements are slower. The prognosis is grave. (This essay received honorable mention by the committee.)

M. Raymond (*Hospt. Med. Society*) gave the history of a case of chorea, followed by paralysis and atrophy of muscles. In this disease paralysis is rare and atrophy still more rare. It is probable that the condition of nerve cells that produce chorea is responsible for the atrophy. M. Ollivier said he divided choreics into two classes, rheumatic and nervous. In the first are heart disease and trophic troubles, in the other, not. The trophic troubles are to be attributed to rheumatism.

CASE OF CHOREA ATTENDED WITH MULTIPLE NEURITIS.

This was the case of a girl who had had successive attacks of chorea. With the third attack she had multiple neuritis, involving all the limbs. Location of degeneration in hands, feet, arms, legs, and thighs. Muscles atrophied, reflexes gone. Muscles of face, neck, and trunk not involved. Although the paralysis was profound and extensive, recovery was complete. The neuritis could not be attributed to alcohol, nor to exposure to cold or wet, nor to rheumatism. The brother of this patient had been treated by

Dr. Fry for chorea, and he suggests that possibly the extensive multiple neuritis in the first case was due to an infectious cause.—DR. F. R. FRY, (*Jour. Nervous and Ment. Dis.*, June).

THERAPEUTICS.

Dr. E. C. Seguin (*N. Y. Med. Jour.*, *May 10, et. seq.*) has delivered some lectures to the Medical Society of the University of Toronto, on the treatment and management of neuroses.   The following is a brief summary: He divides epilepsy into two groups, organic and idiopathic.   In regard to the treatment of idiopathic epilepsy, he is a pessimist as to its curability and has not yet published any case as cured.   Concerning the treatment of epilepsy by bromides, it sometimes requires one or two months of experimentation to find the right dose for a given patient.   Small children bear larger doses of bromide proportionally than adults.   Many epileptic children from two to six years need from forty to sixty grains a day to arrest the attacks.   There is a certain proportion in adults between the size and weight of the patient and his capacity for resisting drugs.   The existence of organic heart disease, or of a feeble heart with a sluggish relaxed state of the circulation, generally decreases the ability to withstand bromides. Hence the necessity of examining the patient's heart and arteries and of occasionally combining digitalis and bromides.   Acne should never serve as a guide to the doses of bromide, as its occurrence depends upon the peculiarities of the patient.   In the treatment of idiopathic epilepsy the aim should be to keep up a slight degree of bromism and this requires extreme care in the first doses and in subsequent variations.   It is one of the most delicate tasks in medicine to keep a patient steadily at the point of therapeutic bromism for several years, avoiding truly toxic effects and not allowing the nervous apparatus to re-acquire enough excitability to permit of an attack.   He says there is no difference between the anti-epileptic action of the different bromides and that there is no advantage in combining them.   In regard to the time of administration his first general rule is to give as few doses *per diem* as possible.   Another rule is to give most of the bromide at a time, and four or six hours before the attack should occur, if they recur with any regularity.   Where the fits occur at night he gives but one dose and that in the evening.   In other cases where the fit is liable to occur just before waking the patient should be roused at two or four A. M., and take part or whole of the dose.   The first dose of the day should be given on waking if the stomach will stand it.   In regard to the uniformity of dosage from day to day and week to week, in a few cases of *grand mal* in which attacks occur only at night, when the proper dose has once been discovered it is not necessary to make any change for months or even years. In cases where the attacks occur at intervals, which may be quasi-regular, as pre-menstrual, the daily dose may be increased just before the dangerous period and then decreased in a few days. The reason for increasing the dose may be increasing age and size of young patients, particularly the approach of the menstrual function, the exposure of the patient to unusual excitement or fatigue. When the patient has been three years without any manifestation of the disease, systematic reduction of the dose may begin. The seasons of the year require some variation in the dosage. Larger doses will be borne in autumn and winter. During the attacks of temporary illness the dosage should be reduced. In some cases of epilepsy he combines chloral with bromide in the proportion of five grains of chloral to ten grains of bromide of sodium. Good results are obtained from combining with bromide a free use of strychnine and atropine or belladonna, giving usually the sulphate of strychnine dissolved, and diluted nitro-muriatic acid. Digitalis and ergot have seemed to succeed in some cases of *petit mal.* Digitalis, strophanthus and caffein are given when the heart is weak, or peripheral circulation sluggish. Iron and cod-liver oil are useful in some cases of epilepsy. In regard to the treatment of chorea our main stay is arsenic. He advises giving fifteen drops or more of Fowler's solution three times a day. The most important factor in the successful treatment of chorea, especially in its chronic form, is absolute rest, which should be mental as well as physical. In chorea, eye strain is a secondary cause of much importance and some choreics should have ocular defects cured. The true pathological cause or condition of chorea is, however, deeper than ocular defects or phimosis or cardiac disease, that is, in all cases there is a fundamental defect in cerebral power. He says that exercise in the ordinary sense of the word is not beneficial in chorea, but that during convalescence the practice of a few gymnastic movements under a teacher's or parent's guidance, and without anyone else present, may prove an advantage. In regard to migraine, he calls attention to its frequent association with ocular defects, and recommends ophthalmoscopic examination in all cases. For the treatment of this disorder cannabis indica, belladonna, atropine, and hyoscyamine are advised. They do good by reducing the accommodative effort of the eye and relieving the strain.

As extract of cannabis indica is of very uncertain strength either Squibb's or Herring's should be used, as they are reliable. The commencing dose should be ⅙ grain combined with 1-60 of arsenious acid, or with iron or digitallis, according to indications, in pillular form three times a day before meals. Each week the dose of cannabis indica should be increased ⅙ grain until it produces a light headed, semi-drowsy dreamy state. Male patients can usually take a grain three times a day. Women can seldom take more than half that amount. This maximum dose should be kept up many months or a year or longer. Diet intended to correct lithæmia should be adopted, as this is often associated with migraine. During the paroxysm of migraine the regular medicine should be suspended. The two chief remedies for migraine are antipyrine and caffein. The former should be given in a single large dose before the paroxysm comes on, if possible. Fifteen or twenty grains to the female patient, twenty or thirty to the male. Caffein is good in cases having optic aura or premonitory symptoms, hemianopsia, or a hazy vision a few minutes before the pain appears. He does not adopt the usual classification of angio-spastic and angio-paralytic. He holds that in all attacks there is a spasm of the arteries at first and dilatation later. Neither morphine nor opium should ever be given for this affection.

For *trigeminal neuralgia* he recommends Duquesnel's crystallized aconitine. A pill of 1-200 of a grain is given twice a day to women, three times a day to men, gradually increased until two pills every three hours are given and physiological effects are obtained. These are tingling and numbness in the face, tongue, and extremities, with a sense of chilliness, usually marked along the spine. Having found the dose of toleration it should be kept up daily for several weeks after the pain has ceased. With this remedy red iodide of mercury should be given, dose gradually increased from 1-20 to 1-5 or 1-6 of a grain combined with iodide of potassium. As a case of this affection approaches cure there are sometimes spots of exquisite sensitiveness on the face and head, which, if irritated, will produce a return of the neuralgia. Blistering these spots will cause them to disappear. Cod liver oil should also be given in this disease.

For the treatment of *exophthalmic goitre* he also recommends aconitine in doses to produce tingling of the lips and extremities to be continued for days or weeks, occasionally stopping for a few days. This reduces the pulse rate and arterial tension. In some cases bandaging the eyes reduces the ophthalmia. A soft pad of cotton is placed over each eye and a flannel bandage is applied to produce gentle pressure. At first this should only be used for an hour or two twice a day, later for periods of two to four hours. Sometimes it is left on all night.

Concerning the diet and hygiene of nervous patients, he says that lithemia and oxaluria frequently accompany neurasthenia, and that the excessive use of fatty and starchy foods is a potent cause of this condition. In regard to diet for nervous patients generally, they should drink large quantities of water, and as they usually avoid a diet of fatty foods, physicians should insist on their eating pork, fat roast beef, butter, cream, and using cod liver oil. These should be used in abundance and persistently. Sweet and starchy foods should be eaten very sparingly, as they require to be transformed into food by complicated chemical action within the body, while oil, butter, cream, and fat are already prepared for emulsion. Oranges should not be eaten before breakfast as the acid they contain checks the flow of gastric juice. All green foods of the spinach group are of special value. Soups are considered indigestible and often harmful. Stimulants are injurious to neurasthenics. For persons of delicate digestion milk should not be taken with solid food. The notion that animal food is bad in convulsive disorders is a mere fad. Phosphorus, when given, ought to be administered in alcohol and glycerine, or disolved in oil.

The *rest cure* for neurasthenia should never be attempted at home. Stimulants in neuroses are generally objectionable, as they weaken the patient's power of resistance. They should only be prescribed on the clearest indications. Morphine should not be given in hysteria, and should always be given with the greatest caution. Bromides should not be given for insomnia, as they have no tendency to produce sleep. They should not be given in traumatic neuroses, as symptoms of this disease may be aggravated by their use. Dr. Seguin gives a cup of black coffee to dyspeptics in the morning, and believes it a valuable remedy in nervous disorders.

In *The Lancet* of March 8th, is published a lecture by H. C. Bastion, on *Some Points in the Prognosis and Treatment of Cases of Hemiplegia.* He said the cases of hemiplegia might be divided into three groups, those of abrupt onset due to hemorrhage, embolism and thrombosis; those due to abscesses of the brain, hydatid cysts and pachymeningitis, or tumors; and lastly, those cases known as functional or hysterical paralysis. In the functional hemiplegia the face as a rule escapes, and with the motor paralysis there is anæsthesia. The pathology of this form of paralysis is not known; the prognosis is good. The prognosis in cases of sudden onset, due to hemorrhage or vascular effusion, is a many-sided question. One must estimate the depth of the coma, the degree of flaccidity of the limbs, and the amount of stertor. If soon after the onset of apoplexy, the temperature drops to 96, the case is to be considered serious. In all cases of apoplexy the prognosis should be guarded, because at any time there may be a fresh and fatal bleeding. In regard to the treatment the patient should be put in a recumbent position in a cool room, and if there is much heat and throbbing about the head, cold should be applied and hot bottles placed at the feet. If the face is flushed bromide of potassium should be given. Speaking of the difficulty of diagnosis in these cases, he said he had seen apoplexy simulated by thrombosis of the basilar in one case, and of the middle cerebral in another. The treatment of this latter condition should be exactly the reverse of that of apoplexy. The diagnosis being difficult between the two conditions, one should be very cautious in adopting active treatment, as the risk of doing harm is very great. In some of these cases of paralysis, the aphasia which accompanied it was soon recovered from; in other cases it remained for a long time, or even permanently.

He mentioned as an occasional sequence of the apoplectic condition convulsive seizures, which might be either general or confined to the paralytic side. In some cases hallucinations and delusions might follow the paralysis, and sometimes blindness occurs on the same side as the hemiplegia. In regard to prognosis he said if the patient began to improve in five or six days, and the improvement seemed to be progressive, he would probably entirely recover. If the paralysis remained absolute for two or three weeks, the chances were against complete recovery. Nothing should be done for the paralyzed muscles for two or three weeks; after that, faradization and tonics should be used. In regard to rigidity of paralyzed limbs he could recommend no treatment.

SYMPTOMATOLOGY AND THERAPEUTICS OF MIGRAINE.

The author adopts the usual division of this disease into angio-spastic and angio-paralytic. He has observed that a prolonged series of attacks of migraine sometimes precedes severe and incurable nervous disease. He has found that on the affected side the galvanic excitability is greatly increased and electrical resistance greatly diminished. He has also noticed this in other nervous diseases. In a case of epilepsy treated by galvanism he was able to predict an attack by the change of resistance several days before. The reduction of resistance is relative, not absolute. The treatment in the intervals of the attack should be directed first to the relief of the constipation which usually exists. For this he recommends the bitter waters and faradization of the abdomen by means of the author's method of increasing and decreasing currents. An important adjunct is active life in the open air. He has his patients drink several glasses of warm water after walking to promote perspiration and wash out the stomach. Electricity is the chief remedy during the attack as well as during the interval. Some are relieved by galvanism, some by the induced current. He generally tries galvanism first. His method of applying it is as follows: the cathode is placed to the nucha, the anode to the eye of the affected side. After a few moments the latter is moved to the temple, then to the forehead, then to the sound side along the auriculo-maxillary groove across the throat to the temple of the affected side. The current strength is gently increased during the application. This process can be repeated several times, the current each time being somewhat stronger, ending with the current strength that was used in the beginning. The limit of current strength is the endurance of the patient. When a patient improves for a time under the use of one current, and then ceases to gain, the other current should be used. During the attack he uses ergotin in plethoric persons and salycilate of sodium in anæmic.—DR. W. B. NEFTEL, (*Archive f. Psychiatrie*, Band 21, Heft 1.)

"The Rational Treatment of Sciatica" is the subject of a paper by Dr. G. M. Hammond (*Jour. Ner. and Ment. Dis.*, May). He considers the disease an inflammation of the nerve sheath or interstitial tissue. To relieve pain he advises phenacetine in doses of 15 grains and also antipyrine and acetanelide. He gives morphine if pain is severe. For the cure of neuritis he uses rest, cold, and electricity. The continuous current alone should be used. The negative pole should be large and strapped to the foot, and the positive pole should be applied to the nerve where it emerges from the pelvis. It should be applied twice a day.

———

Dr. Haig (*Brit. Med. Jour.*, May 31), writes on the causation of reduced arterial tension by mercury. He calls attention to the diuretic effect of mercury, and that it increases the amount of urea and lowers the arterial tension. He mentions his previous contention that uric acid in the blood increases arterial tension. Various experiments which he has performed, show that calomel reduces the amount of uric acid in the urine and also in the blood, and that it thus lowers arterial tension.

———

Dr. Stockton (*Am. Jour. Med. Sci.*, July), writes on *Clinical Results of Gastric Faradization.* He has devised an electrode which is passed into the stomach. This electrode is used as the positive pole of the battery, the negative pole being applied to the spine. He reports highly satisfactory results from this method of treatment.

———

W. J. Morton (*N. Y. Med. Rec.*, May 31), writes on *The Place of Static Electricity in Medicine.* He antagonizes the view of Dr. Starr that the static current is superficial, not going deeper than the skin. He claims that as it produces muscular contractions which are of sufficieut energy to lift bodies of several pounds weight, it must penetrate to the deeper tissues. There is no form of current that penetrates the tissues more deeply than the static, and in proper hands there is no form of electricity equal to it in curative effect.

———

Semple reports (*Lancet*, June 14) a case of chorea treated with arsenic followed by pigmentation of the skin and paralysis of the legs. Knee jerks were absent and the reaction of degeneration was very marked with atrophy of the leg muscles. The patient recovered in about two months. The pigmentation and paralysis were attributed to arsenic.

———

Dr. A. Dobisch, of Zwittau, (*Lancet*, June 14) has used for the purpose of procuring local anaesthesia, a spray with Dr. Richardson's spray apparatus, composed of ten parts of chloroform, fifteen parts of sulphuric ether, and one part of menthol. After one minute's application of this spray complete anæsthesia of the skin and neighboring tissues is obtained, which lasts for from two to six minutes, and suffices for the performance of such minor operations as opening abcesses of the cervical glands, incising a deeply seated whitlow, and the excision of an epithelioma of the nose.

Dr. Mortimer Granville gives iodine in gout and is of the beilef that it decomposes the urates, bringing away iodide of sodium and leaving uric acid in an unirritating condition. He either gives the tincture or iodoform or hydriodic acid.

———

Dr. Collier (*Lancet*, June 21) reports a case of Spasmodic Torticollis treated by nerve ligature followed by complete and permanent recovery. He placed a loop of silver wire around the spinal accessory nerve and twisted it so as to produce slight compression. The wound closed over the wire suture which was allowed to remain.

———

Dr. Gordon (*Brit. Med. Jour.*, Mar. 29), publishes a contribution to the study of sulphonal. The following are his conclusions:

1. It reduced the excitability of the reflex function of the spinal cord.

2. It diminished peripheral sensation.

3. Clinical observations showed that large doses slowed respiratory acts.

4. It did not affect the pulse rate.

5. That it slowly destroyed the conductivity of motor nerves, and that subsequent washing with salt solution, tended to restore it.

6. That small doses, 5 to 10 grains, increased excretion of urea.

7. That large doses diminished excretion of urea.

8. That under the influence of the drug the excretion of phosphates was diminished.

9. That it had no marked influence on the excretion of the fluid constituents of urine.

10. That in none of these experiments was any cutaneous eruption observed.

11. There was no flushing or perspiration markedly observed.

12. That its influence on temperature was negative.

13. That it occasionally caused vomiting, but there was no marked loss of appetite.

14. That diarrhœa was noticed occasionally.

15. That in good health a hypnotic effect was distinctly produced.

16. That in cases of insomnia it was reliable.

17. That the sleep that followed its administration was generally tranquil and refreshing.

18. That sometimes the patient awoke with a feeling of confusion.

19. That inco-ordination of the upper extremities occurred occasionally.

20. That inco-ordination of the lower extremities occurred occasionally.
21. That a feeling of depression occasionally supervened.
22. That giddiness was observed.

———

Dr. Jastrowitz spoke on the treatment of insomnia at the recent meeting of the Berlin Medical Society. He said that in senile insomnia opium acted better than morphine and that he also used it in delirium tremens. He regarded chloral hydrate as the most powerful and certain of all hypnotics. Sulphonal had the advantage of not affecting the heart, though its action is too slow to be suitable for acute affections. He said that it was especially advantageous in conditions of motor restlessness. In mania he advises giving from seven and one-half to fifteen grains once during the day and a second dose at night.

———

At the 7th of June meeting of the "Soc. of Biology," Mairet and Bosc presented a communication on the physiological action of chloralamide. Until now this remedy has been said to have no action upon the heart. These authors say it lowers arterial tension and predisposes to congestions. Otherwise their conclusions coincide with those of other observers.

———

PREVENTION OF THE TOXIC EFFECTS OF COCAINE.

Dr. Gluck (*N. Y. Med. Rec.*, June) writes on this subject. He uses a combination of cocaine and carbolic acid, dissolving two drops of the acid in one dram of distilled water, and then adding ten grains of cocaine hydrochlorate. Carbolic acid is itself an anæsthetic, and as it coagulates albumen and forms a superficial eshcar, it prevents the rapid absorption of cocaine. He claims that the anæsthetic effects of cocaine are increased by this mixture, that its toxic effects are prevented, and that the solution is rendered aseptic.

———

ON THE ANALGESIC ACTION OF METHYLENE BLUE.

Prof. Ehrlich and Dr. Leppman have used this chemical in doses of $1\frac{1}{2}$ to $7\frac{1}{2}$ grains (15 grains per day). Hypodermically, they gave up to $1\frac{1}{4}$ grains per dose. They report excellent effects in all neuritic process, and rheumatic affections of muscles, joints and tendons.

Its action begins about 2 hours after administration, and gradually increases in intensity to a total relief of pain. It has no action on pulse or respiration. It acted well in 2 cases of angiospastic migraine, relieving one, curing the other. It had no effect in ulcer of the stomach, syphilitic headache, in states of excitement or in insomnia.

It is highly important that the preparation used is free from chloride of zinc or other impurities.—(*Deutsche Medizin Wochenschrift*, No. 23, 1890.)

ON HYPNAL.

Hypnal is a colorless and tasteless crystalline body, produced by mixing solutions of chloral and antipyrine. It coniains 53 % of the former and 45 % of the latter.

Bardet has found its action as a hypnotic, to be stronger than that of an equivalent quantity of chloral. It is free from all the after effects of chloral and Bardet has used it with success as an anodyne and hypnotic in neuralgias in children, as well as in insomnia. Dose is not stated.—(*Wiener Medizin Blaetter*, No. 15, 1890.)

ON A CASE OF GRAVE'S DISEASE TREATED BY STRUMECTOMY.

Patient aged 24; symptoms observed first 1½ years before; pulse 140–160. Eyes very prominent. Moderate goitre, tremor and insomnia, etc. Removal of all, but part of left lobe of gland; 45 days after the subjective sensations had disappeared; pulse steady 70 to 80. Exopthalmos was less.— (*Kuemmel Deutsche Medizin Wochenschrift*, No. 20, 1890.)

ON CODEIN

Dr. M. Lœwenmeyer reports as the result of extensive trials. Codein is similar in its action, though weaker than morphia. It has no bad incidental effects.

In functional visceral neuroses, codein will replace morphia, but not in organic troubles.

It cannot replace morphia in the pains due to organic affections of the brain, cord, or peripheral nerves.

Its hypnotic effect is stronger than its anodyne, and can be beneficially utilized in conditions of excitement, in insomnia, and in the treatment of the morphine habit.

The dose is from ½ to ¾ grain. If the latter quantity is in action a larger dose will probably be found to have no effect. The initial dose does not need increasing.—(*Deutsche Medizin Wochenschrift*, No. 20, 1890.)

Russell and Taylor (*Lancet*, May 17) give their results of treatment of eyilepsy with borax. Their usual dose is thirty grains in twenty-four hours. They have found that this remedy succeeds when bromide fails. Some of the unfavorable results of the administration of the remedy are as follows: A vesicular eruption on the lips, slight exema upon the mouth, and in one case erythematous condition of the hand. It sometimes produces disturbance of the stomach and emaciation.

Dr. Stewart (*Lancet*, Apr. 26), reports a number of cases of nocturnal epilepsy treated by borax. He concludes that this remedy has a peculiar influence over nocturnal seizures, and when the fits occur both during the day and night a combination of bromide and borax promises the best results.

———

Reynolds (*Lancet*, May 17) makes a contribution to the *Clinical History of Exophthalmic Goitre.* Forty-eight of the reported cases were women, and one a man. In one case there was no goitre. His principal treatment is the administration of iodine, bromine, and iron in combination. Other remedies are used for special indications. The results of treatment have been highly satisfactory.

———

Dr. Christian (*Bul. Gen. Ther.*, May 8th) uses hypodermic injections of ergotin in the epileptiform convulsions of general paralysis. At the asylum at Charenton the death rate of paralytics has been diminished by this remedy.

———

M. Bardet (*Bul Gen Therapeutics*, Apr. 23) has used hypnol in stridulous laryngitis of a child, and in the insomnia of neuralgia, tuberculosis, and other diseases. He finds a gramme is usually sufficient to produce sleep. It is especially useful in spasmodic conditions. It does not irritate the stomach.

———

Dr. Dresch (*Bul. Gen. Ther.*, Apr. 30) uses salicylate soda in chorea. He considers the disease has a microbic origin. He keeps his patients in bed for some time with milk diet, soda baths, and salycilate soda in fifteen grain doses. As the patient improves solid food is given and muscular exercise in the way of mild gymnastics. In the latter part of the disease he gives arsenic.

———

M. Mascaral (*Meeting Society Biology* of May 21st) presented a paper on the *Treatment of Peripheral Facial Paralysis by Faradic Electro Puncture.* Rheumatic facial paralysis may be cured by acupuncture combined with a weak faradic current. It should only be applied four or five minutes once a day. Should be begun within eight or ten days of the onset of the paralysis.

———

Dr. Mays (*N. Y. Med. Jour.*, April 12) recommends a hypodermic treatment of asthma with strychnine and atropine. He says that asthma is essentially a spasmodic neurosis of the pneumogastric. He begins with 1-50 of a grain of strychnine and 1-150 of a grain of atropine daily, gradually increasing the former to 1-20 or 1-25 grain and the latter to 1-100 of a grain. After a thorough impression is made on the disease the drugs are administered every other day and as the patient improves are gradually abandoned. While some cases get well under this treatment alone, in others their influence must be fortified by efforts which seek to control the cause of the attack, and by measures which tend to build up the general system.

The *Brit. Med. Jour.* reports two recent investigations concerning the therapeutic action of exalgine. In doses of ½ to 2 grains Dr. Frazier found that it relieved pain in suitable cases. Dujardine-Beaumetz and Bardet give large doses, 4 to 8 grains. Pain is relieved or diminished in one-half an hour or an hour after it is given. In some cases the relief is permanent, in others the pain returns. In neuralgias of all kinds due to cold and influenza it is especially applicable, as facial, dental, intercostal neuralgia, and to sciatica. It also gives relief in the pains of locomotor ataxia, cardiac disease. and pleurisy.

Pope (*Deutsche Med. Zeitsch.*) has used exalgine in various neuralgias, dysmenorrhœa and headache, in doses of 4 or 5 grains repeated several times in 24 hours. In some cases there is a tendency to syncope after taking the medicine. One great advantage is, that it relieves pain without producing narcotism. It is without smell or taste and is best administered in weak alcoholic solutions.

Dr. John Aulde (*N. Y. Med. Jour.*) has delivered a series of lectures at the ''Medico Chirugical College'' of Philadelphia, on the clinical application of drugs. In regard to belladonna he says it is a remedy of great power in the treatment of affections accompanied by spasms. In the case of injuries, spasm or pain, and stiffness of muscles, as in the case of torticollis, belladonna acts as a paralyzer of the terminal filaments of the motor nerves and thus allays pain. In small doses it is useful in conditions of sub-normal temperature. He recommends atropine as a heart stimulant, and tablets of atropine he recommends for sore throat and cold from exposure. He gives rhus toxicodendron for rheumatism in ½ drop doses of the tincture. He recommends strophanthus as a heart tonic in doses of 5 drops of the tincture. He considers that it is preferable to digitalis. In regard to nux vomica he gives it in conditions where people complain that they are too tired to sleep. It is an excellent stimulant in all low conditions of the system. He advises gelsemium in the treatment of neuralgia connected with dysmenorrhœa. He recommends cannabis indica when gelsemium fails, and in cases of supra-orbital neuralgia, especially those cases due to impoverished blood.

An article on *The Action of Caffein on Tissues and Heat Phenomena*, by Dr. Reichert, is published in *The N. Y. Med. Jour.* for April 26. He concludes that caffein increases heat production and therefore increases destructive tissue metamorphosis. He believes that coffee cannot to any extent replace food, and that it does not add to capacity for work. Dr. Reichert writes on the "Action of Caffein on the Circulation," in the *Therapeutic Gazette* for May. He concludes that caffein diminishes the heart's efficiency for work, arrests it in diastole, sometimes induces sudden paralysis, and is, therefore, a cardiac depressant. The asserted stimulant action upon the circulation is, doubtless, subjective, and dependent upon an excitation of the cerebral centres.

Germain See's investigations on the action of caffein on the motor and respiratory functions, are published in the *Bul. Genl. Ther.*, April 25. Caffein in small and repeated doses facilitates muscular work in augmenting the activity of the motor nervous system. It prevents breathlessness and palpitation consecutive to effort. It augments the waste of carbon, particularly of the muscles, and does not restrain nitrogenous waste. It is, therefore, not in the strict sense of the word, a means of saving. Caffein has not the property of replacing food, it only replaces the tonic excitation that the ingestion of food produces. Caffein, far from sparing the reserves, will place a fasting man in a position to undertake his work only by attacking those reserves, the destruction of which it hastens by the excitation of the nervous system and that of the muscles. The organism will soon use up its nutritive supply and the caffein will not prevent it.

Dr. William Goodell (*Medical News*) writes an article on the abuse of uterine treatment through mistaken diagnosis. He closes his article with the following pertinent suggestions:

"From a large experience I humbly offer to the reader the following watchwords as broad helps to diagnosis. In the first place, always bear in mind what another has pithily said, that ' woman has some organs outside of the pelvis.' *Secondly*, Each neurotic case will usually have a tale of fret or grief, of cark and care, of wear and tear. *Thirdly*, Scant, or delayed or suppressed menstruation, is far more frequently the result of nerve-exhaustion than of uterine diseases. *Fourthly*, Anteflexion, *per se*, is not a pathological condition. It is so when associated with sterility or with painful menstruation, and only then does it need treatment. *Fifthly*, An irritable bladder is more often a nerve symptom than a uterine one. *Sixthly*, In a large number of cases of supposed or of actual uterine disease which display marked gastric disturbance, if the tongue be clean, the essential disease will be found to be neurotic; and it must be treated so. *Seventhly*, Almost every supposed uterine case, characterized by excess of sensibility and by scantiness of will-power, is essentially a neurosis, *Eighthly*, In the vast majority of cases in which the woman takes to her bed and stays there indefinitely, from some supposed uterine lesion, she is bedridden from her brain and not from her womb. I will go further, and assert that this will be the rule, even when the womb itself is displaced, or is disordered by disease, or by a lesion that is not in itself exacting or dangerous to life. *Finally*, Uterine symptoms are not *always* present in cases of uterine disease. Nor when present, and even urgent, do they *necessarily* come from uterine disease, for they may be merely nerve-counterfeits of uterine disease."

_____

Horsley (*Brit. Med. Jour.*) writes on a possible means of arresting the progress of myxœdema and allied diseases. He considers that it is generally accepted by pathologists of the present day, that the diseases known as myx-cretinism and cachexia strumipriva are the result of want of function of the thyroid gland, and calls especial attention to Schiff's observation of the effects of transplantation of the thyroid gland into the peritoneal cavity. Schiff found that in dogs "thyroidectomy loses its danger and an essential amount of its effect, if one previously introduces and fixes in the abdominal cavity other thyroid glands from an animal of the same species." He speaks of experiments recently performed by Dr. Von Eiselberg, in which in a number of animals after extirpation of the thyroid gland, it is transplanted into the peritoneal tissue. In all cases where the transplanted gland failed to grow the animals died of typical symptoms of loss of the thyroid. In the cases where the animals showed no symptoms, it was found on post mortem examination made some months subsequently, that the transplanted gland had grown and was functionally active. These results show that the transplanted gland will, if it can be induced to grow, provide for the needs of the body, just as well as though it were in the neck. In discussing the question as to what kind of gland would be best for the purpose Horsley suggests that of the sheep. In this animal the gland is somewhat like that of man in its anatomical structure.

_____

At a meeting of the "New York Academy of Medicine" (*New York Medical Journal*) Dr. Hubbard presented a report of a case of *Pott's Paraplegia Treated by Suspension.* He employed daily suspension for a few minutes with decided improvement in a month; however, up to time of report complete recovery had not taken place. Dr. Putzel said that he believed the majority of cases of Pott's paraplegia were not due to pressure but to transverse myelitis. His experience with suspension had taught him to consider it a method of only temporary relief. Dr. Birdsall thought that where Pott's paraplegia was due to myelitis the disease was fatal. He believed that many cases were due to irritation or a pressure upon the roots of the spinal nerves in their passage through the foramina. Dr. L. C. Gray, said that excluding cases which were complicated by organic lesions of the cord, he thought that the ætiology of Pott's paraplegia could be explained by reflex causes. Nerve stretching in these diseases was a very different thing from what it was in ataxia. The latter disease had a very complicated pathology aud embraced several distinct varieties. It was a singular fact that the results alleged by Charcot had not been obtained by other observers.

Rosenbaum presented an article to the "Berlin Med. Soc.," April 24, on *Treatment of Ataxia by Suspension.* In 61 cases, 25 were improved by suspension, 9 cases were doubtful, the others were not improved. In the cases benefited, appetite and health improved, pains lessened, crises more rare, with lessening of inco-ordination.

In 4 cases of myelitis, and in 3 of paralysis agitans, results of suspension were negative. Dr. Guttman stated that he had obtained benefit from suspension in a small proportion of cases of ataxia, and also in 4 cases of sciatica there was improvement, and where antipyrine was given 3 cases were cured. Leyden denied that suspension was beneficial in the treatment of any form of nervous disease, except that owing to psychic influence there was some temporary benefit.

ON A RATIONAL METHOD OF OBTAINING EXTENSIONS OF THE SPINAL CORD AND COLUMN.

Dr. Stillman, of Chicago, read an article with the above title before the "Chicago Medical Society." The lecture was illustrated with pictures of the extension apparatus which Dr. Stillman uses, and which he claims is superior to the ordinary methods adopted. The extension can be made with the patient in a horizontal position, or hanging from the neck with the back resting upon a curved board, which can be properly compared to one half of a wagon wheel, the back lying against the convex surface of the wheel. It seems to have the advantage over the ordinary methods of pressing comfortably against the spine, and in this way relieving the neck of a large part of the weight of the body. The position of the patient can also be reversed when required, that is, the anterior surface of the body being placed against this convex surface, the patient being allowed to assist in the suspension by the use of his hands.

Russell and Taylor (*Brain*) give their experience of treatment by suspension. They treated 45 cases of which 32 were tabes. Of the 45 cases, 9 improved, 5 became distinctly worse, and in the rest there was no change. They seem to think that the improvement which has been noted in suspension is due to the natural intermittence of the disease and to the psychic effect upon the patient. Their experience with the treatment is certainly not encouraging.

Mr. Lucas (*Guy's Hospital Reports*, Vol. 46) reports two cases of division of the musculo-spiral nerve with entire restoration of function after suture. The first was the case of a boy who was stabbed through the arm with a chisel, producing a complete division of the nerve. The damaged portion of the nerve was cut away and the ends united by catgut sutures.

The second was the case of a man who was stabbed in the back part of the arm. There was anæsthesia of all the back parts of the arm and hand, and paralysis of muscle supplied by the nerve. The cut ends of the nerve were united and the patient made a complete recovery.

SURGERY OF THE BRAIN AND CORD.

A CASE OF SPINA BIFIDA SUCCESSFULLY TREATED BY OPERATION.

The tumor was split, the cord reduced and the coverings sutured by the *etagen-suture*, the erector spinæ muscles being drawn together in the median line. Operation on the 14th day after birth.—DR. BAYER in *Wiener Med. Presse*, No. 19.

Dr. Keller reports (*Brit. Med Jour.*, Mar. 29) *Trephining in a Case of Actinomycosis of the Brain*. The patient had paralysis and left hemiplegia, which involved the face, with headache and vomiting, and at the time of the operation was unconscious from the effects of the disease. In the right ascending parietal convolution was found an abscess which was evacuated, and the patient awoke. The patient made a good recovery with, however, some remaining weakness of the left arm. About a year afterwards he had a relapse, with a return of the symptoms associated with convulsions, of which he died. The middle third of the right frontal and parietal convolution was found occupied by a large abscess.

In the treatment of cerebral hemorrhage, Dr. Ludwig Schwarz recommends arteriotomy of the anterior branch of the temporal artery, on both sides if necessary. He claims that this will cause a reduction of the blood pressure in the carotids with the loss of less blood than follows venesection. —(*Wiener Med. Blaetter*, No. 26.)

ON THE TREATMENT OF HYDRENCEPHALOCELE.

Prof. Wolfler treated a case of Hydrencephalocele, after pressure and iodine injections had failed, by incision, reduction of the protruding brain, and ligature of the sac. The hernia was a double one, two different protrusions through one cranial defect. Only one was operated upon. The patient died of sepsis. He prefers reduction of the brain mass, even if the bony defect must be enlarged, to Bergmann's process of ablation which exposes the patient to the danger of opening a ventricle.—(*Wiener Med. Presse,* No. 15.)

CLINICAL CONTRIBUTION TO CEREBRAL SURGERY.

Dr. H. Oppenheim reports a case of the removal of a cystic glioma from the right motor region, with prompt recovery and complete relief from all symptoms.

The second case was one of hemiplegia spastica infantilis in a girl 12 years old caused by a meningo-encephalitis at the age of 4. Athetosis of right side, paresis and contracture of right half of body, several epileptic attacks daily.

Trephining over fissure of Rolando on left side showed a cyst with clear colorless contents, whose anterior wall was excised. Recovery was undisturbed. The athetosis became less marked, contractures less and epileptic attacks lighter and with longer intervals.—(*Deutsche Med. Wochenschr.,* No. 27.)

Prof. Kraske, of Freiburg, has trephined the vertebral canal in 4 cases of spondylitic paralysis. The removal of the granulations with the sharp spoon was followed by relief, which however, was of short duration.

The operation is justified in cases where the process has its seat in the arches, and where the process in the body is stationary. In those cases where the paralysis rapidly invades the bladder it is justified as a last resort.— (*Deutsche Med. Wochenschr.,* No. 22.)

Dr. F. X. Dercum reported at the meeting of the "American Neurological Society" in Philadelphia, a case of *Paraplegia Cured by Operation.* The patient was completely paralyzed and suffered much pain with paralysis of both sphincters. The spines and arches from the first to the fifth dorsal vertebræ were removed. The patient completely recovered.

Lane (*Lancet,* July 5) reports a second case of recovery after operation for paraplegia, due to angular curvature. In his first case, which was published a year ago, the cord was found to be compressed between the body of one vertebra and the margin of the laminæ of another vertebra. The patient completely recovered In the present case the pressure upon the cord resulted from a mass of tubercular material, situated between the laminæ and dura.   The tubercular mass was removed and the patient entirely recovered.

Frank and Church (*Amer. Jour. of Med. Sci.*) make a contribution to *Brain Surgery*, the article consisting of a history of a series of cases of operations for various cerebral pathological conditions.   One was a case of terminal dementia of alleged traumatic origin.   The skull was trephined with great improvement which lasted nine months.   The patient relapsed and a second operation was performed with marked and increasing improvement.

The second case was one of Jacksonian epilepsy, in which a sarcoma was removed with improvement of all the symptoms.   Subsequent relapses from probable recurrence of neoplasm.

The third case was one of idiocy, with generalized and continuous choreoid movements of seven years duration.   Complete cessation of movements followed operation, but patient died the third day.

An article by Dr. W. G. Thompson on *Successful Brain Grafting*, is published in *The N. Y. Med. Jour.*, June 28.   He performed a number of experiments on cats and dogs, in which pieces of brain were transferred from the brain of a cat to that of a dog and *vice versa*.   Autopsies were made at times varying from a few days to several weeks, and in every case it was found that the transplanted brain had united firmly with the surrounding tissues and still maintained its original brain' structure.   The transplanted tissue did not take on the function of the part removed.

Dr. Battle (*Lectures on Injury to the Head, Lancet*, July 5, *et seq.*) says among the important symptoms of fracture of the base are hemorrhage from the nose and mouth, vomiting of blood, and either hemorrhage or watery discharge from the ear, and a collection of blood under the conjunctiva or over the mastoid process.   In cases of skull fracture optic neuritis sometimes results.   It most frequently occurs in connection with fracture of the base.   Injury of the cranial nerve with accompanying paralysis may result from fracture.   The alteration of temperature following cranial fracture does not differ from that of injuries in other parts of the body of corresponding magnitude.   If there is meningitis, or the brain substance is lacerated, high fever is apt to follow.   Loss of consciousness does not always follow fracture.

Dr. W. F. Morgan (*N. Y. Med. Rec., April 19*) reports a case of *Fracture of the Condyle of the Occipital Bone.*   The patient was a miner who was injured by a mass of rock falling upon him, the accident having happened to him twice before under similar circumstances.   The patient had motor and sensory paralysis of upper and lower extremities and occasional "clonic spasms." The patient died suddenly, and on autopsy it was found that the right transverse process of the atlas was destroyed by caries. The articular surface of its lateral mass half absorbed. The right condyle of the occipital bone had been fractured and was carious. The ligaments of the neck were disorganized.

TREATMENT OF FRACTURES OF THE BASE OF THE SKULL.

(Jour. Am. Med. Assn., Aug., by Hal C. Wyman, M. D.) Dr. Wyman has examined 34 cases of fracture of the skull *post-mortem,* of which 26 were fractures of the base. In the majority the clot was in the space between the bone and the dura, the middle meningeal artery having been wounded. In the minority the clot was in the substance of the brain from rupture of the pia or of the cerebral arteries. In diagnosis the most important fact is, that 80% of all fractures of the skull involve the base. Orbital ecchymosis, watery or bloody discharges from the mouth and nose, and paralysis of cranial nerves are important signs of basal fracture. The most important point in diagnosis of fracture of the base is the existence of an "interval of sense" following the injury. The patient meets with a fall or receives a blow on the head, becomes unconscious for a time, then regains consciousness; he may walk a distance, becomes faint, vomits, and again becomes unconscious. The injury has probably caused a rupture of a blood vessel, and the blood being slowly poured out produces the unconsciousness. Dr. Wyman has devised a scoop made out of a watch spring, which can be bent in any direction and allows of the removal of the blood clot which adheres so firmly to the membranes. His views summarized are: trephine all cases that have had a sense interval, trephine the basal portion and remove the clot with the properly constructed scoop.

Watson-Cheyne *(Brit. Med. Jour.)* reports a case of *Abscess of Left Tempero-Sphenoidal Lobe with Operation and Recovery.* The patient had an old standing disease of the ear in which acute symptoms had recently developed. He had acute pain in left temporal region with dizziness when standing. The pain later shifted to the occipital region and became more intense. The patient had rigors followed by fever and boisterous delirium. He was trephined one inch and a quarter behind the centre of auditory meatus. The abscess was found in the brain substance. Its contents measuring an ounce, was evacuated. The patient recovered.

Milligan and Hare report *(Brit. Med. Jour.)* a case of *Abscess of the Cerebellum* following chronic otitis media with operation and death. The patient had had chronic discharge from both ears, but for a few months previous, the discharge from the left ear had ceased. Recently the discharge from the right ear had suddenly stopped and there developed intense pain on the right side of the head. There was paresis of the soft palate and uvula of the right side, and right optic neuritis. The patient became delirious; subnormal temperature, vomiting, irregular pulse. The skull was trephined over the cerebellum on a line below the lateral sinus, and an abscess was found from which several drachms of pus were evacuated. The patient died about twenty-four hours after the operation.

Dr. W. F. Mittendorf (*N. Y. Med. Rec.*, April 5,) reports the successful removal of an *Anterior Encephalocele*. The patient was a male child four days old. The growth was located at the root of the nose, and had left the skull between the cribriform plate of the ethmoid and the frontal bone, and appeared anteriorly between the frontal and nasal bones. The tumor was removed and the child entirely recovered.

Drs. Hale White and A. Lane report to the "Clinical Society" of London, a case of trephining for old *Hemiplegia with Epilepsy*. The patient was a man who had left hemiplegia after a fall on the head. He also had fits which were generally limited to the left side of the body, with paralysis of that side, the left arm being flexed and muscles wasted. He had persist- · ent headache. He was trephined on the right side over the middle of the fissure of Rolando. The bone in this region was three quarters of an inch thick and the dura was also thickened. During convalescence the patient had two fits. Since then he had had none though seventeen months had elapsed. The leg had improved very much but the arm had remained paralyzed as before.

J. Mitchell Clark (*The Lancet*) reports a case of *Syphilitic Growth in Dura* with operation, and death from septicæmia. At the time of operation the condition was as follows: right hemiplegia, slight motor aphasia, frequent occurrence of fits preceded by aura in the right great toe. The patient was trephined in the left motor region. A tumor about an inch thick and of cartilagenous substance was partly removed from the left dura. A small tumor was found attached to the upper part of the falx cerebri which had pressed upon the paracentral lobule. It should have been mentioned that there was distinct deficiency of sensation of the right side of the body.

Dr. Clay Shaw and Mr. Harrison Cripps report (*Brit. Med. Jour.*, June 14,) a case of *Trephining in General Paralysis*. The patient was a man who gave a history of a blow received on the left side of the head 15 months previously. He developed general paralysis, the symptoms being well marked at the time of the operation. A portion of all the membranes of the brain was removed through the trephine opening. The patient was very much improved after the operation. Returned to his home and took a position as ticket collector for a railroad company. Five months after the operation his condition was about the same as before it was performed.

Dr. Shaw reports the final result in a case of trephining done in July, 1889. After the operation the patient improved very much mentally, but the motor symptoms were not modified. Six months after the operation he died. Dr. Shaw claims that the operation was justified as it afforded temporary improvement.

Dr. G. L. Richards (*Medical Record*, April 19,)reports a case of *Trephining for Fractured Skull*. The patient was a boy 16 years old who had a depressed fracture of the left parietal bone with partial paralysis of the right side with partial unconsciousness. The skull was trephined and the depressed bone raised. The patient gradually recovered from his paralysis and in about six weeks was entirely well.

M. Routier (*Paris Acad. Med.*) reported case of *Trephining for Hemiplegia·* There was loss of consciousness with right hemiplegia some time after a fall. He trephined and raised several pieces of depressed bone, and placed an iodoform tampon in the wound. Four hours after operation patient regained consciousness and hemiplegia disappeared. On the next day there was aphasia which disappeared on removing tampon which M. Routier thought caused the aphasic condition.

At "19th Cong. German Surgeons," Schmidt reported the case of patient who had epilepsy following injury to skull. An abscess of brain (location not stated) was discovered and evacuated with complete cure of epilepsy. (*8th–12th April, Jour. des Soc. Scientifiques.*)

Ballance reports (*Lancet*, May 17 and 24) operations for removal of *Thrombi from Lateral Sinus.* One patient who had had chronic ear disease for fifteen years, was admitted with all the indications of thrombosis. After other treatment failed the mastoid was trephined. The sinus contained an offensive blood clot. The internal jugular vein was ligated in two places and divided between the ligatures. Subsequently the patient had pyæmia with abscesses in various parts of the body, but ultimately completely recovered. In the second and third cases, which were very similar to this, death took place after the operation. The fourth case, almost identical with the preceding, entirely recovered after the operation.

In regard to diagnosis the following signs are to be looked for: First, the history of purulent discharge from the ear, for a period of more than a year; second, sudden onset of illness with headache, vomiting, rigor and pain in the affected ear; third, an oscillatory temperature, reaching from

103 to 105 F., and then dropping below 100; fourth, repeated vomiting; fifth, two, three or more rigors; sixth, local œdema and tenderness over the mastoid, or in the course of the internal jugular; seventh, tenderness on deep pressure, at the posterior portion of the mastoid and below the external occipital protuberance; eighth, stiffness of the muscles of the back and side of the neck; ninth, optic neuritis in pyæmia occurring from disease of the temporal bone independently of thrombosis of the lateral sinus. The diseased bone should be freely exposed and aseptisized with sublimate, and the vein ligatured in the neck.

———.

At the Thirty-ninth Congress of Swiss Physicians, M. Krœnlein presented a case of rupture of the middle meningeal artery resulting from injury, but without fracture of the skull. There was complete paralysis with all the symptoms of compression. The skull was trephined and a clot was found in the parieto-occipital region. The wound at point of rupture of vessels was tamponned with iodoform gauze. In two hours patient regained consciousness, and in six weeks was well. *Jour. des Soc Scientifiques, Apr.*

———

M. Lucas Championniere communicated to the "Academie de Medicine" the result of an operation for removal of an old clot from the brain in a case of cerebral hemorrhage. The patient was a man of 53 who, after an attack of apoplexy had paresis of the right leg, marked contraction of the right hand, and epileptiform attacks which increased in frequency and severity. The operation was performed and an encysted clot was found in the ascending frontal convolution and removed. On the following day the contraction of the hand had ceased. When the patient left his bed he found he could walk with ease. The patient had a convulsion about two months after the operation, but this had not been repeated up to the date of the publication of the report 4 months subsequent to the trephining.

Dr. C. B. Porter (*Boston Med. and Surg. Jour.*, April *10*) reports several cases of *Fracture of the Skull* with recovery. The first was a man who had extensive fracture of the skull from being struck by a moving train. Part of the orbital arch, external angular process of frontal bone, and roof of the orbit, were torn away, while the great wing of the sphenoid, squamous portion of the temporal, and part of the parietal and frontal were depressed. Some fragments of bone were lost. He removed several fragments, raised the part that was depressed, the orbital arch was returned to its place and held by a wire suture. There was no paralysis following the injury and the patient recovered.

The second was the case of a man who fell down an elevator, producing depressed fracture of the right parietal bone without paralysis, but with partial unconsciousness. The depressed portions of bone were removed and the patient entirely recovered.

He quotes an interesting case, which is apparently by Walsam of St. Bartholomew's Hospital, in which the patient after a blow on the head lost all recollection of his previous life. He was not otherwise affected, resumed his occupation, and married. Subsequently he became epileptic and was trephined. His health was restored by the operation, as well as his memory of events *prior to the blow*. The whole of his life, however, between the injury and the operation was a total blank. He did not know his wife and could hardly be convinced that he was married. Dr. Porter advocates early trephining in any skull fracture.

PSYCHOLOGICAL.

MENTAL PHYSIOLOGY.

Mr. Sully read before the London Neurological Society *(Brain)* an article on the *Psycho-Physical Process of Attention.* The completed process of attention is the fixation of a particular sensation. Attention is detention in consciousness. The first step in the complication of conditions is where the sensation is pleasurable or painful. The next stage is the partial determination of the direction of attention to pre-existent lines of interest. As we grow older the more vigorous sort of attention tends to confine itself within certain limits, necessary to our special pursuits and tastes. We each see what we are prepared to assimilate or understand. The next stage in the process consists in voluntary or consciously selective attention. He refers to the fact that attention stands in a particularly close relation to the process of motor innervation. Dr. Bain seeks to account for the whole process of that control by help of the motor process. According to him every idea is compounded of an element of passive and of muscular sensation. Mr. Sully holds that attention, though always accompanied by motor processes, cannot wholly be resolved into a motor phenomenon.

Dr. Henry Maudsley contributes an article to the April number of *Mind*, on the *Cerebral Cortex and its Work.* He begins with a study of reflex action in its simplest form, and then passes to a consideration of the higher cerebral centres. He says reason is essentially reflex action; this is the fact that it is important to note, though, of course, there are differences between an act of reason and a simple reflex act. The plan of the entire nervous system is the nervous mechanism of a reflex act, and the development of such a system in its highest parts is the complication of such mechanism.

The motor region so-called is not motor, but centres of motor abstracts, that is, efferent aspects of cortical reflexes, which all our thoughts are. Consciousness and will are not original agents, they are derivative exponents of what is being done, not doers of it.   The nervous system can work without them; they are, when present, imperfect indicators of it.

AN INFANT PRODIGY.

Dr. S. V. Clevenger gives (*Alien. and Neurol.*, July) an account of the wonderful memory of a three year old child, Oscar Moore.   Oscar is a colored boy who was born blind.   He has mastered an appalling array of statistics, such as the areas in square miles of hundreds of countries, the population of the world's principal cities, the birthdays of the presidents, names of the cities of the United States, etc.   His mental development seems to be symmetrical, and he desires to understand the meaning of all that he is taught.   Of the seven brothers and sisters of Oscar all of them died probably of tuberculosis, excepting his two sisters, who are living. While idiots sometimes show a wonderful memory in certain lines there is no reason to suppose that Oscar is in any degree idiotic.   One of the conspicuous faculties of many great men is a wonderful memory.

PATHOLOGY AND SYMPTOMATOLOGY.

Dr. C. F. Folsom ("Trans. of Assn. of Amer. Physicians," Vol. 4) has an able paper on the Earlier Stage of General Paralysis.   The article is based on notes of cases that have been under Dr. Folsom's care in private practice.   His study shows that the earlier symptoms of general paralysis as set forth in text books, indicate a comparatively advanced stage of the disease, and that antedating these more marked symptoms is a stage in which there are more subtle and obscure changes of character and yet which strictly belong to the natural history of the disease.   He says: "The earliest signs of general paralysis are of the slightest possible brain failure; if, for instance, a strong, healthy man, in or near the prime of life, distinctly not of the nervous, neurotic or neurasthenic type, shows some loss of interest in his affairs, or impaired faculty of attending to them; if he becomes varyingly absent-minded, heedless, indifferent, negligent, apathetic, inconsiderate, and though able to follow his routine duties, his ability to take up new work is, no matter how little, diminished; if he can less well command mental attention and concentration, conception, perception, reflection, judgment; if there is an unwonted lack of initiative, and if exertion causes mental and physical fatigue; if the emotions are intensified and easily change, or are excited readily from trifling causes; if the sexual instinct is not reasonably controlled; if the finer feelings are even slightly blunted; if the person in question regards with a placid apathy his own acts of indifference and irritability and their consequences, and especially if at times he sees himself in his true light and suddenly fails again to do so; if any symptoms of cerebral vaso-motor disturbance are noticed, however vague and variable. Naturally there may be many or few of these indications in a given person. This group of symptoms seems very striking, but may be compatible with the performance of usual duties."

In the discussion of the paper Dr. Pepper made some interesting observations on the probable causes of general paralysis. He holds that the process of general paralysis may be initiated by many disturbing and irritating causes, and that it is often preceded by symptoms largely lithæmic, and that in a large proportion of cases these symptoms may be dispelled and the establishment of the process averted.

Dr. G. H. Savage writes (*Brit. Med. Jour.*, Apr. 5) on the *Warnings of General Paralysis of the Insane.* This disease is a degeneration, a premature decay most generally met with in middle-aged married men, inhabitants of cities, and who are flesh eaters and drinkers of alcohol. There are two prevailing forms of onset, a gradual and a sudden. In the first the highest faculties show the earliest signs of change, and special attainments fail before general. The finer social and muscular adaptations fail first. In the other form it may be ushered in with a convulsive seizure and an attack of mania, and in both there are probably many unnoticed changes. Among the many motor symptoms are a feeling of fatigue on slight exertion, sometimes associated with slight ataxic symptoms. Sometimes temporary aphasia or slight difficulty of articulation, or dropping of letters in writing, or fatigue after writing, are among the symptoms of the disease. Special accomplishments are early impaired; an artist, for instance, loses his skill and his ability to blend colors properly. There is also loss of power of social accommodation. The patient's habits and tastes insidiously change, and he falls out of step with his fellows. There is loss of power of attention, want of persistency, restlessness, and weakened memory. There is tremor of the finer muscles, uncertainty of gait, tendency to fall, and he is easily affected by drugs, and especially by intoxicants. Change of temper, irritability, and sometimes depression, occur.

Dr. G. M. Robertson writes an article (*Jour. Ment. Sci.*, July) with the title, "Does Mania Include Two Distinct Varieties of Insanity, or Should it be Subdivided?" He considers that the basis of mania is an emotional state and that there are two principal varieties. The first is the furious or raging maniac, who is destructive, boisterous, irritable and dangerous to those around him. The other is the hilarious maniac who is boisterous and rollicking, his mental condition being one of friendliness and familiarity. He says: "If the above rather brief description of these forms of mania be true to nature, we maintain that the fact is established that mania includes two distinct symptomatological varieties of insanity."

Dr. J. G. Kiernan (*Alien. and Neu.*, July,) writes on the *Mental Symptoms of Phthisis as Illustrated by Keats and Emily Bronte.* Keats was irritable and suspicious from his youth and predisposed to consumption. Keats became noted for his moody taciturnity, which at times amounted to almost panaphobia. He became misanthropic and the companionship of friends was tedious. He was subject to attacks of emotional exaltation accompanied with irritability and symptoms of phthisical insanity. Taking into account his lack of insufficiency of purpose, his suspicious tendencies, his emotional tendencies, his irascibility and pugnacity, as well as his heredity, the influence of phthisis on his mental state is clearly manifest.

The father of Emily Bronté was semi-insane, the mother consumptive, and the history of the two Bronté sisters certainly affords abundant evidence of the mental symptoms associated with phthisis.

The mental condition of phthisis needs as much attention as the physical. To calm the restlessness is to increase the chances of recovery.

Suckling (*Am. Jour. of Med. Sci.*, May) writes on *Agoraphobia and Allied Morbid Fears.* He states that this is only a symptom and not a disease of itself, and that any exhausting disease, habit or occupation may be the exciting cause. The proper treatment is to remove any faulty habit of depression, influence or work, and to prescribe a generous diet, cheerful society, change of air and tonics.

MENTAL DERANGEMENT AND MULTIPLE NEURITIS.

The author has observed a number of cases in which the most prominent symptoms were loss of memory and confusion of ideas. In the severe cases loss of memory was complete; in milder ones it took the form of forgetfulness. This was most conspicuous for time, place, and recent events. The disturbance resembles that of alcoholic paralysis.—DR. KORSAKOW, *Archiv. Psychiatrie*, Band 21, Heft 3.

CHOREA IN THE ADULT AMONG THE INSANE.

The author finds that there is 1 choreic in 425 of the population of insane hospitals. In long standing cases there is a tendency to mental deterioration. Adults are sometimes, though rarely, attacked with chorea while suffering from rheumatism. Chorea and epilepsy are intimately related to each other. Epileptic convulsions (Jacksonian) may be confined to a single member. The same is true of choreic convulsions.—(DR. T. DILLER, *Amer. Jour. of Med. Sci. Apr.*)

MANIA FOLLOWING ETHER.

Dr. Gorton, Superintendent of the Butler Hospital for the Insane, publishes a history of two cases. One was the case of a boy fourteen years of age, who took ether before having teeth extracted. He had an attack of subacute mania followed by dementia. The second was the case of a young lady who took ether to have a number of teeth extracted. The administration was tedious and she was several hours recovering from the influence of the anæsthetic. In a day or two she returned to her work, but her friends noticed there was a marked change in her mental condition. She afterwards became insane. The patient was improving at the time of the report. Dr. C. L. Dana has stated that several of his neurasthenic patients have dated the beginning of their nervous troubles to the administration of ether.—(*Am. Jour. of Insanity*, April.)

Joffroy (*Annales Medico-Psychologiques*, May) reported to the March meeting of the "Med. Psychological Soc. of Paris," the history of a case of *Insanity Occurring with Exophthalmic Goitre.* The mental condition was that of depression. In exophthalmic goitre there are certain psychic modifications frequently observed, which are, restlessness, irritability, incapacity for close mental application, and depression. In other cases there is excitement. The first tend to become melancholiacs, the latter maniacs. In a certain proportion of cases these slight deviations from the normal mental state become still more marked and pass into some form of insanity.

The following abstracts of papers read before the sixth Italian Psychiatric Congress are taken from the reports of the meetings in the *Revista Sperimentale*, XVI, I and II.

PARANOIA, BY VENTURI.—The author begins with a brief history of paranoia recognized as monomania by Esquirol and developed by Griesinger into his secondary delirium from whence the modern conception, while the French pass from the idea of Esquirol in a more analytical manner to the insanity of persecution of Lasegue and Le Grand du Saulle, then through a unification of this delirium with that of grandure to the chronic delirium of Magnan. Thus according to the French, in paranoia we have the delusions of grandure following logically those of persecution. and the whole evolution of the insanity is connected with a certain degree of dementia. The author accepts this last view but denies the logical connection between the two insanities, and calls attention to the analogous genesis of religious delusion, sometimes consecutive to those of an erotic nature, and in ordinary life we frequently find the erotic sentiment hidden by those of a religious nature.

With this preface the author proposes the following different types of paranoia:

1.   A paranoia of memory, consisting essentially of an insanity of recollections, which are either fantastic creations of the imagination or distorted memories of occurrences.

2.   A paranoia of jealousy, different from that of persecution.

3.   A paranoia of miserliness.

4.   A paranoia of anger, and one of grief, due to systematized emotional conditions manifested without any connection with the delusions.

5.   A paranoia motoria, based on rythmic and systematized movements and an altered morül, of which Morselli has treated in his Turin lectures.

6.   A paranoia that may be called an eccentric and objective insanity, consisting of delusions nourished by superstitions or current notions in such a way as to induce what are to the subject, coherent acts. Such are the states of demonomania and the zeanthropy of the middle ages.

The author adds that the diverse methods of our judgments of these various morbid mental states are due largely to the way in which the observer collects and recognizes the different psychic symptoms, therefore he recommends taking down all that the patient says, avoiding influencing him as far as possible, and on the basis of the material thus obtained to construct the clinical inductions.

Finally he proposes a last form consisting in the use of certain words or groups of phrases without any delusional idea or dementia, a condition that may be called a species of verbal paranoia.

___

THE CEREBRO-SPINAL FLUID IN THE INSANE.

GONZALES AND G. B. VERGA.—The authors report some researches on the quantity and density of the cerebro-spinal fluid in 152 cadavers of the insane (56 women and 96 men), the results of which are summed up as follows:

1.   The value of the observation of the cerebro-spinal liquid examined in the cadaver, is not absolute, since it is impossible to collect it in totality.

2.   As regards insanity in general it is more abundant than in the normal man.

3.   It is more abundant in paralytic, pellagrous, demented and melancholic insane.

7

4. It seems that mental disturbances in their commencement and first manifestations, tend to increase the quantity of the cerebro-spinal fluid, and that it later decreases, to increase again in the chronic condition, in approaching dementia, and in terminal demented states.

5. It is especially scanty in the epileptic insane.

6. It is in notable quantity in inflammatory conditions of the brain.

7. It abounds with intercurrent diseases of a sub-acute course, especially in those, the progress of which is slow and apyretic, such as marasmus.

8. The quantity is least with cardiopathics.

9. In acute visceral disorders and infectious diseases in general, the quantity is small.

10. The density varies between 1010 and 1017.

11. In most cases of mental disorder the density is above the normal.

12. The density increases most in the simple insanities, and in alcoholism when not complicated with dementia, in epilepsy, periencephalitis and pellagra.

13. The grade of density is also influenced by the condition of the nervous centres, by intercurrent affections or the cause of death, and by the lapse of time between the decease and the autopsy, increasing from the hour of death.

ON NERVOUS AND PSYCHICAL DISTURBANCES FOLLOWING BILATERAL CASTRATION IN THE MALE.

Dr. M. Weiss, of Prague, has observed in a male, with a nervous heredity, various nervous troubles following extirpation of both testicles.

These consisted of flashes of heat, sweatings, unpleasant sensations in abdomen and head, vertigo, oppression of the chest, melancholia, and tremblings.—*Wiener Med. Presse*, No. 22, 1890.

INSANITY FROM BRIGHT'S DISEASE.

Dr. G. Vassali, *Rivista Sperimentale* 1890, pp. 95–103, reports four cases analogous to those published by Bremer (*Jour. Nerv. and Ment. Dis.*, June, 1888) of insanity developing in the course of chronic renal disorder in which the principle rational symptoms of the somatic affection were masked largely by the mental disease constituting thus according to the author's views a psychic equivalent of the uræmic attack. He also refers to a recent memoir by Hoessin (*Munchener Med. Wchnschr.*, Dec., 1889,) as confirming this view. He says, however, that it is still to be determined whether this mental derangement is to be attributed to deficient elimination of urea or to a leucomaine. Nowadays, he says, the etiology of chronic nephritis is generally attributed to some chemical or infective toxic agent. Hence, we should consider the possibility of the psychic disturbances being due to defective elimination on the part of the kidney diseased by a ptomaine elaborated from the germ which either directly by its local action, or indirectly by the passage of its venom, is the origin of the nephritis itself.

FIBRE-ATROPHY IN THE CEREBELLAR CORTEX.

While engaged in verifying the statements of Tuczek, in regard to the atrophy of nerve fibres in the cerebral cortex in dementia paralytica and allied disorders, the attention of the author was directed to the cerebellum. He found in this organ also, various grades of atrophy, down to almost complete disappearance of the nerve fibres.   In the adult the nerve-fibres of the cerebellar cortex present themselves as a dense interlacing network, while in the new born child the network is much simpler, consisting mainly of a minimal number of longitudinal fibres.   The author compares the microscopical picture to that obtained by laying three or more nets over each other.

In the first degree of atrophy the network is larger and coarser and composed of fewer fibres, as if one of the nets had been lifted off.   This change is most marked at the inner boundary of the gray substance.   In the second degrees of severity, the network is still coarser, showing numerous gaps, and the nerve fibres are often broken.

In the most severe degree all traces of a network have disappeared, only a few isolated fibres running at random across the field, these persisting longest around the layer of ganglion cells.

The second grade of atrophy was found in 3 cases of progressive paralysis, 1 of chronic paranoia and 1 of melancholia.

A case of acquired idiocy showed similar changes which were referred to arrested development.

The severe grade of atrophy was found in 7 cases of progressive paralysis and 1 of senile dementia.

Chronic inflammatory changes in the meninges or cortex, or combinations of them, so frequent in the cerebrum, are rarely found in the cerebellum.   The histological elements of the cerebellum, other than the nerve fibres, were normal, or showed only slight changes, in a few cases.

He considers the process to be a primary degenerative atrophic sclerosis allied to that found in the cerebrum.   In its distribution it is allied to the system diseases, affecting those fibres entering the cerebellum through the peduncles.—DR. A. MEYER, *Archiv. f. Psychiatrie*, XXI Band, Heft 1.

ON CYSTIC DEGENERATION OF THE BRAIN.

The author has examined 8 brains showing this change.   Three were from cases of progressive paralysis, 2 from tabetic paralytics, 1 from a case of melancholia, the other 2 showed no brain symptoms during life.

The size of the cysts varied from a point barely visible up to the size of a pea. Their shape was, in the greatest number of instances, round or elliptical, although irregular compound cysts were often found, formed by the coalescence of smaller ones. Some, where the section cut them in the proper diameter, were seen to be long cylinders, which might be conical at one or both ends. They were rarely completely closed, but were usually connected with each other by a minute canal, often visible to the unaided eye, and frequently containing a vessel. They were usually arranged in rows corresponding to the course of the vessels, the smallest ones being situated nearest the surface. They were found not only in the cortex but also in the white matter of the convolutions, and some were found in the large basal ganglia.

The contour was in most cases sharp and clear, although in those situated near the surface of the brain, the wall was often rough.

The cyst wall was in all cases composed of the surrounding tissues, no lining membrane being detectable. Some of the cysts often showed ridges of tissue in their interior, which contained a vessel.

The tissue composing the cyst wall was normal in appearance, only rarely showing a crowding of its elements or a concentric arrangement of the ganglion cells around the cavity.

A vessel was often found running parallel to the cyst wall separated from it by a layer of neuroglia. Previous investigations have always described the vessels as projecting into the cyst, but not running through it. Serial sections and embedding in celloidine have shown this to be an error. The vessel, usually an artery of small or smallest size, was, upon entering the cavity, deflected along one wall. In the greater number of cases the vessels were pervious, although in some they were seen to be obliterated. Most authors agree in referring these cysts to dilatations of the perivascular lymph channels.

This is due to occlusion of them at some point, or to an increased difficulty of flow of the sub-arachnoid fluid, or both. This view is borne out by the fact that the cysts were largest and most numerous in a case in which there was a development of sub-arachnoid cysts in the sulci and adhesion of the arachnoid to the summit of the convolutions.—Prof. A. Pieck, *Archiv. f. Psychiatrie*, Band XXI, H. 3.

ASPECTS AND OUTLOOK OF INSANITY IN AMERICA.

Dr. Godding (*Jour. of Insanity, July.*) in considering the philanthropic aspect of those engaged in the care of the insane in America, pays a beautiful tribute to Miss Dix, and to Cook, Metcalf, Gray, and Sawyer who were martyrs to the cause. He laments the difficulty of classification and of individual treatment due to the large size and overcrowding of insane hospital wards. He properly denounces the interference of politics in the management of hospitals. The majority of those admitted to our hospitals will probably remain permanent inmates, as statistics show that only about one-third recover.

THE INSANE IN THE STATE OF MOHILEFF, RUSSIA.

Dr. Ignatieff (*Kowalewswskij's Archives of Psychi. and Neurol.*, No. 2,) says that in this state there is but one hospital for the insane. A new building is soon to be erected. The present building is divided into two wings which separate the sexes. The corridors between the rooms serve as dining rooms. The patients are only allowed to walk in the hospital grounds during the summer time, and even then only occasionally. The only ventilation of the building is obtained through the windows. There is also no suitable place for bathing, except the Russian bath, and the supervision of this is so poor that it does more harm than good. There is one physician who lives outside the building and receives a salary of 600 roubles a year. The number of attendants is small, and they are usually incompetent. There are 1,361 insane in the state of Mohileff, which is about one insane person to every 1,000 of the population. The most frequent form of mental disease is dementia in its various degrees. Out of 260 cases treated in this hospital during the last 3 years 24% were discharged as cured or improved. The death rate was 19%. When we consider the insufficient food, the imperfect ventilation, the limited amount of exercise allowed the patients, and the entire absence of any mental diversion, the wonder is that the mortality is not greater.

THE PROPAGATION OF INSANITY AND ALLIED NEUROSES.

Dr. Strahan, (*Jour. Ment. Sci.*, July), claims that statistics show that there is a great increase of insanity and nervous diseases in Great Britain, and he suggests that there are two ways by which this increase may be checked. First, by educating the people, and second, by legislative enactment limiting the marriage of those who inherit tendency to insanity or nervous disease. He says: "Ultimately, I fear, the latter course must be adopted, for the reason that many of those who aid in this propagation are actuated by sordid motives, as rank or wealth, rather than ignorance, while many of those bearing the insane diathesis are so impulsive and ill-balanced that they are at most semi-responsible, and with these the teaching of science, however convincing to the thoughtful, can never have any great weight."

"Of course the old cry of 'interference with the freedom of the subject' will arise like a specter to bar the path of legislation, but this ghost has been laid before and will be again. These wretched creatures far down in the scale of degeneration, with just sufficient intelligence to keep them from outraging the usages of society, who create nothing, add nothing to the commonwealth, but are instead a charge upon the community; these have no more right to claim freedom of action as to procreation than has the leper to mingle with the populace."

LATE CURE OF INSANITY.

Ventra (*Revista Sperimentale*) reported a study of 23 cases of recovery from insanity of long duration; 12 of these psycho-neuroses (3 mania, 6 melancholia, 3 sensorial insanity) recovered after five, seven, nine, and 1 after eleven years, and 11 of degenerative psychoses (9 primary paranoia, 1 hysterical insanity and 1 secondary paranoia, recovering after five, nine, eleven and 2 after the lapse of twenty years). His conclusions are: (1) That normally in the psycho-neurotic forms, the recovery, when it occurs, is earlier than in paranoia, and that it is rare after the eighth year. (2) That in the psycho-neuroses the uncomplicated forms may have a longer duration than those with simple hereditary pre-disposition. (3) That the degenerated forms, taking the special cases in which recovery occurs, appertain to the class of superior degenerations (Magnan). (4) That in the degenerated forms as well as in the psycho-neuroses, a recovery may occur in spite of the long duration, with the return of the individual to his former mental state. (5) That in these recoveries the good influence of mental and physical treatment cannot be ignored. Therefore, care should be taken in the asylums to study more carefully the chronic patients and to do all that is possible to facilitate recovery and to determine those cases in which it seems most probable.

THERAPEUTICS.

MASSAGE IN INSANITY.

Guicciardi (*Revista Sperimeutale*) gives the following points on massage, employed in the cure of mental disorders.

1. Massage accelerates the lymphatic circulation and has an influence both local and general on the circulation of the blood.

2. It increases the nutrition of the muscles.

3. Under the form of *tapotment*, either light or heavy, repeated for a considerable time, it exhausts the motor and sensory nerves (producing fatigue and anæsthesia).

4. *Effluerage* and *petrissage*, equivalent in their effects, when practiced lightly, momentarily increase the irritability of the nerves; when used vigorously, decidedly and rapidly decrease this irritability. The practical point of his communication was, that massage is indicated in depressive forms of insanity. It is quickly useful in simple melancholia associated with bodily wasting. That in some cases of stuporous melancholia it produced, together with decided improvement in the physical condition, a change in the mental symptoms to a more or less decided exaltation, and that its benefits in most all cases are, increase in weight of the body, improved muscular power and digestion, etc.

URALIUM.

Bernardni (*Revista Sperimentale*) has studied the action of this new hypnotic, which is a body obtained by the mixture of chloral and urethan in the quantities corresponding to their respective molecular weights. He obtained excellent results from its use in various mental disorders. The drug was administered in various forms as powder, or dissolved in alcohol and water. In whatever form it was taken the results were always the same. Sleep was generally produced by a dose of two grains, and only in a few cases was there need of more. Hypnotic effect followed the administration in a quarter to half an hour, rarely in an hour to an hour and a half. He did not observe any action on the temperature, the heart, or the blood pressure. There was no disturbance of stomach or headache, such as is sometimes produced by chloral, and it did not appear that the patients became habituated to the remedy.

ANTIPYRINE.

Agostina and Berarducci (*Revista Sperimentale*), starting from the notions that a great deal of the agitation amongst the insane was due to painful sensations, experimented with antipyrine in 8 cases, 2 of simple mania and 6 of maniacal excitement, symptomatic of cerebral lesion (partial meningitis, etc). They found that while antipyrine did not relieve in the psycho-neurotic forms, in the other cases the quantity of one to two grains had rather a lasting quieting effect. They propose to carry their experiments further.

Roscioli at the same meeting reported that he had employed antipyrine as a sedative and tonic to the nervous system in 14 insane in quantities of 8 grains per diem (without any inconvenience by the gastro-enteric or hypodermic methods.) By this method he had no abscesses, but only pain at the point of puncture, the administration by the mouth was therefore preferable. From these experiments it follows that antipyrine is a good remedy in epilepsy, diminishing or arresting the lighter convulsive attacks and the consecutive conditions of agitation, while in other forms of insanity the drug has given no results other than to aggravate the general condition of the patient.

In still another communication Amadei reported that in his experience while antipyrine is efficacious only in vertigo and petit mal, and that the brom-ides often fail in the more severe forms, these two remedies combined were very effective in many obstinate cases. By the use of this combination he had completely suppressed epileptic attacks in the Cremona asylum for a period of four months. He admitted that the communication was not based on quite sufficient experience but would resume the subject after further experiments.

SULPHONAL AS A HYPNOTIC.

Dr. Owen Copp says (*Am. Jour. In.*, Apr.), "It is probable that sulphonal is a reliable hypnotic in the majority of quiet or mildly agitated cases of insanity; that motor excitement impairs its action; that the dose is variable, especially in women, and should be graduated to suit the individual; that in persons of average strength ten grains for women or fifteen for men are not excessive initial doses; that disagreeable symptoms may follow twenty grains in women or thirty grains in men, and that amounts much beyond thirty grains are of doubtful utility."

HYDROTHERAPY IN INSANITY.

Maccabruni *(Revista Sperimentale)* reported on observations made in con-tinuance of those reported by him at the congress in Sienna, on the effects of hydrotherapy in the Insane. He found that frontal irrigation gave better results, that hydrotherapy was of little or no use in congenital or acquired mental debility, and that it was injurious to paralytics. That it had some advantages in cases of mental exhaltation, and that while it aided the de-pressed forms, excepting those of the anxious type, that in severe alcoholism it was of little use, while it was of benefit in a few pellagrous cases. The beneficial action of hydrotherapy consists in its functional excitation dis-posing the organism to a gradual improvement. That its action is directly on the nervous system is shown by its very decided perturbing effects, and he disagreed with the view of Bernheim that suggestion had any thing to do with it. He recommended that these methods of treatment be more largely used in asylums.

Dr. Goodall (*Lancet*, May 24), writes on ihe *Treatment of the Insane.* He considers that asylums must remain in large measure simply places of refuge. The treatment consists in the adoption of measures for the preven-tion of mishap to the patient and his neighbors, the medical treatment being merely to relieve passing bodily ailments. The only exception in medical treatment is the use of sedatives, which are good and certainly use-ful. In speaking of restraint, he says it is disallowed in public, but practiced in private, and though we stand at present in terror of the name, the present fashion of alienists will probably change as fashions do. He advised employment for the insane, and massage for melancholia.

Dr. Kinnier (*Med. Rec.*, July 12) writes on the use of *Chloralamid in the Treatment of Insanity*. It has a slightly bitter taste and can be given in powder or dissolved in alcohol or water. In his experience it acts too slowly to be of much value in cases of insanity with maniacal excitement. In cases of melancholia it acts very well. In beginning its use 25 grains should be the largest dose given.

Cycling for the Insane is the title of an article by Dr. Ewart in *The Jour. of Ment. Sci.*, for July. Advocating as he does outdoor exercise for insane people, he thinks that the use of the bicycle for a certain class of insane persons would be especially beneficial.

HYPNOTISM.

At a meeting of the "Islington Med. S." (*Lancet*, May 31) an interesting discussion took place on the therapeutic uses of hypnotism. Sir Andrew Clark thought its use should be discouraged and said that the habitual practice of hypnotism upon women is injurious, morally and intellectually.

Dr. Funkhouser (*Alien. and Neurol.*, July) writes on *Hypnotism*. He says that hypnotism is not a suspension of will power, and that some people can be hypnotized against their will. People who are hypnotized are not necessarily weak and hysterical. Hypnotism sometimes has the effect of brightening the faculties. He has seen no deleterious results in his experience, though irresponsible agents might develop neuroses in using it.

THE ETHICS OF HYPNOTISM.

Dr. Geo. M. Gould contributes an article to a recent number of *The Open Court* on this subject. He says that physiologically the "hypnotic state of the somnambulic type" is a diseased sleep and morbid perversion of the attention. Psychologically it is a ruthless interruption of the normal activities of the mind. Medically it is disease, and the induction by the physician of this condition for any purpose runs counter to all our therapeutic ideals. It is a wanton playing upon the diseased personality of another by one who has no right to the power.

Dr. H. C. Wood (*The Lancet*, January 11,) contributes an article on *Hypnotism in Therapeutics Without Suggestion*. He states that in a recent visit to Paris he devoted considerable time to the observation of the prac- tice of physicians in the treatment of diseases by hypnotism. He found that all the physicians believed that the cause of the cure was suggestion. On his return to Philadelphia he selected two cases upon which he desired to experiment. The first was that of a man affected with tremor. He hypnotized the patient four times with the result of a complete cure, although he used no suggestion. The second was one of complete paraly- sis of the legs with some weakness and numbness of the limbs. The case had been diagnosed by Dr. Dercum as hysterical. After eight treatments she was able to walk long distances. No suggestion was employed in this case.

THE NATURE OF THE SOMATIC PHENOMENA OF HYPNOTISM.

The following are the conclusions of an extended memoir by Prof. Tamburini on this subject, in the *Rivista Sperimentale*, 1890, pages 147 to 174.

1. The somatic phenomena of hypnotism described as belonging to the several stages of grand hypnotism, namely the lethargic, cataleptic and somnambulistic (neuro-muscular hyper-excitability, cataleptic plasticity and musculo-cutaneous hyper-excitability) may be found in a restricted number of cases of grand hysteria, independently of any suggestions.

2. The said somatic phenomena, therefore, do not justify a nosographic division into three distinct stages, ("three different nervous states varying one from another, and each provided with its own set of symptoms," Char- cot), because these conditions are found mixed and confused, and only represent the altering manifestations of exaggerated reflex excitability, the changes of which are caused solely by the diverse nature, intensity and duration of the stimuli.

3. The said somatic phenomena are not characteristic of grand hypno- tism, since they are met with independently of suggestion and in the waking state, in cases of grand hysteria, in which they appear as so many hysteri- cal stigmata.

4. These then, in the few cases in which we meet them during hypnotism, are not the effect of it, but only represent the symptoms proper to hysteria, that are awakened in the hypnotic condition, either through increase of reflex excitability, or by stimuli operating directly upon them (acting under the guise of trauma or some other agent revealing the latent hysterical diathesis).

5. Hypnotism is therefore not a provoked neurosis, since in the few cases in which it has this appearance, we have really to do with pathological phenomena, appertaining to the hysterical condition, and the hypnotic sleep acts in these cases only as an exquisite revelation of the latent hysteria.

6. Hypnotism is only a simple condition of induced sleep. It has no pathological significance, but it has a double power of producing a certain increase of reflex excitability and a notable increase of suggestability, which furnishes the key to all the somatic and psychic phenomena of hypnotism.

7. The phenomena of hypnotism may vary infinitely, according as we have to do with healthy and robust, or debilitated, diseased, or neuropathic individuals or hysterical cases of the simple or of the major variety, but in all these cases the complicated phenomena should not be considered as appertaining to hypnotism *per se*, but as pre-existing morbid conditions that it only calls out.

8. The innumerable apparent forms that have given rise to the distinctions of major and minor hypnotism, are therefore (aside from the various possible degrees of somnolence) only to be understood as the artificial suggestion of pre-existing or superimposed pathological conditions.

MEDICO—LEGAL.

Dr. Knapp (*Bos. Med. and Sur. Jour.*, April 17, 24) writes on *Accidents From the Electric Current*. His conclusions are as follows:

"1. Currents of high potential may produce no permanent effect upon the human organism, or they may cause severe burns without other effects, or they may give rise to nervous symptoms of various kinds, similar to those seen after other injuries,—the so-called " traumatic neuroses."

2. Currents of high potential may prove fatal immediately, or they may give rise to burns which later cause death.

3. The limit of safety from death or injury from currents of high potential has not yet been determined, and is probably variable.

4. The alternating current is probably more dangerous than a continuous current of equal electromotive force."

Dr. Miller (*Medico-Legal Jour.*) publishes results of experiments in regard to the *Post Mortem Absorption of Strychnine*. A solution of strychnine was injected into a rabbit eighteen hours after death. After having been buried twelve days its organs were examined. The presence of strychnine was verified in the spinal cord, brain and liver.

In the same journal Mott states it has been shown that a poison injected into the body after death will, by the process of osmosis, disseminate itself through every organ of the body.

DELIRIUM TREMENS AS A DEFENSE OF CRIMINALITY.

T. Crisp Poole, Esq., writes on this subject in the June number of *The Medico-Legal Jour.* He states that the law exempts from punishment for crime persons who are afflicted with delirium tremens, because the delirium is not the result which could ordinarily be expected from indulgence. Another reason for this exemption is, that the disease may result from abstaining from drink, and may therefore be the accompaniment of an honest effort at reformation. He suggests that when a homicide has been committed under the frenzy of delirium tremens the person should be kept under custody for a period not less than two years.

Dr. T. L. Wright; (*Alienist and Neurologist*, Jan.,) writes of the state of *Inebriate Responsibility*, in which he says in viewing the effects of alcohol upon human nature, a great fact should always enter into consideration, that alcohol in appreciable quantities invariably produces paralysis in some degree, great or small, in every organ and function of the body and mind. He calls special attention to the fact that in drunkenness there is a disturbance in the normal relationship of the various parts of the human organism. There is a sort of general bodily and mental inco-ordination. His conclusion is, that the morbid self-exaltation of the drunken mind impels to a line of conduct out of relation to the fitness of things, and beyond the capacity of the person to control. He says alcoholic paralysis impairs not only the sense of feeling, but the co-ordinating power of the body, so that community of action between the organs is difficult or impossible.

Dr. H. N. Moyer (*Medico-Legal Jour.*, Mar.) contributes an article on the *Medico-Legal Relations of Shock.* It is manifest that injuries must of necessity have two results in causing death, either by immediate shock, or by their remote and secondary consequences, such as hemorrhage, inflammation, abscess, etc. The different indications upon which diagnosis of shock can be based are the following: First, duration, if the injuries have been inflicted more than twenty-four hours, we are scarcely justified in attributing death to this cause; second, the condition of the heart; if empty and contracted it is strong negative evidence against the theory of shock; third, are the injuries in their nature and extent sufficient to cause fatal shock? An error is apt to be made in establishing shock as an absolute cause of death when it is only one of several concurrent causes, the relative importance of which varies in different cases.

Clark Bell (*Medico-Legal Jour.*, June) writes on *Electricity in its Medico-Legal Relations*. He says that electrical engineers should be called upon to pass upon the questions:

1. What precautions could be taken to reduce the danger of injury to persons from the use of high pressure wires, whether submerged in great cities or suspended in air outside city limits?

2. What precaution could be adopted in the distribution of the lights, by transformers, so as to make the consumer safe as to personal injury?

3. What quantity can be carried over each class of conductors, and what regulations and limitations as to quantity should be placed on each class and kind of conductors?

M. Louis Proval (*Annales Medico-Psychologiques*, Jul. and Aug.) contributes an article on the *Legal Responsibility of the Insane*. He holds that the court should consider all the circumstances of the life of the accused, especially those that relate to his childhood and education. If the person has had good advantages, has been well educated, has had opportunities for acquiring correct moral principles, he should be dealt with more rigidly than one who has been born of ignorant and degenerate parents, and who is without education, and has spent his early life amid conditions that tend to moral and intellectual debasement. Dr. Ball has also proposed to adopt a principle of limited responsibility for those who, having grown up and lived in the midst of vice and unfavorable conditions generally, are accused of crime.

Dr. H. H. Smith of Philadelphia, publishes a paper on *Concussion of the Spine in its Medico-Legal Aspect*, in the *Jour. of the Amer. Med. Assn.*, June. His conclusions were summarized as follows:

1. Concussion of the spine may sometimes occur from various forms of violence, there being nothing peculiar in the application of the force to the body, as the result of derailment or collision of railroad trains.

2. The pathological changes in the molecular structure of the cord as the result of shaking, jarring, or so-called concussion of the cord, where attended by paralytic symptoms, may be due to hemorrhagic effusions, or be shown post-mortem in softening, or localized or limited atrophy. In cases due to hemorrhage, the symptoms may be improved by judicious treatment and permanent disability prevented.

3. The possibility of pre-existing neurasthenia, or hysteria, or fraud, on the part of the claimant should be carefully noted in forming a diagnosis in these cases.

4. As the question of permanent disability, justifying exemplary damages, is frequently raised in claims of the kind alluded to, it should be recollected in forming a prognosis that numerous cases are reported of

recovery, or marked improvement, even after the occurrence of paralysis.

5. No physician should go into court and swear that a plaintiff had had a concussion of the spinal cord or of ts nerves, unless he has proved the disturbance of the normal functions of the cord, as shown in sensation or motion, or both, and that the symptoms appeared soon after the injury.

———

The London Med. Rec. says that steps have been taken in Hungary for the formation of a senate of medico-legal experts. It will consist of fifteen members and it will be its duty to report to the Minister of Justice on all cases on which the courts require the aid of medical science.

NEWS AND COMMENTS.

The editor of this journal does not flatter himself that the profession is waiting breathlessly for the appearance of a new periodical. It is already afflicted with so many that the issue of a new one should be accompanied with an apology rather than with the common assumption that it will "fill a long-felt want."

If there is any special reason why an apology is not demanded in this case, it is, that the Review, being a compendium of the current literature of an important specialty, will occupy a unique position in Medical Journalism. In this form it is believed that it will find a place in the medical world.

At the outset a few words of explanation seem needful. Original articles will occasionally appear in the REVIEW, and will be written by specialists, but it is the intention to have not more than one in any issue. While this periodical will practically mirror the literature of the specialty, there must of necessity be a choice made in selecting articles, for not all can be noticed. Preference will, therefore, be shown to literature that has an immediate practical bearing, to the neglect, if necessary, of technical and speculative productions, however meritorious. It may be well to state that the REVIEW will not depend for existence upon its subscription list, and that it is not an experiment, but from the first may be considered a permanency.

Prof. William Goodell in the *Medical News*, protests against the habit of physicians of attributing woman's ailments to uterine disease, and probing, cutting, cauterizing, and stitching these poor victims of a theory. The protest of Dr. Goodell carries the more weight from the fact that he is a

gynæcologist, and considering his large experience he ought
to know whereof he speaks.   This is not the first time that
the profession has been warned against this *utero-mania* that
affects many physicians.   Neurologists have frequently called
attention to the unnecessary interference in treatment with
the female pelvic organs by physicians, and now that gynæcol-
ogists have joined in the protest we may hope that there will
be a reformation in this direction.   Every neurologist knows
that frequently symptoms referable to the uterus occur in
cases of nerve failure, symptoms which are but local expres-
sions of the general condition of the nervous system.

If women have head-ache, spine-ache, indigestion and cardiac palpitation in consequence of nervous exhaustion, why
should they not have uterine-ache, ovarian pain, menstrual
irregularities from the same cause?   The sexual apparatus of
woman as the agency of reproduction embodies the supreme
function of her being, and it is of necessity complexly related
to the nervous system.   There being this intimate structural
and functional relationship it could not be otherwise than that
the disorders of one should be reflected upon the other.   It is
an unhappy day for any woman who has nerve failure with
false uterine symptoms when she is told that she has uterine
disease.   Being already morbid and despondent there is a
special vividness in the phrase that impresses her, and this
ghost of "womb trouble" haunts her with its nasty associations, aggravating the original trouble and making its cure
more difficult.   What physicians ought to bear in mind in
this connection is, that in any condition of poor health in
women there are almost certain to be symptoms of uterine
derangement.   The thing to do is to treat the primary condition and not its counterfeit.

———————

Some months ago there appeared in the "Medical News"
a symposium by several prominent neurologists and gynæcolo-
gists on co-education and the higher education of woman.
The articles were contributed by Prof. William Goodell, Dr.

S. Weir Mitchell, Dr. T. Gaillard Thomas, Dr. J. R. Chadwick, Dr. M. Allen Starr, and Dr. J. J. Putnam. They were written in response to a series of questions addressed to these gentlemen by the editor of the *News*. These gentlemen were requested to give their opinion as to whether co-education is advisable; whether the higher education of woman tends to produce disease and reduce capacity for child bearing.

Dr. Starr calls attention to the fact that women have not the power that men have of concentrating the attention. Dr. Putnam says, "It is a noteworthy fact that the question is seldom asked how the health of men is affected by college life," and he can see no reason why the health of women should suffer from study more than that of men. Dr. Thomas sees no danger from co-education or from the higher education of woman, and Dr. J. R. Chadwick expresses a similar opinion.

Dr. Goodell thinks that as colleges are now conducted the higher education of woman is injurious, though he suggests the fault is in the mode of education rather than in the simple fact itself. He considers that woman's dress has much to do with crippling her health.

Dr. Mitchell believes that constant hard study is more injurious to women than to men. He says he considers co-education abominable, foolish and absurd.

In regard to the higher education of woman it is worthy of remark that the world is coming to recognize that she is fit for something else beside child bearing and domestic service. In the earlier and ruder stages of human existence, fecundity was not unnaturally regarded as the chief measure of woman's usefulness, but if as her mental capacity is increased by education, the race gains thereby, we can certainly regard a reduction of her fertility without alarm. The reply of the lion in the fable that her solitary cub was a *lion* emphasizes the point. What this world needs is talent, and the educated mother who can give birth to a Bacon or a Spencer can afford to be indifferent to the capacity of her more prolific neighbor.

8

Much of the impaired health that educated women suffer from, and that seems to have its origin in student life, can be avoided by recognizing that education should be a physical, as well as a mental discipline. Every teacher should keep constantly in view the interdependence of mind and body, or to state it more narrowly, brain and muscle. The accepted cerebral localization of the function, of the various organs, the relative size of the heart and brain, the demonstrated vital filiation of the entire organism, show the necessity of combined mental and physical training in order to accomplish the best results.

The achievements of modern manual training show that the effectiveness of mental drill is vastly increased by associating it with physical culture.

It is greatly to the credit of modern pedagogy that it recognizes the disciplinary effects of education aside from the mere imparting of information. In education *what* one learns is not so important as *how* he learns, that is, the mental *habits* that study develops are of vital importance, for these habits must serve in working out the problems of daily life. Our facts may fail us utterly, but the mental discipline that comes from the struggle for their acquisition, serves us ever after. Not only school life, but all life, is discipline, having for its object the development of character, that consummate flower of our existence, and the facts that we so eagerly seek and so soon revise are the steps of the ladder by which we climb. The overcoming of obstacles, the incessant struggle of life, builds up character, and though we may often fail, part of the gain is in the vigor and self-confidence that effort brings. The infant grasps at the moon, and though it fails in its immediate object, yet by this and a thousand like efforts it learns the uses of its senses and its muscles; and we infants of a larger growth reach our hands of thought for much that eludes our grasp, but the effort contributes to our development and adds to the mentality of the race.

The literature of insanity records no warmer discussions, nor greater differences of opinion than in the matter of mechanical restraint. Especially have these differences existed between English and American alienists. In Great Britain, from the time of Connolly little or no mechanical restraint has been used in the management of the insane until very recently. Lately, however, there has been a change in the views of many British superintendents in regard to the use of it, the tendency being to use it in certain cases. On the other hand, American superintendents have used it less than formerly, and there are now several hospitals for insane in the United States where it is not used. The new Lunacy Law of Great Britain recognizes for the first time in the history of lunacy the right of superintendents to use mechanical restraint in the management of the insane. As might have been expected there has been much discussion of the subject among English alienists. The *Journal of Mental Science* contains a correspondence which passed between Dr. Alex. Robertson and Dr. D. Yellowlees, that fairly illustrates the position of the two sides. Dr. Robertson, who is opposed to the use of restraint, claims that if it is beneficial this should be shown in fewer suicides, homicides, and all accidents the result of violence. Dr. Yellowlees uses restraint in cases where the suicidal impulse is intensely strong, in cases of extreme and exceptional violence, in extremely destructive cases, and the helpless and incessantly restless patients. He also says: "The question whether the use of restraint is ever beneficial, and therefore right, in the treatment of the insane, might surely in these days be considered on its own merits, and apart from traditional authority or personal bias. There is no other question of medical treatment about which physicians may not legitimately differ, and agree to differ; but let any one dare to think or act independently as regards this particular treatment, let him dare to say that restraint prescribed by a humane and experienced physician is totally different from restraint inflicted by cruel and unen lightened men in bygone days, and he at once encounters reproach and blame, as if non-restraint were a rule imposed from heaven, whose universal obligation and absolute wisdom it was little less than sacrilege to question."

In regard to the advisability of the use of mechanical restraint in insanity we do not believe that statistics or experience however large will decide it, and it will probably forever remain an open question. That a given insane person shall or shall not be restrained will depend quite as much upon the mental bias of the physician, as upon the mental condition of the patient. One physician, believing that restraint will prevent violence, and considering it humane, will impose it. Another physician, believing all mechanical restraint unnecessary or cruel, will refuse to use it under any circumstances.

We fancy we see the human race still existing upon this planet thousands of years from now and wrestling with many new problems, but with some of the old ones still on its hands. Among the latter the question of restraint for the insane will be one; Alienists will still be discussing it and using, too, the same arguments for and against, that were urged by their predecessors of the semi-civilized nineteenth century.

In his last annual report of the Royal Edinburgh Asylum, Dr. Clouston observes that the types of the mental symptoms of the educated are more differentiated and distinct, while the lower we go in the social scale the less distinct and complex are the manifestations of insanity. In savages the insanity is more apt to assume the form of acute delirium. Among the educated classes we find more melancholia and monomania, which indicate more subtle reasoning and more vivid picturing of the feelings. Dr. Clouston's experience leads him to think that the mental and moral causes, such as grief, worry, disappointment in love, etc., are more common among the educated, while the physical causes, such as intemperance, bodily disease, and excesses, are more common among the uneducated. This is rather a fascinating generalization, but we believe the exceptions are so numerous that it can have but little value. Surely the uneducated man does not escape the consuming effects of worry, anxiety, and disappointment; on the contrary, from the very conditions of his existence they must be a constant element in his life. In him, too, as they operate upon a mind less developed and with fewer resources they will more easily impair the mental vigor and develope the morbid mental habits that readily pass into insanity. The uneducated man is exempt from some of the causes of insanity that beset his more educated neighbor, but this exception is dearly bought, for he is subject to other causes more numerous and more potent. The uncertainties of employment, the insufficiency of remuneration, the petty cares and grinding anxieties that are the partners of ignorance and poverty, are influences that depress vitality, weaken the resisting powers, and make the occurrence of insanity an easy matter.

Dr. Clouston says that in Scotland the death rate is one-third greater, and in England one-half greater among the uneducated, and that a smaller proportion of the uneducated recover from insanity. His explanation of this is, that among the uneducated insane there is·more organic brain disease, and hence less insanity that is recoverable. We doubt if this accounts for all the difference, for it is probably a larger and more complex problem than this would seem to indicate. There is another element in the problem that is important, and that is, the greater mental strength and larger mental resources that are part of the equipment of intelligence. The weaker mind and habitually narrow range of ideas of the ignorant man, make insanity a more serious thing for him than for the educated man, whose stronger mind and greater resources find various forms of diversion by which he can more easily escape from the tyrannizing and depressing power of his delusions. We have had frequent opportunities of observing how difficult it is to divert an uneducated

insane man, who cares for no amusement and whose life has
been narrowed to an occupation that has made a minimum
of demand upon the reasoning faculties.   On the contrary, we
have observed the greater ease with which an intelligent
insane person can be entertained and diverted, and drawn
away from his morbid self.   These mental therapeutics that
are thus more applicable to the intelligent, are of vital importance in insanity, for they rest the brain that is exhausted
with the dull round of monotonous and harassing delusions
and afford an opportunity for recuperation.

We most emphatically dissent from the editorial opinion of
an eastern medical journal, that women physicians have seldom
been successful, and that experience shows they are, with
few exceptions, unadapted to the profession.   We believe
as large a portion of women physicians succeed as men, and
have no doubt so far as talent is concerned they equally merit
their suceess.

Considering the opposition they have had to overcome in
the profession and out of it, their success is surprising, and
the fact that they have conquered many of these prejudices,
and have thus far won, is the best of evidence that they
deserve to win.   The argument that woman unsexes herself
by entering a profession is puerile.   There is no reason why
any occupation woman now engages in should interfere with
any womanly duty or function any more than those duties she
has always performed.   Any man who had his choice would
doubtless as willingly choose a practicing woman physician
for his mother as one engaged in ordinary domestic work.
Ten years ago a prominent Englishman said he did not
believe there was a woman in the United Kingdom who
understood the principle of the rule of three, and yet recently
a young girl of twenty-two won the highest academic honors
in the realm, showing the highest proficiency in mathematics.

There are doubtless women physicians who are not adapted
to the profession, but the law of survival of the fittest will
sooner or later lay its firm hand upon them.

Many women are, however, specially adapted to the practice of medicine, and we hail their entrance into the profession
as an evidence of that progress that everywhere is enlarging
the opportunities and aims of life.   The women of this generation are beginning to feel that in place of the mere drudg-
ery of life they should have a part in the world's work that
educates and elevates.

The address in Medicine delivered by Dr. N. S. Davis
before the "American Medical Association" is able and timely.
His views on the use of antipyretics we believe to be sound.
The wild rush for these remedies that has been observed the
past few years is another illustration of the fact that medicine,
like society, has its fashions, and fashions, too, that it sometimes looks back upon with regret.   Considering that we know
so little of the nature of fever that we are every year framing
new theories to account for it, it would seem that men would
use with caution a remedy that, whatever it does, certainly in
fever conceals from view the real condition of the patient.
To attempt to cure a fever by giving an antipyretic is like
throwing a cloth over a severed artery and assuming the
bleeding has been checked because we do not see it.

Dr. Davis reiterates his well known views of the medicinal
use of alcoholic liquors.   His statistics show that in hospitals
where they are not used the death rate is much smaller than
in hospitals where they are used.   A few years ago the regular ration of beer was discontinued in some of the English
insane asylums with the result that the patients were more
quiet and orderly.

The death of Dr. C. H. Nichols of the Bloomingdale Asylum removed from among alienists an amiable and most lov-
able man.   The doctor had been connected with insane hospitals from early manhood and had long had a high standing
in the specialty.   Although his years of service entitled him to

retirement, he continued to work with his usual enthusiasm, and at the time of his death was preparing to make an addition to the asylum.

Dr. Nichols was a man of exceptional ability, and with this was combined a mental poise and cheerfulness that made him superior to the habits of worry and irritation that sap the strength of many able men. He was modest and retiring, of a kindly disposition and unvarying good nature, yet a very positive man and of strong convictions. He may never have a monument built to his memory, but in his life time he erected one in the hearts of those who knew him and who will not lose the moral that attaches to his useful life.

The helpfulness of the useful life often lies in the history of past struggle that brought out and developed the higher qualities, the self-confidence that gives momentum, the conquered adversities, the perseverance that has passed all obstacles and from the coarse material of experience has woven the fine fabric of character. The study of such a history is valuable to every one; it is an object lesson that is not forgotten. From the time of Dr. Nichols' first connection with hospitals his life was one of ceaseless care and grave responsibility. He was beset by trials that cut deep into the sensibilities, and sometimes break mens' spirit, and yet he was always hopeful and serene, and he had the success that all brave and patient men deserve.

It is profitable to consider also, that such a person, in addition to fulfilling the duties of daily life, has upon others an enduring influence that may be said to represent the immortal part of human nature. It is one of the happy uses of noble natures that the higher qualities which they exhibit do not die with the possessor, but like the energy of the sunlight that is knit into the fibre of the growing tree, they enter into the lives of others and reappear in character.

———

Though it does not belong to our special field we are glad to commend the work of Prof. and Mrs. Geo. W. Peckham of Milwaukee, on the family antidæ of spiders. There is very much in such studies that interests the psychologist, for many of the problems of the human mind have been elucidated by facts of the psychic life of the lower animals. In the spider, as elsewhere in insect life, we see a mimicry of many of the mental characteristics of mankind, as though this clear consciousness of ours had sought itself in animals and found itself in man.

How near the instincts of human and insect life approach each other is illustrated in this interesting essay. Our sympathy is excited by the hard lot of the poor male spider who must become a parasite upon his spouse or, after a tedious and vexing courtship, submit to be eaten by the unfeeling female. We see in this the ancient origin and hopeless fixity of the "female will" by which such hard tyranny is exercised over the so-called sterner sex, for it appears much older than the brief period of human existence and spans the centuries of the past to the time, at least, when the parent family antidæ spun its first web on a carboniferous island.

The Regents of the "University of Wisconsin" have done a wise thing in creating the chair of " Experimental and Comparative Psychology," and Prof. Jastrow, who has been selected to fill it, appears to be a man of special fitness for the place. His articles on " Psychological Research " in *The Open Court*, and his descriptive sketches of the present state of psychology in Europe, furnish the best of evidence of his qualifications for the position he occupies.

Dr. H. M. Hurd, for many years Superintendent of the "Eastern Michigan Asylnm for the Insane," at Pontiac, has been appointed Superintendent of the Johns Hopkins Hospital. In addition to the possession of superior executive capacity, Dr. Hurd has shown exceptional ability in the literary field. His contributions to the literature of Psychiatry have been of a high order, and we trust his interest in the specialty will not abate by reason of the change he has made.

———

Dr. Clark Gapen, formerly professor of jurisprudence in the University of Wisconsin, is practicing in Omaha, and is health officer of that city.   Dr· Gapen stood high in Wisconsin and we predict for him a successful career in Nebraska.

———

Dr. E. H. Van Deusen, who for many years was Superintendent of the "Michigan Asylum for Insane," is living in retirement at Kalamazoo.   Dr. Van Deusen is not yet an old man and takes a lively interest in all that pertains to his former specialty.

———

Dr. D. R. Brower, of Chicago, is spending the summer among the Neurologists of Europe.

———

NEW BOOKS, ETC.

Brief notices of new books on Neurology and Psychiatry will be given in this department, but detailed reviews will not be attempted.

———

THE ANATOMY OF THE CENTRAL NERVOUS ORGANS IN HEALTH AND DISEASE.—By Heinrich Obersteiner.

Translated by Alex. Hill, M. R. C. S.

This work treats of the Anatomy of the Central Nervous Organs, and is complete and unquestionably reliable.   A special and valuable feature of the work is, that after considering the anatomy of a part, the pathology is also given.   The high standing of both the author and the translator is sufficient guaranty of its accuracy.   No progressive physician can afford to be without it.   The American edition is issued by P. Blackiston, Son & Co., Philadelphia, and is an exact reproduction of the English edition.

A RESUME OF LECTURES UPON THE STRUCTURE AND PHYSIOLOGY OF THE CENTRAL NERVOUS SYSTEM WITH LOCALIZATION OF ITS LESIONS.

By C. EUGENE RIGGS, M. D., OF ST. PAUL.

This is a pamphlet of 30 pages which, with additional lectures soon to be published, as we are informed, will represent the course of lectures delivered to the medical class of "The University of Minnesota."

The anatomy of the pons, medulla, and cranial nerves is contained in the first part with minute directions for diagnosis of their various lesions. The second part is devoted to the structure of the "Cura Cerebri and its Lesions," and the last part to "The Internal Capsule and its Lesions." When the remaining part is issued it will form a convenient sized volume, and with its numerous illustrations and concise descriptions will be one of the best aids to diagnosis of central nerve lesions in existence.

The following works have either appeared in German or are advertised to appear. Those advertised are so designated.

Expert Evidence in Insanity.—Leppman.

Clinical Lectures on Psychiatry.—Meynert.

Manual of Mental Medicine.—Scholz.

The Emotional Life in its Principal Phenomena and Relations.—Nahlowsky.

Cyclopædia of the Practice of Medicine.—Eulenberg.   Contains articles by neurologists.

Clinical Instruction in Psychiatry.—Krafft-Ebing (adv).

History of the Care of the Insane in Germany.—Kirchoff (adv).

Visual Impressions and their Analysis.—Kries (adv).

Functions of the Cerebral Cortex.—Munk (adv).

Mechanism of the Cerebral Circulation.—Geigel (adv).

Electro Therapeutics.—Rosenthal and Bernhard, third edition (adv.)

Electricity in Medicine.—Ziemssen, fifth edition (adv).

Electricity and its Application to Practical Medicine.—Meyer, fourth edition (adv).

Compression-Myelitis Following Vertebral Cares: Schuman's Contributions to Cerebral Surgery.—Navratil.

A Text Book of Brain Diseases.—Wernicke, three Vols. (adv).

Craniometry and Cephalometry.—Benedik.

Syphilis of the Central Nervous System.—Oppenheim (adv).

The Journal of Psychology, and Physiology of the Organs of Sense.   To be edited by Ebbinghaus and Kœnig.

Pathology of the Brain, two Vols.   Prof. Henschen of Upsala, written in German.

BRITISH AND AMERICAN.

Nerves of the Human Body.—A. W. Hughes.

Diagnosis of Diseases of Brain, Spinal Cord, and Nerves.—C. W. Suckling.

Cure of the Morphia Habit.—Jennings.

Psycho-Therapeutics.—Liebault.   Translated from the French.

Heredity, with Reference to Disease.—Lithgow.

FRENCH.

Epilepsy and Epileptics.—Ch. Féré.

Diseases of the Nervous System; Muscular Atrophies and Amyotropic Diseases.—F. Raymond.

Regicides.—Régis.

Dreams, their Physiology and Pathology—Tissiè.

Clinical Studies in Mental and Nervous Diseases—Dr. Falret.

No. 2.   NOVEMBER, 1890.   Vol. 1.

THE REVIEW

OF

INSANITY ᴬⁿᵈ NERVOUS DISEASE

A QUARTERLY COMPENDIUM OF THE CURRENT LITERATURE OF

NEUROLOGY AND PSYCHIATRY.

EDITED BY

JAMES H. McBRIDE, M. D.,

Superintendent Milwaukee Sanitarium for Nervous Disease.

ASSOCIATE EDITORS :

LANDON CARTER GRAY, M. D.   C. K. MILLS, M. D.

NEW YORK CITY.   PHILADELPHIA

C. EUGENE RIGGS, M. D.   W. A. JONES, M. D.

ST. PAUL, MINN.   MINNEAPOLIS, MINN.

H. M. BANNISTER, M. D., KANKAKEE, ILL.

$2.00 PER YEAR.   SINGLE NUMBERS, 50ᶜ·

SWAIN & TATE,
BOOK AND JOB PRINTERS,
MILWAUKEE.

INDEX.

REVIEW

OF

INSANITY AND NERVOUS DISEASE.

ORIGINAL ARTICLE.

INFANTILE PARALYSIS.

D. R. BROWER, M. D. ·

(Professor of Diseases of the Nervous System in the Woman's Medical College Chicago, and Professor of Mental Diseases in Rush Medical College, Chicago.)

Synonyms: Infantile Paralysis; Polio-Myelitis; Anterior Acuta; Acute Atrophic Paralysis; Essential Paralysis of Infancy.

Definition: A form of paralysis characterized by a primary febrile stage, a secondary paralysis and a tertiary atrophy without loss of sensibility.

Ætiology: This disease occurs usually between the ages of six months and four years, but why, is a mystery. The period of dentition has been supposed to be an active factor in its production, but surely this physiological process is not sufficient to account for it. It does not seem to be hereditary and it frequently attacks children who seem to be in the most robust health. It is much more frequent in summer than in winter, therefore cold, which has been supposed to be one of the important factors in the production, can hardly be charged with it. It frequently follows the exanthemata, but it can hardly be charged to the patho-

2

logical processes upon which these diseases depend. So we may safely say we do not know the cause of Infantile Paralysis.

Morbid Anatomy and Pathology: Acute inflammation of the cells in the anterior horns of the spinal cord is the pathological condition upon which the symptoms depend, and whether this inflammation begins in the cells or in the neuroglia is a matter of question, but sooner or later it certainly becomes both parenchymatous and interstitial, and the result, be the pathogenesis of the process what it may, is the destruction of the cells in the anterior horns of the spinal cord that are concerned in giving rise to motor impulses and trophic influences.

Symptoms: The disease is usually ushered in by a more or less intense fever with or without convulsions. The temperature is usually about 101°, although occasionally a temperature of 104° has been observed. This fever may last for two days but it usually terminates in a few hours. There is nothing pathognomonic about this fever, it resembles in all its leading symptom the ordinary febricula of childhood, and will be almost invariably mistaken for some gastro-intestinal disorder, or for the beginning of one of the exanthemata. The fever is accompanied, as is usual, with headache, prostration, loss of appetite and restlessness.

It soon subsides, the general health improves and the child is apparently well, when it will be discovered that one or all of the limbs are paralyzed, and then the paralyzed limbs will be found to have undergone rapid atrophy. The temperature of the paralyzed part is always below the normal side, the difference in temperature may be from five to ten degrees. The sensibility of the paralyzed limb, or limbs, is not altered.

This paralysis reaches its maximum of extent and severity within a comparatively brief space of time, and it shows no progressive tendency, on the contrary it soon begins to recede, and after its recession the distribution of the paralysis is found to be exceedingly variable, having, however, a special tendency to manifest itself in the lower extremities, and then confined in its effects to a few of the muscles, especially those muscles that have been the latest in development. Frequently it is hemiplegic in its distribution, and then we will sometimes be in doubt as to whether the paralysis is spinal or cerebral. But the differential diagnosis is usually easily made when the atrophy becomes manifest, and along with this atrophy the diminished reflexes of spinal paralysis, rather than the exaggerated reflexes of cerebral paralysis, will aid in the differentiation. The functions of the bladder and rectum are rarely if ever disturbed.

After a few days, or it may be a few weeks, a gradual improvement in the paralysis takes place; the improvement may affect a greater or smaller number of the muscles involved, and some of the paralyzed muscles may completely recover. The paralyzed muscles as a rule will not contract to a faradic current, but will usually contract to a galvanic current of the proper degree of intensity, and this electrical condition of the muscles is one of the most valuable means of prognosis, as its use is one of the most valuable means of treatment. If the paralyzed muscles will contract to any electrical current, such as the child will tolerate, then there is a prospect of restoration, the chances of recovery being in inverse ratio to the strength of the current required. If the muscles respond to a mild current the prospect of improvement and possible cure is good.

These cells of the anterior horns are the trophic centre of the bones as well as of the muscles, and, therefore, atrophy of bones is also a part of the clinical history of this disease. The skin which derives its trophic function from another portion of the spinal cord is not disturbed in its nutrition, so that bed sores and ulceration of the skin are no part of the phenomena of the disease.

Diagnosis: As already stated, it is impossible to diagnose the febrile stage of this disease from other fevers. The differentiation of the paralysis must depend upon the absence of cerebral symptoms, the rapid onset of the paralysis, the speedy manifestation of the atrophy, the lowered temperature of the paralyzed parts and the absence of sensory disturbances.

Treatment: The treatment of the initial fever will usually be that employed for the treatment of the febricula of children generally, and for this purpose we know of no better remedy than small doses of *hydrargyri chloridum mitis*, say the one-twentieth of a grain given once an hour until its effects are produced upon the intestinal tract. This remedy is not only indicated in cases of gastro-intestinal disorder, but it must be beneficial because of its alterative power upon the inflammatory disturbances in the spinal cord, upon which the fever depends. Were it possible to differentiate this febrile state from the ordinary fevers of childhood, then *ergotin* might be of advantage administered subcutaneously in doses of from one-fourth to one-third of a grain, and *belladonna* might be given with advantage, because of its power in diminishing the blood supply of the spinal cord.

As soon as the diagnosis is certainly made then alteratives should be administered, and the best of this class is iodide of potassium, mild counter-irritation used over the spine, and careful attention given to elimination by bowels, skin, and kidneys.

As soon as we have reason to suppose that the inflammatory products have been removed by this alterative process we should begin the use of electricity. This can usually be done in from a week to ten days after the febrile stage has passed away. The mildest electrical current should be used which is sufficient to produce muscular contraction. The electrical current that does not produce muscular contraction in the paralyzed muscles is of no possible service in the disease. One large electrode, the positive, should be placed at some indifferent point, such as the back, and the negative electrode, a small one, should be placed on the motor points of the paralyzed muscles, and one after another these paralyzed muscles should be made to contract, limiting at each *seance* the contracting of each muscle to three times; many contractions at a *seance*, by exhausting the muscle and the centre, will probably do very much more harm than good, so that *seances* should always be short, and very few contractions of muscles made at any one sitting. It will be found in favorable cases that the strength of the current necessary to produce the contractions will gradually become less and less, and after a time the muscles will be found to respond to the faradic currents. Along with this electrical treatment appropriate gymnastic exercise of the muscles, and massage should be employed. Frequently shampooing the limb with a spray of hot and cold water alternately is sometimes of service.

After the iodide of potassium has been given to the

extent that seems desirable, then remedies should be administered that are stimulating to the nerve centers, and phosphorus, strychnia, and arsenic are the three best remedies for the purpose. These remedies should be administered as freely as is possible without deranging the general nutrition of the patient.

The child should be encouraged to use the paralyzed limbs to the utmost that is possible. The exercise of the will power possible over the action of the spinal centres will improve their nutrition. I think it is unwise, certainly so in the earlier stage of the disease, to use any mechanical contrivances to supplement the action of the muscles. The child wearing such appliances will have less occasion to exercise the will power in arousing into activity the dormant centres of the spinal cord, hence they should be avoided as far as possible. Later in the case they may be necessary, but as long as there is any prospect of cure the use of mechanical contrivances should be avoided.

The electrical treatment of this affection should continue through many months, the progress toward recovery is exceedingly slow in the great majority of cases, but a prolonged, faithful, and persistent use of electricity, along with the tonics indicated, will sometimes bring about surprising results. The difficulty about the treatment of such cases is, that the use of electricity is more or less painful to the child, its frequent application is troublesome and annoying to the mother, and the results as manifested from day to day are so trifling, or indeed so unappreciable, that there does not seem to be sufficient inducement to continue the disagreeable process to the extent that is absolutely necessary. But I am sure that a more prolonged and persistent use of this remedy would be followed in a great many cases by success. But I must condemn the use of electricity for this purpose by the laity. It is easy for the physician to order a battery to be used by the mother or some member of the family, but as a rule such applications of electricity are worse than useless, the current either being too strong and thereby producing exhaustion, or too weak so that no beneficial effects are experienced. The use of electricity should always be in the hands of the physician, and in the beginning of the treatment a daily *seance* is necessary. It is no uncommon experience with me to have cases of infantile paralysis come into my office and when I recommend electricity the friends reply that they have been using it for a long time, possibly for a year, and when I come to ascertain the kind of electricity they have been using I find it has been the faradic current, though the muscles will only respond to the strongest possible galvanic current. Such a use of electricity is deplorable.

70 STATE STREET, CHICAGO.

TRANSLATORS.

ITALIAN.

H. M. BANISTER, M. D., Asst. Physician Eastern Illinois
Hospital for Insane, Kankakee.

GERMAN.

G. J. KAUMHEIMER, M. D., Milwaukee; I. LANGE, M. D.,
Chicago.

RUSSIAN.

T. KACZOROUSKI–PORAY, Chicago.

SCANDINAVIAN AND FRENCH.

C. FRITHIOF LARSON, M. D., Chicago.

SPANISH.

HORACE M. BROWN, M. D., Milwaukee.

NEUROLOGICAL.

ANATOMY AND PHYSIOLOGY.

DEVELOPMENTAL ANOMALIES OF THE CORD.

The specimen was from the cord of a patient who died of progressive paralysis, complicated with degeneration of the lateral columns.

On a section about ¼ inch below the medulla, the anterior gray horn on the right side is seen to be composed of two triangular masses, connected with each other and to the posterior horn, which is normal in shape, by a narrow neck of gray matter. A little further down these masses become oval in shape, and send prolongations almost to the surface of the cord. About an inch below, the masses have united to a single triangular body, which decreases in size as it proceeds downward, while still another inch down it forms a thin arch with its concavity forward.

The white columns showed a normal location in relation to the gray matter.

Both parts of the horn sent fibres to the anterior roots.— (Dr. Buchholz, *Archiv. f. Psychiatrie*, Band XXII, P. 231.)

G. J. K.

In his address on *Cerebral Anatomy* before the "British Med. Assn.," Dr. Cunningham discussed the theories in regard to the origination of cerebral fissures and convolutions. He adopts the explanation of Jelgersma. The gray cortex of the brain, which in members of the same species maintains a tolerably consistent thickness, increases by surface extension. With every advance in the growth of the gray matter there must be a relative increase of the adjacent white matter. Now, if we extend the surface of a smooth brained animal, say four times, and at the same time desire to keep the surface even, we must provide eight times as much white matter to fill the interior of the gray capsule. Or, to put it in different terms, if we lengthen out the radius of the cerebrum, say ten times, we acquire a surface extension one hundred times greater, and an internal capacity one thousand times greater. The geometrical law involved is, that in the growth of a body the surface increases with the second, and the interior with the third power of the radius. Thus it is evident, that if the proportion of gray and white matter is uniform, that in the evolution of a large animal from a small one, a disproportion between the gray capsule and the white core of the cerebrum must result. This is compensated for by the extended cortex pressing itself in folds and so reducing the capacity of the capsule to a degree which brings it into correspondence with the white contents. Hence, the formation of the convolutions is the result of the tendency on the part of the superficial layer to increase by a surface extension, and of a mutual space accommodation of the gray substance and of the white conducting paths. Jelgersma points out that the extension of the cerebral surface depends, first, upon the absolute quantity of the gray matter, second, the thickness with which this is spread over the surface. The absolute quantity of gray matter is determined by the size, or by the mental power of the animal, or by both. Although the thickness of the gray matter is much the same in the same species, it differs considerably in different animal groups. It follows from the theory that the more thinly the gray substance is spread over the surface of the white matter, the more convoluted the brain will be. In the cetacean cerebrum the gray cortex is very thin and hence the surface is extremely convoluted. An explanation similar to Jelgersma's was offered by Baillarger many years ago.—(*Brit. Med. Jour.*, Aug. 2.)

THE RELATION OF THE CONVOLUTIONS TO THE POSTERIOR COLUMNS.

In 1881, Flechsig called attention to bundles of fibres entering the posterior central and adjoining convolutions, which fibres are connected through the tegmental radiations, the stratum lemnisci, etc., with the posterior columns of the cord. He did not then succeed in showing how much of the posterior columns, inter-olivary layer, stratum lemnisci, etc., passed to the cerebral convolutions. A recent pathological case has shown the anatomical relationship of the parts in question, the case being one of parencephalic defect of the central convolutions with secondary degeneration of the loop of Henle and the pyramidal tract. The subject was a woman who died at 54, having had right sided paralysis from her second year. The disease involved exclusively the substance of the mantle. (Cortex and adjacent medullary area, as far as the ependyma of the lateral ventricle, and not involving the internal capsule or basal ganglia.) The left posterior central convolution seemed to be absent, and a portion of the paracentral lobe was destroyed. The defect of the medullary substance was found below the upper third of the anterior central convolution, and below the most anterior part of the superior parietal lobule. Externally these convolutions showed no abnormality. The medullary defect was so situated that a portion of the corona radiata and the anterior convolutions, as well as of the superior parietal lobule, were interrupted. Hence parts of the motor zone had alone suffered, while the parietal convolutions remained almost entirely intact. Secondary degeneration was found as follows: left pyramidal tract traced from hemisphere to lower portion of cord markedly degenerated, the degeneration of the crus involving $\frac{3}{4}$ of the crusta from within outwards, and hence destroying that part in which Flechsig had located the pyramidal tract. The authors concluded that $\frac{5}{6}$ of the fibres that passed from the nuclei of the posterior columns through the inter olivary layer to the ' cerebrum ultimately reached the central convolutions, probably the posterior central and paracentral lobule. The so-called motor zone is also a sensory centre. There exists a connection between the central convolutions and cerebellar hemispheres, through which the cerebellum influences the motor tract independently of the pyramidal path.—(Prof. P. FLECHSIG and HÆSEL, *Neurol. Centralblatt,* July 15, 1890.)   I. L.

SURGICAL ANATOMY OF THE MASTOID REGION.

Birmingham reported to the " Brit. Med. Assn." investigations on this subject. He says the usual accounts of the lateral sinus, which describe it as running horizontally downward and forward from the occiptital protuberance to a point one inch behind the meatus where it is said to turn down behind the posterior margin of the mastoid process is not correct. Nor is the statement correct that the transverse fissure is marked on the surface by the portion of Reed's base line behind the meatus. The sinus in its course forward from the occiptital protuberance is arched. From the region of the protuberance it ascends above the base line to a point 1½ inches behind, and nearly ¾ of an inch above the centre of the bony meatus; here it bends and runs down in front of the posterior margin of the mastoid process about ½ inch behind the meatus. It turns into the jugular foramen at a point about ¼ inch below the level of the floor of the meatus. The floor of the cranium is usually about ¼ of an inch above the roof of the meatus, but here there is some variation. In trephining for cerebellar disease a point should be selected two inches behind the centre of the meatus and one inch below the base line.—(*Brit. Med. Jour.*, Sept. 20.)

TISSUE CHANGE IN SLEEP.

The following are the conclusions of an experimental investigation by Dr. A. Marro, as reported in the *Archivio Italiano* XXVII., V., 1890.

1.   That in the nocturnal repose the amount of the chlorides eliminated in the urine is diminished.

2.   Likewise the amount of sulphuric acid is lessened.

3.   There is an increase in the total amount of phosphoric anhydride.

4.   The increase of the phosphoric anhydride occurs in a peculiar manner, combined with earthy alkaloids.

5. Reversing the periods of activity and mental repose so that the night is given to work and the day to sleep, we still preserve the same modifications of elimination of phosphoric anhydride, but in a less pronounced degree. The same occurs with the elimination of chlorides.

6. The maximum elimination of phosphoric anhydride occurs in the early hours of sleep.

7. The greatest elimination of chlorides occurs in the morning hours.

Dr. Marro conjectures that the comparative inactivity of the intestinal movements during sleep may explain the diminished elimination of the chlorides, possibly also the horizontal position favoring the circulation in the cartilagineous tissues which are rich in chlorides, may favor their absorption from the blood and thus render the quantity less in the urine.

The greater elimination of phosphoric acid is not so readily explained. Marro holds it as probable that it depends somewhat on the condition of the digestion.

ELECTROTONUS OF THE MOTOR NERVES.

The following are the conclusions of a paper by Dr. Brugia read before the Italian Freniatrical Society in Novara *Archivio Italiano* XXVI., I and II, 1890:

1. The time required by the An. O. contraction is much more considerable than that by Ka. Cl. contraction.

2. The time required by the Ka. Cl. contraction is longer than that of the Ka. O. contraction.

3. In a series of moderate galvanic and faradic excitations of opening and closing, the rhythm of the excitation is constant; the time of reaction is sensibly equal, but after a certain time presents a slight but progressive increase.

4. The time of reaction is in every case inversely proportional to the degree of excitation.

5. In debilitated individuals whose muscular nutrition is deficient, the plantar reaction time is longer than normal and

in, as it were, an identical manner, follow the reactions in healthy individuals whose muscles have been subjected to long exertions.

6. We have not noticed any marked difference in the symmetrical muscles of the same subject, nor have had the opportunity to estimate the influence upon the reaction time of toxic substances that increase or diminish the neuro-muscular excitability.

As regards the effect of polarization, the following has been observed:

1. Both catalectrotonus and analectrotonus, but more particularly the latter, cause a notable retardation of the velocity of transmission.

2. Progressively increasing the polarization causes an equal increase of the reaction time; but while the anodic polarization to a certain degree completely hinders transmission, the catalectrotonus may attain a very high intensity before exhausting the conductability of a nerve.

3. While with cessation of the catalectrotonus the retardation of the muscular action ceases almost immediately, a rather longer time is required for a nerve that has passed into an analectrotonic state to regain its full power of transmission.

4. Increasing the stimulus, which is without effect on an analectronized nerve, in catalectrotonus sufficiently compensates for the difficulty of conduction.

In nerves offering the reaction of degeneration the following was observed:

1. In the first stage in which the faradic and galvanic excitability is only diminished, the electrotonic effects are more marked and prompt than in the normal condition.

2. In the stage in which only the galvanic muscular excitability persists, the direct stimulation of the muscle is characterized by an extreme slowness of reaction to which a little later is added the electrotonic condition.

3. If the nerve under examination is moderately chilled by an ether atomization, the time of reaction diminishes, the pain

prevents continuance of the experiment and determining if, in a later period, the contraction time is lengthened.

The electrotonic modifications follow as in the normal condition.

MALE AND FEMALE RESPIRATORY MOVEMENTS.

Dr. W. Smith concludes after elaborate investigations that the teaching that there is an actual difference in the respiratory movements of the sexes, is incorrect, and that the difference, when it exists, is wholly due to the effects of woman's mode of dress.—(*Brit. Med. Jour.*, Oct. 11.)

INTRACRANIAL PRESSURE.

The old theory of so-called "brain pressure" considered it an axiom that the substance of the brain was incompressible, and taught that pathological and space diminishing areas of the cranial cavity were due to displacement of the cerebro-spinal fluid. In stating the mechanism of cerebral compression, the author performed experiments upon animals, using laminaria with which to produce compression of the brain. At the same time he measured the arterial pressure. He concluded that the brain substance can be compressed, especially by a slow developing pathological area. The lessening volume of the brain, being attained by the cerebral fluids leaving the cranial cavity through the blood and lymph vessels, there is less condensation of the nerve tissue proper. The "intracranial pathological area" grows at the expense of the brain substance, and not at the expense of the cerebro-spinal fluid, which is not displaced. Measurements of blood pressure do not show disturbances of circulation in the cranium under the influence of an increasing "intracranial pathological area." In regard to anæmia of the brain, he proves by examination of compressed brains, that the development of an "intracranial area" has no influence upon the lumen of the cerebral capillaries. He found that they not only did not diminish in

size, but dilated.   Long continued pressure of the brain with
laminaria, resulted in the development of small blood vessels
even in the parts compressed.   The blood circulation of the
brain is increased under these circumstances.   If there is no
brain compression and if "intracranial areas" have no tendency to produce anæmia, then the old theory of symptoms
of brain compression must be false.   The following phenomena
of compression were observed: Attacks of contra-lateral
spasms with disturbances of consciousness; trophic disturbances of the eye, and disorders of motor innervation of the
same; spasms with increased tendon reflexes, and tremor of
the half of the body opposite compressed hemisphere; finally
paraplegia.   Removal of the compressing area from the
cranial cavity causes disappearance of these symptoms.   The
author has studied the relation of the arterial and venous
curves, while he forced a 6% solution of salt with a certain
varying pressure into the cranium.   He finds that there is no
consistent relation between the amount of pressure and the
reaction of the animal, death often resulting before the so-
called symptoms of compression occurred, the pressure on
the injected fluid being less than the carotid pressure.   Death
also resulted when the pressure of the solution was far greater
than the carotid pressure.   The injections do not mechanically
close the cerebral capillaries, but do so by excitation and subsequent paralysis of the vaso-motor centres.   The pressure in
the veins of the neck is increased by intracranial injections,
but neither phenomena is produced before the arterial curve
is disturbed.   This venous pressure takes place at first very
gradually; later, however, with great rapidity, when death
ensues with phenomena of irritation.   Any pathological fluid
that may accumulate inside the cranial cavity has a tendency
to enter the venous sinuses, when the pressure is greater than
that of the intracranial veins, and at first this will increase the
venous outflow.

　　If the veins are incapable of returning the accumulated
fluid, it is forced into the lymphatic vessels.   If the latter

are incapable of containing it, it enters the interstices of
nervous tissue and œdema of the brain results.   This brings
the œdematous fluid in contact with nerve elements proper,
producing an excitement which may be followed by paralysis
and death.   These investigations explain also the physiological
uses of the cerebro-spinal fluid.   This fluid transudes from
the capillaries into the arachnoid space, and from here
is taken up by the venous sinuses.   Hence there cannot be
formed more fluid than corresponds to the intracranial spaces,
a regulation tending to prevent cerebral pressure.—(Dr. ADAM-
KIEWICZ in *Wiener Med. Presse*, Aug. 10.)   I. L.

THE NERVES OF TASTE.

SYNOPSIS OF LECTURE BY PROF. E. GLEY.

The nerves of taste are of two kinds, *accessory* and *essential.*
In the first group come the *hypoglossal*, the *facial*, and the
pneumogastric, that is, the motor nerves of the tongue.   Prof.
Gley has shown by experiments, that *movements* of the
tongue are often necessary to a correct appreciation of taste.
The action of the hypoglossal and the facial are well known.
The pneumogastric sends filaments through the superior
laryngeal nerve to the posterior part of the tongue and thus
establishes a connection (of a reflex nature) between the
tongue and stomach, which plays an important part in the act
of swallowing.

The *essential nerves of taste* are the *glossopharyngeal, lingual*
and the *chorda tympani.*   These are indispensable in the function of taste.

The glossopharyngeal distributes taste fibres to the base of
the tongue and velum palati; it is the principal nerve of taste.

The lingual nerve supplies the taste function on the margin
of the tongue.

In regard to the chorda tympani, after reviewing the experiments which have been made to prove its gustatory function,
he gives the hypotheses, which have been advanced in regard

3

to the gustatory filaments, which pass through the chorda. Finally he gives the following, which he thinks has the greatest probability. The sensory filaments, leaving the tongue through the lingual, reach the inferior maxillary; further, the otic ganglion, from there, passes through the small superficial and deep petrosal nerves to Jacobson's nerve and so to the trunk of the glossopharyngeal. The glossopharyngeal is thus the principal nerve of taste. Prof. Gley closes his lecture with the remark, that as yet the nerves of taste are not demonstrated beyond question, and that there is room for investigation and discoveries.—(*La Tribune Medicale,* July 17th.)

THE BULBO-CAVERNOUS REFLEX.

Under this name Dr. Onanoff designates a sudden contraction of the ischial and bulbo-cavernous muscles determined by the mechanical excitation of the glans penis in the normal man. To obtain this it is necessary to proceed in the following manner: The left index finger being placed on the bulbous portion of the penis, the right hand strokes lightly and rapidly with a piece of paper the dorsal surface of the gland, or pinches lightly the mucous membrane. The index finger then in normal conditions perceives a more or less intense shock ("secousse") due to the sudden contraction of the ischial and bulbo-cavernous muscles.

The results of his researches are as follows:

In sixty-two cases, adults considered normal, the reflex never failed to appear.

In aged, impotent subjects the reflex is abolished or at least hardly perceptible.

In three cases of hemiplegia ("hemiplegie vulgaire") when the genital functions were not affected, the reflex was normal and without exaggeration. In two cases of transverse myelitis in the superior lumbar region, the reflex was exaggerated; in these two cases an erection followed unknown to

the patient. In progressive locomoter ataxia the urinary
troubles do not seem to have any connection with the reflex;
when the patients show the reflex, they have maintained their
sexual functions intact or exaggerated.

It may, however, appear that certain tabetic subjects have
lost their sexual functions although the reflex is present, but
then the impotency is not permanent and will disappear under
proper treatment.   Should the reflex also be absent, it is to
be inferred that the impotency is permanent and treatment
unavailing.

In urinary troubles, hemorrhoids and certain neuropathic
conditions with impotency, the reflex is of great importance in
forming the prognosis.   Dr. Onanoff formulates the following
conclusions:

1.   The normal man has a reflex which he terms "bulbo-
cavernous."

2.   When the sexual function is impaired the presence of
this reflex indicates a dynamic origin and leads us to give a
favorable prognosis.   The absence of the same is the sign of
an organic lesion and the prognosis is accordingly grave.—
Bulletin gen. de Thirapeute, Aug. 8, 1890.

PATHOLOGY AND SYMPTOMATOLOGY.

CORTICAL VISUAL CENTRES.

A man of 61 had an apoplectic seizure, followed by hemianopsia; death occurred seven years afterwards.   A focus of
softening as large as a hazel nut was found in the cuneus.
Lingual lobe and gyrus hyppocampus also involved.   There
was secondary degeneration of corpus genic. ext. and the pul-
vinár.   Optic tract atrophied.   In another case there was
incomplete hemianopsia with immunity of the fixation point.
Autopsy showed integrity of cuneus with disease of gyrus

angularis and precuneus.   The incomplete hemianopsia was
therefore due to the partial involvement of the occipital lobe.
The immunity of the fixation point due to non-involvement of
the cuneus.—(Dr. von Monakow in *Neurol. Central.*, Aug. 15).

I. L.

CHANGES IN THE OPTIC TRACT AND NERVE, FOLLOWING LESIONS OF THE OCCIPITAL LOBE.

Dr. C. Moeli reports three cases of this kind.   The first
was the case of an idiot, 17 years old, with an extensive fissure in the right cerebral hemisphere, extending into the ventricle. The left optic tract and right nerve were greatly
atrophied.   Case two was a case of hydrocephalus internal,
either congenital or beginning soon after birth, in a male
aged 17.   The entire right parietal lobe was hardly ⅛ inch
thick, the occipital lobe but little more.   The thinning on the
left side was less extensive.   There was very pronounced
atrophy of the left optic tract, and of both nerves, but especially the right one; also of the corpora ˏgeniculata lateral
and the medulla of the testes.

Case three was that of a man 44 years old, who had an
apoplectic attack two years before.   There was almost total
blindness, with almost normal pupillary reaction.   There were
found foci of softening in both occipital lobes, more extensive
on the right side.   Microscopic examination showed granular
degeneration of the posterior limb of the internal capsule,
extending on the left side into the thalamus, with extensive
degeneration in the lateral half of the corpus genic. lat.
There was slight degeneration in both optic tracts and nerves.
These cases, especially the third, are of interest, because the
results of experiments on animals tend to show that atrophy
of the optic tracts does not result from lesions of the occipital
lobes in adult animals, but that they do not result in the new
born.

In three dogs operated on by Munk, Monakow found no degeneration except in one, which was killed two years afterward.

Only six other cases have been reported, in three of which the lesions were not limited to the occipital lobe.—(*Archiv. f. Psychiatrie*, Band XXII, p. 73.) G. J. K.

THEORY OF APHASIA.

A male, age 62, presented on admission, word-deafness and paraphasia. The former improved in the course of time, the paraphasia remained until death, 4 years later.

Autopsy showed a sclerotic focus in the posterior two-thirds of the first, and of the adjacent part of the second temporal convolution as well as of a small part of the gyrus longus of the insula.

The author considers the case as a transition form between cortical and subcortical sensory aphasia.—(DR. K. CRAMER, *Archiv. of Psychiatrie*, Band XXII.) G. J. K.

FUNCTIONAL APHEMIA.—(APHASIA.)

Dr. Jacob publishes notes of several cases. The first was a man of 50 who had fits from infancy, and who, after reverses at the age of 34, showed mental instability, violent temper, excitability, etc. He gradually lost power of speech, and for five years had not spoken. Having suffered dislocation of the arm, ether was administered to effect reduction. After recovering from the effect of the ether he immediately regained speech and remained well. The next case was a man, who eight years previously had epilepsy for a short time. He suddenly lost power of speech and had been aphasic for about a week when he came under observation. Ether was administered and he immediately recovered his ability to talk. The doctor seems to consider the cases as hysterical.—(*Brit. Med. Jour.*, Sept. 13.)

TUMOR OF FOURTH VENTRICLE.

Drs. Noyes and Dana report a case with post mortem examination. The patient had suffered with occasional attacks of severe headache from 1873 to 1888. At first pain was limited to left side; more recently had it on right side. Headache usually occurred in forenoon. He had dizziness and roaring in ears, numbness and tingling in right hand and arm, and slightly in right leg. Weaker on right side and often staggered. His gait, when walking, resembled that of a drunken man, staggering chiefly to the right. Nystagmus of right eye (left eye having been enucleated), smell and taste impaired. Later had pains in ends of fingers, back of leg, knee, and foot of left side. Four days before death had an epileptic fit. There were three tumors in floor of medulla, compressing and partly destroying the left lemniscus, left pyramidal tract, left 8th, 9th, and 10th nerve nuclei, and left, and perhaps right, 3rd nerve, and probably 4th, 6th, and 7th. The first symptoms, vertigo and tinnitus, were probably due to irritation of the nuclei of vestibular and cochlear nerves. Paræsthesia of right side due to pressure and irritation of sensory tracts. Right hemiparesis due to pressure of left motor tract. The forced movements to the right were probably due to involvement of the left middle cerebellar peduncle. Burning sensation in right side of face, and paresis of left masseter and temporal, due to involvement of 5th nerve. There was total decussation of the optic nerve fibres.—(*N. Y. Med. Rec.*, July 26.)

BILATERAL CEREBRAL TUMOR.

Dr. McBride reports an infiltrating sarcomatous tumor, involving the motor region of both hemispheres symmetrically. The patient was affected with a slow progressive paralysis beginning in the legs, afterwards involving the arms and trunk and which finally became complete, and was accompanied by atrophy of optic nerves. No other cranial nerves

were involved.   Anæsthesia coextensive with paralysis.   Infiltrating tumor was found involving upper ⅔ of motor region
on both sides, and posterior ⅓ of first and second frontal,
extending on left side to upper surface of corpus callosum; on
right side not quite so deep.   Middle region of both crura
softened, and both anterior pyramids.—(*Jour. Ner. and Ment.
Dis.*, Aug.)

TUMOR OF MENINGES.

De Grandmaisson reports a case of Jacksonian epilepsy in
which on autopsy there was a fibrous tumor of meninges in
Rolandic region and membranes adherent to cortex.   Ver-
chere said that he had trephined a case of Jacksonian epilepsy.
He does not state that there had been a fracture.   No abnormality of brain or membranes found.   Two months subsequent there had been no return of fits, and he had complete
relief from headache from which he had suffered.—(*Jour.
Societes Scientifiques*, July 16.)

THALAMIC TUMOR.

Dr. F. X. Dercum publishes a case which began with vertigo
and headache and later was accompanied by temporary blindness.   Two years subsequently right leg began to feel stiff and
numb, and occasionally he struck the toes of his right foot in
walking.   At times he would have curious spells, when right
arm was drawn up, which at first occurred only once in a few
days, but became more frequent, and afterwards occurred every
few minutes.   Some weakness of right arm and leg, the right
leg being dragged and slightly spastic.   Knee jerk exaggerated
on right side; muscle and temperature sense on right side
diminished; weakness of right side increased; mind became
very much weakened.   Not long before death there was complete paralysis of right side.   There was right lateral hemianopsia, more marked on left side.   Visual field con-

tracted, hemianopsic pupillary inaction sign (Wernicke) manifest. On autopsy a glioma-sarcoma was found in the pulvinar,
though it had invaded the tubercle and slightly the caudate
nucleus. The sensory disorders, hemianopsic pupil, and
paralysis of arm and leg, were probably due to pressure of
the tumor. The athetotic attacks were probably due to irritation of the motor portion of the capsule.—(*Jour. Ner. and
Ment. Dis.*, Aug.)

NERVOUS SYPHILIS.

Dr. Julius Althaus read an article with this title before the
"Berlin Med. Congress." He considers that the most
important remedy is the periodical and long continued hypodermic injections of small doses of a non-irritant insoluble
preparation of mercury. He recommends what he terms the
"carbolized mercurial cream," a preparation which consists
of metallic mercury rubbed up with lanoline, and mixed with
a certain preparation of carbolized oil. Perfect homogeneity
of the mass, great stability, painless injection, absence of
swellings and abscesses, great efficiency in truly specific
lesions, and absence of the risk of stomatitis and dysentery, if
a certain dose is not exceeded, are among some of the advantages of the preparation.—(*Lancet*, Aug. 16.)

ANATOMICAL ALTERATIONS IN ACUTE BROMIDE POISONING AND THEIR CLINICAL IMPORTANCE.

Dr. Biorchia Nigris (*Bull. delle Scienze Med. di Bologna
Archivio Italiano*, XXVII., V., 1890), publishes the following
results of his experiments on dogs and white mice in the
physiological laboratory of the University at Bologna:

1. Young animals bear much better than older ones and
for a longer time, excessive doses of the bromide (one grain
of bromide per kilo of weight.) Salts of bromide in these

doses cause death in the older animals by multiple hemorrhages occurring in all the organs of the body, and due to a
lesion of the lining of the vessels; in young animals these
hemorrhages are much more limited, or altogether lacking, and
death is due to a diffuse parenchymatous neuritis, and a lobu-
lar pulmonitis, and degenerative processes in the parenchymatous elements of the various organs.

2. The bromide salts containing bromates and bromites in
these continued doses do not produce any greater effects than
the pure bromides, except that the digestive canal may undergo a violent hemorrhagic gastro-enteritis.

3. Doses of 35-40 centigr. of bromide per kilo of body
weight when administered for some consecutive days cause
albuminuria. The urine then contains also red globules and
renal cylinders and cells.

4. From these facts it would follow that equivalent doses .
are contra-indicated in the human subject when there exists a
nephritis or a hemorrhagic disposition or other disease of the
vascular system (syphilis atheroma).

5. Doses of from ten to twenty centigrams per kilo of body
weight of the animal gave rise to no anatomical lesions.

LESIONS OF THE CENTRAL ORGANS OF THE BRAIN.

Under this caption Dr. Andrea Verga (*Archivio Italiano*,
XXVII., Fasc. 1 and 11, Jan. and March, 1890), discusses the
lesions of the corpus callosum, fornix and septum lucidum.
His conclusions are as follows:

1. The central organs of the brain (corpus callosum, fornix
and septum lucidum) may be found lacking in man, either
wholly or in part.

2. This deficiency may depend on a primary defect of
structure, but perhaps more frequently it depends upon pathological processes during fœtal life or shortly after birth, that

may cause laceration or destruction of these organs. This is particularly the case with the fornix and septum lucidum.

3. The disease that generally causes this destruction of the central organs is hydrops of the ventricles, occurring at a time when their walls are exceedingly thin and soft, thus favoring this injury.

4. The lack of the central organs of the brain is not only compatible with organic life, but also with animal life, and it may not be accompanied with any characteristic symptoms by which it can be diagnosed *intra vitam.* In two cases dullness of the olfactory and auditory senses was observed.

5. By this series of curious facts there is demonstrated to a certain degree the independent actions of the cerebral hemispheres, and it becomes desirable to know how far the individuals suffering from this defect are the more developed on one or the other of the two sides of the brain.

It is not possible to assume from these exceptional cases that these organs are without any essential functions. To a certain extent the clinical histories presenting these anomalies, have generally been defective.

CHOREA CORPUSCLES.

Dr. Wollenburg demonstrated (at a society meeting in Berlin) preparations from the brains of two choreic patients. One, fresh, was from a woman of 34, who had died of progressive paresis. The other, from a boy, had been hardened. In both cases the inner limbs of the nucleus lenticularis were found to contain a large number of glistening, strongly refracting bodies, which were mostly arranged around the small vessels, although some were scattered throughout the tissue. The outer limb contained none. The corpuscles resisted all strains and reagents, except strong $H_2 SO_4$. These "chorea corpuscles" were first described by Elischer in 1874. His statements were corroborated by Flechsig and Jakowenko, although later investigators do not seem to have noticed them.

Dr. Sander claimed to have found them in non-choreic brains. No hypothesis as to their nature or importance is offered. Dr. Siemerling claims they do not occur in paresis, or in normal brains.—(*Berlin Klin. Wochensch.*, 1890, p. 877.)

G. J. K.

CHOREA MINOR, AND RHEUMATISM.

Among 18,074 children treated in five years 121 were found to have chorea (0.6 %), 46 were boys and 75 girls. The majority were between 6 and 13 years, one boy and six girls were 5 years old, two girls respectively 4 and 3½ years old. Eleven cases (9.0 %) had been preceded or were accompanied by rheumatism, thirteen cases had a valvular lesion, three cases presented both rheumatism and valvular trouble. The author concludes that chorea may often be developed on a rheumatic basis, but that the rheumatic virus can by no means be made responsible for all cases.—(DR. P. MEYER, *Berlin Klin. Wochensch*, No. 28, 1890, p. 628.)   G. J. K.

EPIDEMIC OF CHOREA.

This epidemic involved twenty-six children aged from 12 to 14, and all pupils of a single school. Seventy-four girls of one class were vaccinated in the spring of 1885. Nine of these children suffered from catarrhal jaundice within six months following the vaccination, but of light degree. Four of these nine girls afterward had chorea. The original case was that of a girl aged 12 who had suffered from chorea minor each spring for several years. In the course of the summer and fall eighteen cases had occurred among girls, of which thirteen were cases of chorea rhythmica and five, cases of chorea minor. Later on, when no new cases appeared among the girls, eight boys of similar age developed chorea rhythmica. The last case began about nine months after the first. In only one case (a girl of 18) could a preceding rheumatism be found. The author considers the cases mostly hysterical, as the children first affected were allowed to attend school. The majority of the cases received no medical treatment; a few were given arsenic or bromides, as well as cold sponge baths. The author made futile use of metallotherapy. He has been able to find only nine epidemics of chorea reported in literature within the last century.—(DR. RALF WICHMAN, *Deutsche Medic. Wochenschrift*, No. 29 and 30, 1890, p. 633.)

G. J. K.

EPILEPSY FOLLOWING ACUTE INFECTION.

The patient, letter carrier, aet. 19, was of healthy ancestry and good habits. He was revaccinated in June, 1888. This was followed by a severe lymphangitis, with a scarlatiniform rash and pains in the joints. He was compelled to continue at work, and about four weeks after the vaccination had his first epileptic attack. Within a year he had an attack at least once a month, sometimes by day, sometimes at night. Just before coming under the author's care, he had fourteen convulsions in one night. He recovered in two months under potassium bromide, arsenic, and hyoscyamus, and has had no convulsions since. Animal vaccine of a reliable brand had been used.

The only explanation the author can offer is that of a secondary bacterial infection of the nervous system.—(DR. JUL. ALTHAUS, *Deutsche Medic. Wochenschrift*, No. 31, 1890, P. 689.)

G. J. K.

THROMBOSIS OF CEREBRAL SINUSES IN AN ADULT.—RECOVERY.

Miss L., aet. 35, relapsing morphia habitue, very anæmic, was under treatment a second time. About two weeks after beginning treatment, a thrombosis of the left saphenous vein developed, subsiding in twelve days. Five days later, a double dry pleurisy was found, and during its decline, a thrombosis of the right femoral vein developed, which lasted ten days.

Five days after its termination the patient complained of intense pain in forehead and vertex, in the middle line, which resisted all therapeutic measures. About a week after the beginning of the headache, patient became soporose, and in the course of 24 hours was in such a state of coma that nothing could arouse her. On the eleventh day œdema of the nose and both temples was found. On the twelfth day, this œdema had spread over the forehead, the lower lids, the malar regions, and the nasal halves of the upper lids and ocular conjunctiva. Pupils did not react to light, corneæ were anæsthetic, patellar reflexes abolished. Anything poured into the mouth was swallowed. No vomiting, convulsions, opisthotonos, gnashing of teeth, or, incontinence. Urine and fæces were discharged into a bedpan, when it was pushed under her. The temperature was slightly raised (100°). On the fifteenth day, she awoke for a short time. From this time on the soporose intervals became shorter, and the facial œdema rapidly subsided. Three weeks after the beginning of the headache, she was convalescent.

The distribution of the external veins of the face and the œdema, would point first to a thrombosis of the longitudinal sinus, followed a day later by thrombosis of the cavernous sinus.

The patient and three sisters, all had varicose veins of the left lower extremities. The author believes that congenital ectasia of the sinuses, combined with the toxic anæmia, was the main etiological factor in this case.

Treatment consisted of heat applied to the head and concentrated liquid nourishment and stimulants.—(DR. ALBRECHT ERLENMEYER, *Deutsche Medic. Wochenschrift*, No. 35, 1890, P. 782.)   G. J. K.

_____

HYSTERICAL ANÆSTHESIA.

Dr. C. L. Dana writes on this subject, the following being his conclusions:

"1. Its frequent presence in the retinal field, and its peculiar distribution here. 2. Its distribution on the skin, affecting first the pain-nerves, and its modification, disappearance, or transfer by metals, or suggestion, or cutaneous irritants. 3. Its peculiar involvement of the auditory nerve, causing deafness to high and even low notes, as well as dulling the hearing generally. 4. The rarity of muscular and articular anæsthesia, except in connection with profound paralysis. 5. The involvement of taste and smell."—(*Amer. Jour. Med. Sciences*, Oct.)

HYSTERICAL MONOPLEGIA.

(From Prof. Erb's Clinic at Heidelberg.)

The patient was a man 42 years of age and a manufacturer. In April, 1890, he suddenly fell to the floor unconscious and remained so for twenty-four hours. After regaining consciousness left leg was immovable and rigid, left foot painful and stiff. Left leg anæsthesic and to inguinal region. Sphincters not involved. Visual field was much diminished. On examination muscles of left leg found to be hard and rigid, though leg was capable of voluntary flexion and extension. Sense of touch, locality, temperature, pressure, and muscular sense, abolished in left leg. Knee jerk in left leg abolished. Left leg cold and bluish. Patient could walk but dragged left foot. Special senses were normal. Diplopia of right eye at 30 cm., at 50 cm. could not count fingers. Field of vision diminished for all colors. After two electrical treatments, the pains the patient had complained of ceased, and after a few more treatments he recovered and left the hospital. The surprising results from the electrical treatment leads Erb to diagnose this case as one of hysterical monoplegia. The psychological influence with the assurance of recovery was of great importance, and the use of very strong currents restored the conducting power of the nerves by reflex action. It may have acted also by removing any inhibition that may have been set up by the cortex.

HYSTERICAL FACIAL PARALYSIS.

Huet reports a case in which there was hysterical paralysis of the right side of the face with hemiparesis of the same side, hemianæsthesia and hemianalgesia. Later the patient also had paralysis of the left side of the body. Paralysis of the face involved muscles of mastication. Later patient had hysterical convulsions.—(*Neurol. Centralblatt*, Sept. 5.)

G. J. K.

DISORDERS OF SLEEP.

Dr. Weir Mitchell has an article upon this subject in which he discusses a little known group of disorders that are associated with sleep. The cases reported are interesting and unusual, . but as it is difficult to condense the article, the reader is referred to the original for the account of some very novel experiences of people whose condition seems to have been nearly akin to insanity. One was the case of a woman who had an injury to the nose which was followed by persistent headache and loss of smell. Two years later she had on going to sleep a sense of horrible odors, to which was added the sound of voices; she finally had melancholia. In some cases the patient, after sleeping a while, awoke with a feeling of intense fear. One patient said that he could always tell upon retiring whether he would have this trouble or not. The condition was often preceded by very vivid and disagreeable dreams, usually dreaming that after a long absence from home he had returned and found that some one of the family had become idiotic. Sometimes on closing his eyes on retiring at night, a feeling of fear came over him which prevented him from sleeping. Some persons awake from sleep with a feeling of numbness, or motor paralysis, affecting a hand or leg, or one side of the body, or the entire body. The author mentions a condition that he calls "sleep pain," which is a kind of pain or distress that involves the legs and always develops during sleep. He also refers to sensory shocks that some people feel on going to sleep, such as loud nòises, sudden flashes of light, sense of odors, etc.—(*Jour. Med. Sciences,* Aug.)

COMBINATION OF ORGANIC AND FUNCTIONAL NERVOUS DISEASE.

A woman had for several years difficulty of locomotion and speech, failure of eyesight, diplopia and vertigo. There was some spastic paralysis of lower extremities, atrophy of optic nerves, nystagmus, partial anæsthesia of lower extremities. Diagnosis of multiple sclerosis was made, and soon after this, rythmical tremor of right arm appeared. This was cured by hypnotism, which showed that it was an hysterical symptom and did not belong to the original disease. A lady, who had suffered from migraine, had cataleptic attacks with hysterical spasms, paralysis and disorders of sensibility. Had partial paralysis of right leg, fingers of right hand could not be extended, paralysis of right arm with atrophy of interossei, diminution of electrical excitability, and temperature sense diminished. Hemiatrophy of right half of tongue, with diminution of electrical excitability, and concentric narrowing of field of vision for colors in left eye. The diagnosis was that of hysteria combined with cerebral gliosis. Several other histories of interesting cases are given, from all of which the author concludes that hemianopsia may be produced by reflex action, as also may functional cerebral disorders.— (Dr. H. OPPENHEIM, in *Neurol. Central.,* Aug. 15.)   I. L.

NEURASTHENIA FOLLOWING INFLUENZA.

Dr. Lehr reports eleven cases of neurasthenic conditions following influenza. These cases presented no noticeable symptoms indicating their etiology, and recovered promptly under the treatment usual in such cases. Six of these cases had antecedent vague nervous troubles.—*Deutsche Medic. Wochenschrift,* 1890, p. 908.   G. J. K.

GRAVE'S DISEASE.

In a lecture on this disease Dr. Mackenzie says it is found associated with epilepsy, hysteria, chorea, diabetes and insanity. It sometimes occurs in several members of the same family. Oestereicher reported a case of a woman who had ten children, eight of whom had the disease, and one of the latter had four grandchildren affected. The thyroid is generally the largest on the right side in this disease. Of the three symptoms the exophthalmos is the least constant. The following is a brief summary of the symptoms, some of which are not usually mentioned in the descriptions of this disease. Tremor may usually be discovered; it is noticeable when the hand is held out, and attacks the whole extremity, sometimes the whole body. Attacks of trembling also occur, the trembling interfering with the act of writing or sewing. Painful cramps in the feet and legs occur on lying down at night; the legs give way at the knee sometimes when patient is standing, and there is found a feebleness of the lower extremities. Weakness of the ocular muscles occurs, and sometimes ptosis. Elevation of temperature may occur; sometimes it is subnormal. There is occasionally subjective sensation of heat with intolerance of heat and great tolerance of cold. Flushing of the head and neck, excessive sweating, pigmentation of the skin, falling out of the hair, œdema of the subcutaneous tissue (especially of the legs), and great diminution of the electrical resistance, epistaxis, sometimes great emaciation, and pigmentation of the face, neck, sides of the chest, abdomen, lumbar region, flexures of arms and thighs. Mentally the patients are irritable, excitable, restless, low-spirited, spiteful, and untruthful. Mania and melancholia are sometimes developed. Dyspnœa, nervous cough, vomiting, intermittent albuminoria, glycosuria and polyuria are observed. The author and White have reported ten deaths out of eighteen cases. Sudden death sometimes occurs. It is not probable that the sympathetic is involved in this disease, the tendency now being to ascribe it to disease of the central nervous sys-

4

tem.   He advises the following remedies: belladonna in
10 minim doses of the tincture, three times a day; arsenic
and bromides are also useful; iron combined with belladonna
is good.   Galvanism and Faradism have both been found
useful.—(*Lancet*, Sept. 13, 20.)

TINNITUS AURIUM.

Dr. H. M. Jones, in discussing this subject before "The
Brit. Med. Assn.," said that he had twice suffered from it
after intense mental application.   His remarks were based on
the histories of 260 cases.   The condition found on examination of ear was as follows: cerumen in 30; inflammation or
abscess of the meatus in. 7; exostosis of the meatus in 7;
catarrhal changes in the tympanum in 102; the same with
closed eustachian tube in 38; perforation of the membrana in
17; polypus in the tympanum in 4; diseases in the labyrinth
in 47; nasal obstructions, deviation of septum, congestion of
the membrane, etc., in 19; enlarged tonsils in 7; both ears
normal, 26; one ear normal, 7.   Causes were: recent sea
bathing, fever, cardiac weakness, nasal obstruction, puerperal
septicæmia, mental strain, Bright's disease, albuminuria,
quinine, neurotic temperament, and tobacco.   Hearing normal
in both ears in 18 cases.   In 187 of the cases the main symptoms were tinnitus and deafness.   Vertigo existed in 22, and
in 9, typical symptoms of Ménière's disease occurred. (Nausea,
vertigo, syncope, and deafness).   In 7, there were ocular
symptoms with retinal changes.   It is an interesting fact that
in the majority of cases of perforation of the drum of the ear
there is no tinnitus.—(*Brit. Med. Jour.*, Sept. 20).

PARALYSIS AGITANS.

Dr. Frederick Peterson publishes a clinical study of 47
cases of this disease.   In the majority of his cases it developed
between the ages of 50 and 60, more men being affected than
women.   Tremor is present in all his cases, though various

others have been observed in which it is absent. Charcot's statement that the head never takes part in the tremor, is proved to be unfounded. In 9 of Dr. Peterson's cases, the head was involved. Paralysis agitans is about the only disease where the tremor continues when the body is at rest. Rigidity of muscles existed in 41 of his cases. Contractures producing the typical position of the disease were present in over 80%. In some, the flexors of the fore arms were so contracted that extensions could not be made. In some cases there is muscular atrophy. In 9 cases the knee, wrist, and elbow jerks were exaggerated. In 6, they were hypertypical, and in all the rest normal. In one case of eight years standing, with disease limited to left side, there was diminished electrical excitability. There was change of voice in 13 of the 47 cases, which was probably due to rigidity of muscles concerned in speech. These changes were, monotony of tone, high pitch, and piping quality, with sometimes a hesitation in starting to speak. In some of the cases there were subjective sensations of heat, in other cases of cold. Hyperidrosis existed in 4 of the cases. In many of the cases there was diminished intelligence. The pathology of the disease is properly to be ascribed to functional disorders of the motor areas of the cortex, due to nutrition changes of a degenerative character. Codein in doses of two grains combined with hydrobromate of hyoscine, doses of 1-100 of a grain, is recommended to be given two or three times a day.—(*New York Med. Jour.*, Oct. 11).

PATHOLOGY OF SYMPATHETIC.

Dr. Hale White contributes an elaborate article on this subject. He has examined the superior cervical ganglion in forty-one human adults and has found the largest of these ganglia to be about an inch in length and the shortest ¼ inch. The cervical ganglion is sometimes absent. Variations in vascularity of these ganglia (including the semilunar) has no significance whatever. Giovanni's observations on the sympathetic are in many respects erroneous. Seventy-three superior cervical ganglia of lower animals were examined, and it was found that the size of the ganglion varies directly as the size of the animal. His investigations lead him to conclude that the nerve cells of the cervical and semilunar ganglia are healthy in childhood, but that they rapidly undergo degeneration as age advances. He found that the nerve fibres passing through degenerated ganglia were healthy. In comparing results of examinations in man and the lower animals, he has found that in man there are by far the largest proportion of granular, pigmented, degenerative cells. Next, in the monkey there are some, but they are few. In carnivora and ungulates there are few, in lower animals none at all. It is therefore probable that in adult man the superior cervical and semilunar. ganglia are degenerated organs, like the coccyx, or appendix cæci, and that as we descend in the animal scale they become functionally important, being most so the lower we go. In man and animals division ot the cervical sympathetic causes dilatation of vessels with rise of temperature on affected side. In man the pupil contracts upon division of the cervical sympathetic and dilates upon its stimulation. On section of this nerve there is retraction of the eye ball. Injury of the cervical sympathetic is not followed by headache and there is no affection of hearing. Migraine is not due to disease of the sympathetic, neither is angina pectoris nor hemiatrophy of the face, nor exophthalmic goitre. Disease of the splanchnics may cause glycosuria, as may also disease and injury of the central nervous system. In diabetes there are no specific changes in sympathetic ganglia. Addison's disease is not due to disease of the suprarenal capsules nor to disease of the semilunar ganglia. The suprarenal capsules in man may be absent, diseased, or extirpated, without producing symptoms.—(*Guy's Hospt. Reports* for 1889.)

LEAD POISONING.

Dr. G. L. Walton reports a case in which the principle symptom was ataxia. The patient had transient numbness in hands, later in left foot which finally involved the whole leg. Patient's gait was so unsteady that he had to lay hold of objects to keep his balance when walking. Some anæsthesia in left foot and loss of muscular sense. There was also marked inco-ordination. It was thought possible that patient had been poisoned by lead from a tin lined copper kettle in which water had been heated which he had been in the habit of drinking. A large amount of lead was found in the urine which confirmed the supposition of lead poisoning. Patient died, but no autopsy was made. Dr. J. J. Putnam has reported three cases from his own practice of pseudo-tabes from lead poisoning.—(*Boston Med. Jour.*, Oct. 30.)

PROGRESSIVE MUSCULAR ATROPHY.

Dr. C. L. Dana read an article upon this subject before "The Practitioner's Soc. of N. Y.," May 2. He presented a man 40 years of age suffering from progressive muscular atrophy. The disease began at 17 years of age with weakness and atrophy of left shoulder, then the other shoulder, afterwards the left thigh, and afterwards the right thigh became involved; later the fore arms and legs and some of the muscles of the back. There are two groups of progressive muscular atrophy; one of spinal, the other of muscular origin. The muscular form is hereditary. In some cases it begins in the shoulder, in others in the face, and sometimes in the lower extremities. Pseudo-muscular hypertrophy is a type of the same affection. The disease begins in early life and progresses slowly. There are no fibrillary twitchings and no reaction of degeneration as in muscular atrophy of spinal origin.—(*N. Y. Med. Rec.*, Aug. 23.)

ACUTE MYELITIS PRECEDED BY OPTIC NEURITIS.

Dr. J. T. Eskridge, of Denver, publishes an interesting case of this disease. The patient was a man of 49 years of age who accidentally discovered that the sight of his left eye was defective. In a day or two the other eye was involved, and in one week he was totally blind. Ophthalmoscopic examination showed optic neuritis with white atrophy. In about a week from the time his eye-sight began to fail he noticed that his right leg was gradually growing weak, and in a day or two both legs were paretic. From this time on the patient developed the symptoms of acute myelitis. He died in five weeks from the time of the first appearance of eye symptoms. On post mortem it was found that there was diffused ascending myelitis which began in the lower portion of the cord. The author gives an interesting resumé of the literature referring to the coincident occurrence of optic neuritis and myelitis.— (*Jour. Nerv. and Ment. Dis.*, Sept.)

COMPRESSION MYELITIS.

Dr. Goldscheider reports the case of a woman, aet. 61, who, in January, complained of pain in both shoulders. Later the pains disappeared in the left arm, persisted in the right, in which an intention-tremor appeared. In May, a spinal paraplegia with anæsthesia, and incontinence of urine and fæces appeared. A diagnoses of compression myelitis was made. Caries was excluded, carcinoma could not be proven, and although the physical symptoms were scanty, aneurism of the aorta was suspected. Autopsy showed carcinoma of the 2nd, 3rd, and 4th, dorsal vertebræ, the body of the third having almost entirely disappeared. This was secondary to a soft carcinoma of the posterior wall of the stomach near the lesser curvature. There was a considerable aneurism of the ascending and transverse aorta, which, however, was not responsible for the cord symptoms.—*Deutsche Medicin. Wochensch.*, 1890,

G. J. K.

PRESSURE PARALYSIS.

In the discussion of Pott's disease before "The American Orthopædic Assn.," Dr. Halsted Myers said:

In 1,570 cases of Pott's disease presented at the New York Orthopædic Dispensary, 218 were known sooner or later to have become paraplegic. Estimating the duration as to the period from the onset of any symptoms until the patient could walk well, the average duration of those paralyses whose duration was known was, in the cervical region twelve months, in the dorsal region above the eighth vertebra nine and one-half, in the lower dorsal six, in the lumbar eight months. This average period was less where the paralysis came on during treatment, as it did in 85 cases, viz., cervical five, upper dorsal seven, lower dorsal five, lumbar three months.— *N. Y. Med. Rec.*, Oct. 4.

LANDRY'S PARALYSIS.

Nanwerck and Barth were the first to find changes in the peripheral nerves, with intact central apparatus, in this disease. The cause of the disease has been generally referred to an acute intoxication. Eisenlohr has observed two cases which differed considerably in regard to the pathological conditions found. Case 1, which died after eight days, presented hæmorrhagic infarcts in lungs and bowels and a greatly swollen spleen. Central nervous system normal. Numerous spinal and cranial nerves, examined in osmic acid, showed a granular degeneration and breaking up of the axis-cylinders, which were still stained with osmic acid, as well as numerous new cells in the endoneurium, in fact a beginning parenchymatous and interstitial neuritis. After hardening, these changes were much less marked. Culture plates, inoculated from the cord, nerves, and spleen, remained sterile. In case 2, tubercular ulcers of the intestine were found. The central nervous system was normal to the naked eye. The anterior roots and nerves showed similar changes under osmic acid as in the other case. The spinal cord at the level of the 11th and 12th dorsal vertebræ showed the following changes, which occupy the major part of the right lateral column: the meshes of the neuroglia are larger than normal, the trabeculæ thicker, as if swollen. The nuclei not increased. The meshes of the neuroglia were filled with swollen axis cylinders, or irregular figures resulting from their degeneration. The gray substance and posterior roots are normal. The lumbar cord and medulla normal. The bacteriological examination of cord and nerves showed the presence of the staphylococcus pyog. aur., and two other varieties of germs. Tubercle bacilli were not found.

The author calls attention to the fact that the nerves should be examined first in osmic acid, as the changes show much less clearly after the use of Mueller's fluid.—DR. C. EISEN-LOHR, *Deutsche Medic. Wochensch.*, 1890.   G. J. K.

PERIODIC PARALYSIS.

Dr. Goldflam reported at the International Congress, the case of a lad of 17, who was subject at certain intervals to attacks of paralysis, lasting one or two days, and gradually disappearing. During the attack the faradic excitability of the nerves of the arms was greatly diminished, and abolished in the lower extremities, as well as in all the muscles. Patellar reflexes were absent in the attack. Sensation was normal, as well as the function of the bladder. Eleven members of the patient's family suffered in an exactly similar way. The author attributes it to auto-intoxication, and locates the seat of the lesion in the muscles and the terminal nervous filaments.—*Deutsche Medic. Wochenschr.*, 1890.

G. J. K.

POLYOMYELITIS AND POLYNEURITIS.

At the meeting of the "Sociéti de Biologie, July 5th," Dr. Paul Blocq presented a paper with the following conclusions:

1. There are certain morbid forms, whose clinical aspect exactly corresponds to the disease described by Duchesne as acute and subacute anterior spinal paralysis of the adult. These morbid forms are really, according to the author, due to lesions of the anterior cornua of the spinal column.

2. Again a symptom complexus is observed more or less analogous to the preceding in regard to the clinical history, and whose evolution may cause confusion in the diagnosis, but where no appreciable change in the gray matter is detectable.

3. It is not possible to absolutely assert in certain of these cases, that the disorders are associated with peripheral neuritis.

4. In these cases, it is to be inferred that central lesions manifested by clinical signs play a pathogenetic role.

5. It is not possible to point out in an absolute manner the connections between amyotrophic paralyses, polyomyelitis and polyneuritis. The most plausible hypothesis is, that a lesion of the neuromuscular arc exists, more or less intensely localized in the different parts of this apparatus.

───

At the same meeting, Dr. Onanoff gave the results of his investigations on the neuro-muscular bundles ("faisceaux musculaires"):

1. The section of the roots (in dogs) was followed by atrophy of a very small number of fibres in these bundles.

2. The destruction of the vertebral ganglion left only few fibres intact.

3. The bundles can be traced without interruption through the length of the muscles, where they are lodged.

4. The penetration of the nerve into the bundle is affected at different points of its length. Although the number of

nerve fibres in a transverse section rarely exceeds 10 or 12, the course of each one being very short in the interior of the bundle, their number is still very large, 50 to 60 for each millimeter.

5. In lateral amyotrophic sclerosis, syringomyelia and kindred affections, where completely atrophied neuro-muscular bundles are found in the muscles, a large number of intact nerve fibres are found.

6. In lateral amyotrophic sclerosis, and after sections of the anterior roots (in animals), the striated muscle fibres of these bundles become atrophied; sometimes they disappear altogether in the first named condition.

7. The greatest number of nerve fibres sent out by the vertebral ganglion to the muscle body are found in the neuro-muscular bundles. From all these facts we conclude that the striated muscular fibres, both trophic and motor, of the neuro-muscular bundles are dependent on the anterior cornua, and that the nerve fibres of these bundles transmit to the spinal column both intrinsic and extrinsic impressions. It is to be noted that these bundles are found in the greatest number in the thenar, hypothenar and interosseous muscles; also in the flexors of the fingers and the quadriceps femoris.—*La Tribune Medical*, July 10.

INFANTILE PARALYSIS.

Dr. O. Medin (Hygiea, Sept. '90) observed an epidemic of this disease in Stockholm, 1887, within a space of five months. He noted 44 cases and gives a schematic table of the symptoms which is of interest, as it shows many complications not generally found or observed in this disease. Thus we find paresis of the abducens, hypoglossus, facialis, oculomotorius, vagus, accessorius and trigeminus, associated with polioencephalitis, polyneuritis and nephritis acuta. Dr. Medin lays special stress on the facial symptoms; he says that his experience during this epidemic proves the fallacy of the generally accepted theory that the cranial nerves are not often affected. Two post mortem examinations were obtained. Besides changes generally observed in acute infectious diseases, he found hyperæmia of the brain and gray matter of the bulb, hyperæmia of the spinal cord and dura, hemorrhagic points in the anterior horns, especially in cervical and lumbar regions. The crural nerve showed a diffuse redness of the sheath. Microscopically an intense acute inflammatory process of the anterior horns with degeneration of ganglion cells was noted.

He concludes that infantile paralysis is without doubt an acute infectious disease which may be epidemic. He does not consider it contagious.

THE CONTRACTION OF THE VISUAL FIELD IN SYRINGOMYELIA.

Dr. Déjérine points out (Societe de Biologie, July 12th), that in syringomyelia the visual field is considerably lessened without any detectable change in the fundus or any suspected encephalic lesion. Dr. Déjérine and his interne have examined seven persons afflicted with syringomyelia and found in all a contraction of the visual field for all colors. The results were as follows: ·

The contraction of the field was for all colors, but not alike, it was most marked for green and least for white.

Dr. Déjérine points to the diagnostic value of this sign, where hysteria and traumatic neuroses can be excluded.—(*La Tribune Medicale*, July 17th).

MORVAN'S DISEASE AND SYRINGOMYELIA.

At a meeting July 5th of "Societe de Biologie," Dr. Déjérne made some remarks in regard to the identity of the two mentioned diseases. A case of gliomatous syringomyelia with changes in the cutaneous nerves, had been reported as a case of Morvan's disease; hence the remarks.

He does not consider the two diseases identical and points out differences as follows: The loss of tactile sense is altogether the rule in analgesic panaritium (felon), while it very seldom is lost in syringomyelia, the exceptions in fact are very few. Panaritium is just as rare in syringomyelia as it is frequent in Morvan's disease, hence the name "panaris analgesique" given to the disease discovered and described by Morvan. The lesions of the peripheral nerves differ also in the two conditions.

Further, all the twenty cases described by Morvan have occurred in the county (canton) where he practices, therefore Déjérne concludes that this disease is a peripheral neuritis of a regional occurrence and due to an infectious or toxic influence not yet determined.—(*La Tribune Medical,* July 10, '90.)

SYRINGOMYELIA.

Paul Blocq of Paris, has an article on this subject in the October number of *Brain.* "*Syringomyelia is a chronic affection of the spinal cord, characterized anatomically by cavities formed pathologically in this organ, and clinically by certain alterations in the sensibility, associated with trophic disorders.*" In regard to the macroscopic appearance, he says that there is rarely some posterior spinal meningitis consisting of patches of thickening of the *pia.* The cervical enlargement of the cord is most frequently affected. On section of the spinal cord one or more central cavities are visible to the naked eye. Sometimes several of these cavities exist which are entirely separate from the central canal, and always found in the gray matter, most frequently in the posterior horns, the lesion usually being bilateral. The size of the cavity varies; sometimes it is a small slit. At other times it occupies the whole extent of the cord, of which only sufficient remains to form a thin wall around the cavity. The affected area may encroach upon the medulla. In regard to the pathogenesis it is believed to be an inflammation of the neuroglia with gliomatous pro-

liferation which is followed by disintegration and the formation of cavities. The disease commences insidiously, usually
by increasing weakness of the lower limbs. Pain is rare, but
sensation of numbness may be present. Muscular atrophy
soon ensues, and also curvature of the spine. In a well
marked case the hands are wasted and there are various
trophic changes in the skin and nails with spontaneous ulcers
and painless whitlows. There is curvature of the spine, loss
of muscular power, tendon reflexes absent or diminished.
Sensibility to *touch* is retained, but abolished to *pain* and *temperature.* The loss of temperature sense is one of the first
symptoms to appear. It is distributed irregularly over the
body, generally in patches. The same may be said in regard
to pain. The special senses are rarely affected. There is
usually paraplegia of the spasmodic kind or inco-ordination of
the lower limbs. Atrophy of muscles involves the hands,
limbs, and finally the trunk, and gives the reaction of degeneration. The joints are sometimes the seat of arthropathies
similar to those seen in tabes.

PERIODIC CONTRACTURES.

Male, aet. 48; no hereditary taint; has had much sickness
during his life. Was first taken in 1874, with peculiar tetanoid seizures, lasting from ½ to 24 hours, at first invading the
left side only, and with no impairment of consciousness.
During the attack the left leg lay fully extended and absolutely
rigid, the left arm flexed to a right angle, all muscles of the
left side being rigidly contracted. Severe trismus rendered
speech impossible, although this soon disappeared. Between
the attacks only slight abduction and adduction are possible
in the left hip. The left knee is rigid to any justifiable force;
in the other joints motion is possible. The right leg is normal,
as are tendon and skin reflexes on both sides; no fibrillary
tremor. The left elbow cannot be brought out of its flexed
position by any force. In the other joints of the arm, passive
motion is possible; active motion only in the shoulder to a

slight degree. The left sterno mastoid and right scaleni are
rigidly contracted, so as to throw the head to the right, where
it resists all efforts to turn it.   Occasionally he could turn his
head with ease.   He could not puff his cheeks, whistle, or
protrude his tongue.   Sensation distinctly reduced on left
side.   Six weeks after discharge he was brought back unconscious, both legs and back rigid as iron bars; head turned to
left, trismus.   This lasted a few hours.   In a third attack the
head was turned to the *right*.   Paresis of the recti extern.,
super., and infer. of both eyes was noticed.   Phonation without articulation, or vice versa, was impossible.   One and a
half years after the first attack, a paresis of the facial nerve
of both sides, except the ocular branches, was noticed,
although he could draw the angles of the mouth back as a
reflex.   The reflex motions of the eye were good, the voluntary ones almost abolished.   Spasm of the pharynx now
accompanied attempts at swallowing.   During the later
attacks, which lasted three to six days, he was unconscious
and had rises in temperature, which were absent in the beginning.   Urination was always impossible for a short time after
each attack.   A small quantity of albumen was found once.
The attacks were always immediately preceded by a sensation
as of a spindle whirling in the head.

Death ten years after beginning of trouble.   Autopsy
showed absolutely nothing, but slight œdema of the pia.
Brain, medulla, and peripheral nerves were absolutely normal
microscopically.

No similar case could be found in literature.   Westphal,
Cavare, and Hartwig have described paralyses, occurring in a
similar way, but neither have been able to suggest or find a
cause.   Tetany can be excluded; hysteria explains nothing.—
(Dr. Gottfried Leuch, *Virchows Archiv.*, Band 121.)

G. J. K.

DISEASES OF THE CORD FOLLOWING INFLUENZA.

Judging from the periodical literature, the late epidemic was followed by severe nervous sequelæ, much more frequently in Germany than in this country. Dr. Bruno Herzog recounts 29 different conditions reported as following this disease. This does not include psychoses, neuralgias, vertigo, etc. He reports two cases of spastic spinal paraplegia in children aged 8 and 11 years, not relatives. No other etiological factor could be found. Neither case improved under treatment.—*Berlin Klin. Wochensch.* 1890.

G. J. K.

,LARYNGEAL PARALYSIS.

Desvemine in a contribution to the study of this subject divides it into three parts. It is considered something of a mystery why the lesion' of a nerve trunk which contains two classes of fibres, viz., those that supply the dilators of the glottis and those that supply the constrictors, always results in paralysis of the dilators and has no effect upon the constrictors. Since the publication of the ideas of Riegel, Rosenbach, Pemzolo, and others, authors have accepted this as a fact, and have accounted for it on the theory of the special susceptibility of the abductors. Desvemine does not accept this theory and cites the fact that there are cases of bilateral paralysis of the abductors from lesion of one of the inferior laryngeal nerves, and that such paralysis should not be considered peripheral, if by this term is meant abolition of the function of the distal extremity of a nerve from interruption of its conductability at the seat of lesion, or section in its continuity. Second, Krause concludes from his experiments that in these cases it is not a paralytic insufficiency of the dilators of the glottis, but a contraction of the constrictors. The author does not accept this explanation. Krause's experiments have suggested that lesions of the recurrent produces a

certain degree of neuritis which in itself is not sufficient to
produce contraction, or cause a certain degree of compression.
Krause also concludes that the contraction in the constrictors
is produced under similar conditions after section of the recurrent nerve, and explains those cases in which the phenomenon
is bilateral after lesion of one side, by the theory of anomalous
co-associated movements of Hitzig. Desvemine does not
accept this explanation. Third, he considers that experiments
show that these cases may be divided into two groups. In
both, the larynx loses its function as a respiratory organ and
the vocal cords occupy the median position, but in one there
is marked disphonia, or complete aphonia. While in the
other form, if the voice is affected at all, the change is only
in tone. The explanation of Krause is only applicable to the
first form and that only when the phenomena are unilateral
and of short duration. If they are bilateral, it is necessary to
accept the theory of contraction, which, however, is always of
reflex origin. The action of the dilators is overcome by the
constrictors in peripheral lesions which tend to increase the
general mobility of the larynx, either directly, or reflexly. If
the lesion is one which produces a centripetal inhibition the
dilators of the first two lose their physiological characters
owing to the inherent dynamic instability of their bulbar
centres.—(*Revista de Med. y Chirugia Practicas*, Aug. 1890).

------

*Intercostal Neuralgia as the only Symptom of Aortic
Aneurism.* Patient, aet. 39, complained of intense pain in
left side. No zoster; no physical signs for four months; then
a pulsating tumor between left scapula and spine. Autopsy
showed pressure atrophy of bodies of 3rd, 4th, and 5th dorsal
vertebræ, with corresponding ribs, by an aneurism of descending aorta. The dura was exposed. The case is of interest
on account of the neuralgia being the only symptom until four
weeks before death, and the absence of pressure on the cord,
although the spinal canal was opened.—(DR. BRASCH, *Berlin
Klin. Wochensch.*, 1890.)   G. J. K.

Mechanism of the Apoplectic Attack in Embolism. Dr. R. Geigel, in accordance with his theory of the cerebral circulation, explains the process as follows: At the moment of obstruction, all patent arteries dilate, in order to occupy the place of the obstructed one and express the blood from it, as their pressure exceeds that of the obstructed vessel. As long as this process lasts, a corresponding quantity of blood is used for their dilatation, and so is lost to the cerebral circulation. The resulting derangement of the circulation is the slighter, the greater the force of the heart and the remaining available arterial tension. The greater the pressure in the veins, the lighter the coma. Therapeutic deductions to increase the force of the heart and the venous pressure, by the recumbent position, although these will be of more practical importance when the differential diagnosis between embolism and hemorrhage is sure.—(*Virchow's Archiv.*, Band 121, p. 432.)   G. J. K.

A new work on *Acromegaly* by Sonza-Leite is just published. Etiologically neither race, sex, nor heredity is of importance. It usually occurs between the ages of 19 and 26. The important symptoms are the large circumference of the hands, feet, and head, also deviation of spinal column, particularly in its upper part. The signs and symptoms may be summarized as follows: morbid enlargement of hands, extremities and head, spinal deviation, thick neck, large larynx, speech slow and gutteral, voice hard, abdomen enlarged. There may be either hypertrophy or atrophy of muscles, or both, joints enlarged, cardiac hypertrophy, polyuria, disturbances of sensibility rare, complexion yellowish brown, hair thick and growing rapidly. Subjective symptoms are: persistent headache, pains in bones, amenorrhœa, voracious appetite, great thirst, sometimes loss of hearing and smell, cardiac palpitation, and intra abdominal pains. There is usually debility with some depression. Prognosis bad; progress of the disease slow. So far seven autop-

sies have been made.   The principal morbid condition found
is, hypertrophy of the hypophysis cerebri found in three cases,
also hypertrophy of the ganglia of cord and sympathetic, per-
sistence of thymus, alteration of thyroid, hypertrophy of
heart and blood vessels.   Enlargement of skull, hypertrophy
of bones of hands, feet, arms, and clavicle.   I. L.

———

Dr. W. E. Forest comments upon the discussion which
took place at the "World's Med. Congress" on *Chronic
Nephritis*, and complains that there is so much disparity in
the views of great men in the profession that the lesser lights
hardly know what to believe, or whether to believe anything.
He says: "In summing up we find that Dr. Lépine favors the
non-albuminous diet and keeps his patients on milk; uses
iodide of potash; condemns strophanthus and the vapor-bath.
Dr. Granger Stewart favors the hot-air and vapor-bath, and
the milk diet; knows no remedy for the disease.   Dr. Rosen-
stein, is against all medicines and not partial to the milk diet.
Dr. Senator agrees as to the inefficacy of drugs, with the sole
exception of iodide.   This is of great value.   Dr. Aufrecht is
certain that iodide is not of the slightest use.

In consulting the above list and eliminating those remedies
that are rejected by at least one of the authorities quoted, we
find we have absolutely nothing left, not even an accepted
dietary.

Surely the recording angel, your reporter, if there were a
spark of humor in his nature, must have smiled as he made
up his report; while the young and earnest seeker after scientific certainties must have reached the conclusion that the
only certainty in the therapeutics of chronic nephritis is that
nothing is certain."—(*Med. Record*, Oct. 4.)

THERAPEUTICS.

MIGRAINE.

Mr. Peake says he has found only two drugs of any effect in this disease, guarana and antipyrin, the first being more certain in its action. A patient who had taken 20 grains of antipyrin without effect, found that 10 grains guarana relieved the pain. His treatment is as follows: when the symptoms of the first stage come on, the patient should lie down, take a dose of 20 grains of guarana, drink beef tea, and if possible, sleep. When the pain comes on a cup of strong tea and perfect quiet are beneficial.—(*Lancet*, Sept. 27.)

HYPODERMIC USE OF ARSENIC. ·

Dr. H. N. Moyer, of Chicago, read a paper before the "Mississippi Valley Medical Assn.," upon this subject. The author says he has carried the use of arsenic given by the mouth to the utmost bounds of prudence, until the eyes were puffed and vomiting almost incessant, and had then continued the use of arsenic in large doses by hypodermic injections with the result of cessation of all gastric symptoms, and the cure of the disorder. In a case of chorea in a girl, the patient had been placed immediately upon the hypodermic use of arsenic beginning with three minims of a 5% solution, and increasing every second day, until three weeks after beginning the treatment she was receiving 13 minims of the solution at each injection, with an amount of arsenic equivalent to about 36 minims of Fowler's solution. At the ninth injection she was discharged cured. · The author considers Fowler's solution valuable in various forms of glandular enlargements. The action of arsenic given hypodermically is certainly better than when taken by the stomach.

ADMINISTRATION OF IRON.

Dr. John Aulde suggests a new preparation of iron called "levulose ferride." It is prepared in the form of tablet triturates and has a pleasant taste. The fact should be constantly kept in view that metals have a poisonous action upon nerves, nerve centres, muscles, and glandular structures. The prolonged use of insoluble preparations of iron may result in great harm. Iron produces slow contraction of the blood vessels and has a tendency to accumulate in the liver. Small doses increase the functional activity of this organ when given in a soluble and non-astringent form. Large doses limit the amount of work that muscular structure is capable of doing, while small doses increase its capacity.—(*Med. Record*, Oct. 11.)

BROMOFORM IN WHOOPING COUGH.

Neumann has used bromoform in this disease in 61 cases. It had no curative effect, but seemed to diminish severity of the seizures, and in few cases, perhaps, limited the disease. The dose is one to three drops in sweetened water, and the taste of the remedy is pleasant.—(*Therapeutische Monatch.*, July.)   I. L.

CANNABIS INDICA.

Dr. J. Aulde writes an article on the uses of this drug. The antagonists for this remedy are about the same as those in use for morphine poison. The bromides with chloral, gelsemium, and ergot increase the sensible effects of cannabis. Small doses give rise to mental and motor activity. As a pain alleviator it is especially indicated in functional derangements, as in tic douloreux, and dysmenorrhea with anæmia. Great value is shown in the irritable conditions, as in delirum tremens, the cough of phthisis, flatulency, or climacteric. Chronic poisoning is attended with œdema of the face, weakening, tremor, mental feebleness, and finally marasmus. As a remedy for supraorbital neuralgia, no article, perhaps, affords better prospects than cannabis given in small doses at short intervals.—(*Ther. Gazette*, Aug.)

M. Germain Sèe reports his conclusions concerning cannabis indica, which are as follows: It relieves pain and all abnormal sensations, and for these conditions is superior to morphine. In gastric atony and dilatation it has no effect, but is useful in vomiting of these conditions. It is the best stomachic sedative with none of the inconveniences of the narcotics.—(*Jour. de Med. de Bordeaux*, Oct.)

MYTHELINE BLUE IN NEURALGIA.

Prof. J. Steiner very highly recommends this chemical for the treatment of idiopathic neuralgias.

He reports seven cases of neuralgias with one failure, as well as one case of pruritus of the perineum in a male, and a traumatic neurosis, with tenderness of the lumbar spine. This treatment was introduced by Prof. Debove, of Paris.

Methyl chloride (C. H3 Cl.) is a gas which liquifies at a pressure of four atmospheres. A stream of the gas is directed upon the painful part, producing primarily, great refrigeration followed by a reduction of cutaneous irritability, and after several applications by desquamation.—(*Deutsche Medicin. Wochenschrift*, No. 29, 1890.)   G. J. K.

TREATMENT OF MORPHINE DISEASE.

Dr. J. B. Mattison writes upon this subject. His method consists in first producing a certain degree of nervous sedation by means of the bromides, and for this purpose he prefers the bromide of sodium. He gives two doses in twenty-four hours at regular intervals, continuing it until the symptoms desired are obtained. In from four to six days he produces a slight condition of bromism. He begins with a dose of thirty grains twice a day, and increases it until he sometimes gives a drachm and one-half, to two drachms, during the twenty-four hours. The amount of opium used is gradually decreased, leaving off one-third to one-fourth of it at first. One result of the opium habit is, to produce an insusceptibility to the actions of other nervines, requiring large doses to produce the usual effect. On the evening of the day when the last dose of opium is taken, he gives the patient a large dose of morphine to obtain sound sleep. When the patient becomes irritable and restless he gives large doses of cocoa and quinine; if these fail, doses of cannabis indica. The author considers that the usual dose of cannabis indica, as given in the books, is useless. He gives 60 minims of the fluid extract at a dose. Galvanism is advised in the neuralgic troubles of the opium habit. Warm baths are advised, and the external use of ether to relieve pain in the limbs. When the last dose of opium has been taken, the patient is put to bed and kept perfectly quiet for several days. He prefers cannabis indica as an hypnotic for this condition.—(*Ther. Gazette*, Sept.)

TREATMENT OF ATAXIA.

Dr. D. R. Brower, of Chicago, read an article before "The World's Med. Congress" with this title. He considers that there are three forms of ataxia according to predominance of symptoms. First, the *cerebral form*, characterized• by cephalic phenomena, pupillary inequalities, disorder of pupillary reflexes, myosis, transitory diplopia, optic atrophy, deafness, laryngeal symptoms, vertigo, apoplectic crises, etc. Second, a *spinal form*, and third, a *peripheral form*, usually due to traumatism. Treatment must vary according to the form. In all, rest in bed for six or eight months with massage is advised.—(*La Tribune Medical*, Sept. 11.

THERAPEUTICS OF SPASTIC PARALYSIS.

Dr. V. P. Gibney publishes a number of cases. The first patient was a girl of seven years, who was unable to stand alone, knees being persistently flexed. The Achilles tendons, adductors of the thighs and hamstrings, were divided. Knees were straightened and the child kept in a wire cuirass for six weeks. She was afterwards able to walk very well. The next was a boy with spastic paralysis. The child stood upon the outer borders of the foot, leg flexures being tense and resistent. The tense bands in the popliteal space were divided, as were also the tendon Achilles. The patient was put into plaster of Paris from hips to toes. He entirely recovered. The next was a case of talipes equinus. Achilles tendons were very short and both divided. Patient recovered. It is necessary after dividing the tendons to keep the foot in normal position for two or three months after the operation.— (*Jour. Nerv. and Ment. Dis.*, August.)

TREATMENT OF EPILEPSY.

Dr. H. C. Wood of Philadelphia, suggested some years ago, the use of antipyrin and bromide of ammonium, in the treatment of epilepsy. Dr. Potts reports results of treatment of 43 cases of idiopathic epilepsy In all these cases there was marked relief of the symptoms, and in some cases where other remedies had failed to give relief. There were no indications of bromism and none of the unfavorable symptoms of antipyrin observed. Six grains of antipyrin were usually given with 20 of the bromide.—(*University Med. Magazine.*)

THERAPEUTIC USES OF ELECTRICITY.

Dr. Morton Prince says: As a means of diagnosis electricity is of great service in demonstrating the different forms of muscular atrophy and paralysis, in deciding whether the atrophy or paralysis is due to neuritis or disuse, to a myopathy, to cord or brain disease, or functional disorders, and to detection of dissimulation. In such investigations it is essential that the electrodes should be of proper size, applied at the proper point, and with current of known strength. In neuralgia electricity is both palliative and curative, sometimes acting like magic. In neuritis it is only palliative. In atrophy and paralysis following infantile paralysis, joint lesions and disuse, electricity is beneficial, and the two latter can be cured by it. In cerebral hemiplegia the muscles recover more quickly under electricity. In muscular rheumatism electricity is palliative. In neurasthenia it is an excellent tonic. Electricity is not of the slightest use in curing locomotor ataxia, disseminated sclerosis, progressive muscular atrophy of the spinal type, or general paralysis. It is of no use in epilepsy or megrim. In using electricity never put the battery in the hands of the patient to use, and never entrust it to a nurse or to a student.—(*Boston Med. and Surg. Jour.*, Oct. 2.)

VERTIGO.

Dr. Landon Carter Gray, of New York, read a paper before the "Med. Soc. of Va." on this subject. He said that an important generic distinction between vertigo of organic disease and that of a functional nature, is that the former is attended by less irritability and apprehension on the part of the patient than the functional form. Kidney, heart, and organic liver troubles cause slight vertigo, lasting only a short time. In vertigo due to kidney disease the headache sometimes disposes to hebetude or coma, occasionally convulsions; there may be œdema also. In vertigo of organic liver disease there is usually some degree of hebetude, or jaundice, or dropsy. Vertigo due to brain lesions will prove pathognomonic. Thus cerebellar vertigo causes titubation, or one-sided staggering gait, or sudden semicircular whirling. According to Dana, temporal lobe lesions may produce like symptoms. Spinal lesions are not apt to cause more than slight vertigo; in locomotor ataxia dizziness may occur when the patient's eyes are closed. Middle-ear and labyrinthal diseases also cause vertigo.

But there is a chronic vertigo, varying from a sudden sensation of loss of equilibrium to dread of going about, often attended by symptoms of other nervous disturbance, such as tingling of extremities, fulness about head, usually felt most at the vertex, with slight aural ringing, mild insomnia, irritability, etc. It is most common in young and middle-aged adults, and is most frequent in Northern climates in the first warm months of the year. In severe cases, even going from a cold into an over-warm room will induce an attack. Neurasthenia is common. Tongue is usually unaffected, but there is generally an excess of uric acid or oxalate of lime in the urine. Hence Murchison gave' to this symptom-group the name lithæmia. But Dr. Gray said many cases do not present evidence of liver trouble; nor is there a standard by which to judge of an excess of uric acid in the system; and again, cholagogues often aggravate the vertigo as well as nervous symptoms. He said that generally this peculiar form of vertigo is due to some chronic and persistent error of digestion— either of the nitrogenized or starchy elements of food, or of both. Constipation without coated tongue or foul breath is common in this form of digestive disease. Predispositions to gout or rheumatism seem to cause such trouble; but the exciting causes of the vertigo are mental or physical overstrain, great anxiety, malaria, very sedentary life, etc.

Therapeutically, in those not having neurasthenia, he gives twenty drops of dilute nitro-muriatic acid, before meals, in a wine-glass of water, and also one drachm of fluid extract, or two grains of solid extract of cascara sagrada three times daily, reducing the dose if it causes more than two feculent stools a day. Interdict all red meat diet. In about ten days the patient will feel better generally, but the vertigo will still be unaltered. Then stop the acid, and, instead, give the best pepsine and pancreatine—pepsine immediately after meals, and pancreatine an hour and a half later. After a time gradually return to meat diet, but only once a day. In neurasthenic cases, in addition to this treatment, require the patient to take absolute rest—even sometimes in bed for two or three weeks. It is better to err on the side of enforcing too much than too little rest in these cases. In so radically differing from others with reference to the treatment of this chronic form of vertigo, he has only to say that his experience has taught him to so differ.—(*Med. Record*, Oct.)

SURGERY AND TRAUMATIC NEUROSES.

FRACTURE OF THE CRANIUM.

Dr. Rubio, of Madrid, reports a case of cranial fracture followed by intra-cranial exostosis, with operation and recovery. The patient was struck on the head by an iron quoit. Immediately after receiving the blow he lost consciousness, but a few minutes afterwards recovered himself and walked some distance. There was some hemorrhage from the wound. Or examination some time afterward there was depression of the frontal bone at point of injury. There was no motor nor sensory disorders of any kind, but for three months following the injury there was marked change in his mental condition. He became morose and apathetic and was indifferent to what was going on about him. His face lost its natural expression indicating a state of mental hebetude. His speech was very slow and hesitating. Several times during each day he would suddenly lose consciousness for a few moments, and later there was observed trembling of the left inferior extremity, which was considered an initiatory symptom of Jacksonian epilepsy. The skull was trephined and there was found a depressed fracture with an exostosis in the region of the right first frontal convolution. The pieces of bone removed were not replaced, but following the method of implantation of decalcified bone recommended by Dr. Senn of Milwaukee, pieces of bone plate were introduced above the replaced dura. The patient made a rapid and complete recovery. This case tends to confirm the conclusions of Ferrier and others that injury to the anterior lobe produces chiefly mental symptoms. —(*Revista Med. y Cirugia Practicas*, July, 1890.)

A CASE OF BRAIN SURGERY.

Dr. W. A. Hammond publishes an interesting case upon which he operated in his institution. In infancy, patient had an attack of cholera infantum, and during course of the disease, arms and legs became flexed and contracted, more pro-

nounced on left side, and after recovering from sickness the
left side remained contracted. During teething had convulsions; at five years of age had another convulsion; at six convulsions became more frequent. When she was about ten
years of age, was noisy and excited just before the access of
the fits. Some of the fits were accompanied by loss of consciousness, others not. Patient was nineteen years of age at
time of operation. During previous five years had shown
marked symptoms of mental derangement, was at times noisy
and disorderly, mind undeveloped. Left hand was in extreme
state of flexure, permanently contracted; elbow also flexed;
muscles about the shoulder joint paralyzed and atrophied;
muscles of left leg in same condition; left side of face much
less mobile than right; physically, patient was undeveloped;
right side of skull smaller than left. At times the epileptic
convulsions were confined to left side of the body, but at other
times were general. The skull was trephined on the right
side and on puncturing the brain a large amount of fluid was
evacuated. A drainage tube was inserted and the wound
closed. The patient died, and on autopsy a cyst was found
occupying nearly the whole right parietal lobe, and the greater
part of the ascending frontal convolution, and the anterior
part of the occiptital lobe. The cyst also involved the superior temporal convolution and one-half of the middle temporal.
The island of Reil was destroyed, and part of the caudate and
of the lenticular nucleus. The internal capsule was atrophied.
The post mortem confirmed the doctor's diagnosis. The
original disease was probably an extensive meningitis.—
(*N. Y. Med. Jour.*, Sept. 27.)

A CLINICAL CONTRIBUTION TO CEREBRAL SURGERY.

Mrs. A., æt. 36, had suffered for about six months with
headache, mental hebetude and irritability, cortical epilepsy,
beginning in the left side of the face and spreading to the left
arm, especially to thumb and index fingers. This was followed

by a left brachio facial paralysis, the left leg being but slightly involved; sensation was reduced in the paralytic area; syphilis was denied.

Although choked disc, slow pulse, somnolence and vomiting were absent, a neoplasm involving the motor area on the right side, was diagnosed.

The patient was put on mercurial inunctions for two weeks, at the end of which period she was decidedly worse. Vomiting and somnolence had appeared, and the paresis of the left leg had increased. Although the patient was in the sixth month of pregnancy it was decided to operate.

On reaching the dura, no pulsations were noticed. On opening it the brain was found to pulsate feebly. Five convolutions presented themselves; the upper and posterior are normal in color, the lower and anterior one of a dark violet color. The identity of this convolution could not be established by faradic excitation. On incision a large cyst filled with fluid and partially coagulated contents, was found. The cyst wall, composed of gliomatous tissue, was extirpated as far as accessible. Immediately after the operation the pulse rose from 53 to 86. All existing symptoms promptly disappeared, with the exception of slight facial paresis and impairment of sensation in the hand, in three weeks. Four weeks after operation a cortical spasm occurred, which was caused by the retention of a small quantity of pus. She was delivered thirteen weeks after operation.—(Dr. H. Oppenheim and Dr. R. Koehler, *Berlin Klinisch. Wochenschr.*, No. 30, 1890, p. 677.)

CRANIECTOMY IN MICROCEPHALY.

Dr. Lannelongue gives an account of a craniectomy which was performed on a 4-year-old idiotic girl with a microcephalic head. The patient was very poorly developed, less than the average at 2 years; could not stand on her feet nor speak and showed very little intelligence. There were no hereditary taints; parents and five brothers and sisters were healthy. There were no contractures nor epileptiform attacks.

The head was very small, cranium narrow and very flat on the sides, coming up to a point at the vertex presenting the scaphoid shape. The forehead was slanting backward and very narrow. Briefly: a microcephalic head with idiocy. Dr. Lannelongue acted on the hypothesis, that the brain had not sufficient room to develop, hence the idiocy and tardy development of the body.

Dr. L. removed a plate of bone from the cranium along the sagittal suture 9 centimeters long and 6 millimeters broad, reaching from the frontal to the occipital suture. The incision was made on the left side (which was most contracted) about a finger's width from the sagittal suture. The dura matter was not interfered with and the wound healed per primam. The results were highly satisfactory; a rapid development took place, the intelligence was greatly improved; the patient laughs and plays and understands what is said. She is learning to talk, stands upright and commences to walk.

Dr. L. adds that the operation is only the first step; the painstaking education of the child must follow.

Dr. Verneuil reported that he had made a similar operation, but as only a week had elapsed, he could only say that the operation was successful from a chirugical standpoint; although the interference here had been more extended along the whole antero-posterior diameter, the outlook was favorable.—*La Tribune Medical*, July 10th.

___

Lawson reports a case of fracture of the skull in a child aged 8. The patient had headache, drowsiness and occasional vomiting. There was a fracture of the skull on the left side, at the junction of the parietal and frontal bones, with cerebral hernia, which was burned through with a paquelin cautery. The patient recovered.—(*Lancet*, Sept. 27.)

___

SPINAL SURGERY.

Dr. Robert Abbé reports eight cases upon which he has operated; three of paraplegia from fracture; one of early curetting of a vertebra for Pott's Disease; two of tumor of the vertebral canal for paraplegia; two of intra-dural section of some of the posterior roots of the brachial plexus for neuralgia. The first was a case of injury to the spine between the 11th and 12th dorsal vertebra. Complete paralysis and paraplegia with incontinence. Intra-dural adhesions were broken up and the dura was sutured, but there was no relief of the paralysis. The next was a case of fracture of the dorsal vertebra, with complete division of the spinal cord, the object of the operation being to unite some of the dorsal and lumbar roots in order to restore sensation. Another case of fracture followed by paralysis was operated upon with benefit. The next case was one of extra-dural tubercular tumor of the spine with complete paralysis. Operation resulted in cure. Operation upon a patient with paraplegia with extra-dural sarcoma resulted fatally.—(*Med. Rec.*, July 26.)

BRAIN TUMOR.

Drs. L. Bremer and N. B. Carson report a case with unusual symptoms. The patient was suddenly seized with a spasm in left arm while at work. In a few minutes the spasm returned. From this time he had two of these spasms every day for a year, after which they increased in frequency. In six weeks the toes of his left foot turned under, his speech was affected, and he had occasional jerks or twitchings of muscles, in various parts of the body. Head was drawn to left side and fixed in this position. In a short time the toes of right foot were turned under, while the whole left side of body was much weaker than right. He vomited every morning for three months after onset of the disease. During the last two years he vomited about twice a week. There were frequent induced and uninterrupted contractions of the muscles which produced great suffering. Mouth was liable to close suddenly; fingers of left hand were clinched; hand was flexed at the wrist and pronated, and it was with consid-

erable effort of the will that the hand was extended and fingers opened.   There was a tendency of the arm to be flexed
at the elbow; leg was frequently extended at the ankle, knee
stiff, foot inverted.   While knees were stiff, legs tended to
flex on abdomen.   Every voluntary movement was at once
opposed by the simultaneous contraction of the antagonists.
Rigidity passed off during sleep.   No sensory disturbances of
any kind.   The skull was trephined on the right side over the
ascending convolutions and a tumor was found and a considerable quantity of reddish brown fluid was evacuated.   The
cortex at this point presented a reddish brown color.   A cavity was found about the size of a small hickory nut.   The
part of diseased brain substance removed was about the size
of a large walnut.   After operation the left fingers, part of the
hand, and ulnar side of arm were anæsthetic, as was also the
platysma region, and the fingers lost their muscular sensibility.   Further reports will be made on the progress of the case.
—(*Amer. Jour. Med. Sciences,* Sept.)

Dr. C. A. Wheaton reports a case of paraplegia due to vertebral disease, upon which he operated.   On examination there
was found a decided prominence of the 8th dorsal spine.   The
lamina of the 8th dorsal was removed, as were also the laminæ
of the 2nd vertebra above and 2nd below.   A few hours after
the operation he went into a profound collapse and died.
Dr. Wheaton thinks the result might have been different had
he, after the operation, given a sufficient support to the spine
in the form of a jacket.   He considers it probable that some
sudden change of position was the cause of the fatal collapse.
—(*Northwestern Lancet,* Sept. 1.)

TUBERCLE OF CORD.

Dr. Herter read an article upon this subject before " The
Amer. Neurological Assn." in which he reported three of his
own cases.   In all there are nine cases of this disease, of
which the records are sufficiently detailed to enable us to form

a picture of their essential features. Solitary tubercle of the cord is a disease of early life. Pain, weakness, and paræsthesia exist early and are at first unilateral, later they become bilateral and loss of power is complete. Anæsthesia and spasm of muscles are also observed. Symptoms attain intensity rapidly, usually reaching maximum of severity in two months. In sixteen cases in which location of tumor was noted, five were in cervical, seven in dorsal, three in lumbar, one at junction of lumbar and dorsal region. In some cases secondary degeneration was found, in others not. Meningitis of the cord with softening about tumor, and sometimes local myelitis have been observed. The rapid progress of symptoms, mildness of irritation phenomena, tubercular disease in other organs, are characters that help to identify this disease.— (*Jour. Nerv. and Ment. Dis.*, Oct.)

NERVE GRAFTING.

Atkinson reported to the "Brit. Med. Assn.," a number of successful cases. It is important to bear in mind that the substitution of a new piece of nerve for a piece that has been lost, (nerve grafting) is a totally different thing from the uniting of a nerve, (nerve suturing). In one case the median nerve had been divided in an operation. Two and one-half inches of the tibial nerve from an arm that happened to be amputated at the same time, was transplanted to the fore arm of his patient. Healing occurred by first intention and sensation began to return in 36 hours, and in five weeks was complete. The muscles also partly recovered their power. A number of other cases were reported with similar results. Strict asepsis is important, though union may occur following suppuration. Return of sensation occurs before motion. Better results are obtained when the grafting is performed immediately after the injury.—(*Brit. Med. Jour.*, Sept. 13.)

TRAUMATIC NEUROSES.

Dr. J. Hoffman reports 24 cases of traumatic neuroses, for the purpose of determining the degree of disability for work and consequent damages.

He divides them into 3 classes: Class 1, includes 10 cases, which he concluded were really sufferers. Class 2, comprises 6 cases, which presented strongly exaggerated symptoms. Class 3, comprises 8 simulators.

He protests against Oppenheim's views that the diagnosis may be made in three-fourths of the cases at once. This he claims would allow a great many frauds to obtain damages. He relates 3 cases, in which the simulators were only detected after several weeks residence in hospital. These men had previously obtained written opinions regarding their trouble, varying from one-half to total disability.

There is no rule which can be followed in detecting malingerers; each case must be treated according to the presenting indications.

He prefers the old nomenclature of traumatic hysteria, cerebro-spinal commotion, or traumatic neurasthenia to the term of traumatic neurosis, which is vague and does not represent any definite grouping or sequence of symptoms.—(*Berlin Klin. Wochensch.*, No. 29, 1890.)

In the same journal, No. 30, Oppenheim protests against Hoffman's assumption that he gives certificates of disability on the first examination. He says: "I have made it a matter of principle never to give an opinion without observation in hospital."   G. J. K.

———

Dr. S. V. Clevenger read an article before the "Amer. Med. Assn.," at Nashville, on *Erichsen's Disease as a form of Traumatic Neurosis.* In stating his reasons for giving the term, "Erichsen's Disease" he said: "It is a serious disturbance of the functions of the spinal cord, without there being demonstrable cord lesions. Such disturbances can be best accounted for by supposing that the spinal sympathetic system

6

has been deranged, secondarily interfering with the blood supply of the cord and its membranes, conjoined with other vaso-motor phenomena, such as emotionalism, flushings, cardiac rapidity, hyperidrosis, sleeplessness, headaches, directly due to the original sympathetic system derangement.'' In the discussion, Dr. Judd expressed the opinion that a large number of persons bringing suits against corporations for injury from accidents were insincere. Every person claiming disability from concussion of the spinal cord, where no object-ive symptoms are to be discovered, is to be considered as planning for your aid in enabling him to rob his employer. Dr. Clark Gapen endorsed the term ''Erichsen's Disease'' in order to fix upon Erichsen the obloquy he deserved. In his opinion, cases of so-called spinal concussion are usually hys-terical or malingering. Dr. Moyer could not agree with Dr. Judd that a concussion case only occurred associated with damage suits. In the past five years he had seen a large number of concussion cases in which there was no suspicion of a suit. Dr. Everts of Cincinnati, and Dr. Kiernan of Chi-cago, also approved of the use of the term, as did Dr. Lydston.

PSYCHOLOGICAL.

TIME-RELATIONS OF MENTAL PHENOMENA.

This is an able and exhaustive paper by Joseph Jastrow, professor of psychology at the '' University of Wisconsin.'' It is difficult to condense the article and make it intelligible, for to be thoroughly understood it should be carefully studied. It has been argued as a proof of the immateriality of thought that its operations were out of relation to time, but modern investigations show that every mental process must occupy time, and must conform to certain well established laws. It being established that so apparently simple a process as sen-sation, involves the passage of an impulse along nerve fibres, it is plain that the rate of traveling of this impulse sets a limit

to the time of the entire process, as well as to all more complicated mental operations, in which sensations are involved.
In 1844, Müller despaired of ever being able to show the
speed of nerve activity, but soon after that Helmholtz showed
the rate in the nerve of a frog, finding it to be 86 feet per
second. In man it is 110 feet per second, much slower than
light, or even than sound, and only a little faster than an
express train. In regard to the analysis of reactions, a great
variety of actions may be viewed as responses to stimuli.
There is a flash of light and we wink; a burning cinder falls
upon the hand and we draw it away; a bell rings and the
engineer starts his train; the clock strikes and we stop work.
In these, and many other actions, the time-relations are more
or less definite and important. In the simple reaction there
is physiological and also a psychological portion. The physiological time elements include: first, the time for the sense
reaction to respond to the impression; second, the time for
the passage of the impulse inward along the nerves; third,
the return passage of the motor impulse from the brain to
nerve and muscle; fourth, the time for the contraction of the
muscle. The time thus left unaccounted for is taken up by
the psychological processes and transformation of the sensory
into the motor impulse. The reflex act takes less time than
the voluntary one. The reaction time differs for the different
senses. Hearing is shortest, touch is intermediate, and sight
longest. Reaction to the sense of temperature is longer than
to contact. Reaction to heat is longer than to cold. When
the stimulus is intense the reaction time is shortened. When
the subject knows the nature of the experiment, and expects
a certain result, the reaction time is shortened. In addition
to the above, Prof. Jastrow considers the "overlapping of
mental processes," the "effects of practice and fatigue," of
miscellaneous and individual variations," of "time associations," and of "unlimited associations." A description is
given of the apparatus used in carrying out these experiments.

PATHOLOGY AND SYMPTOMATOLOGY.

Dr. Gustave Lopez states that there is a marked increase in the amount of insanity in the Island of Cuba. Among men from the northern portion of the island there is more insanity than among those from the cities, while in women the proportion is reversed. Among the natives of Cuba the increase of mental disease seems to be in direct relation to the proximity of their residence to the cities, and nearly twice as many blacks as whites are affected. Among the half breeds, or *Mestizos*, there is no noticeable predominance of insanity among the women, while the number of men affected is three and one half times greater than the women. General paresis and mania from alcohol are very common.—(*Revista de Med. y Cirugia Practicas*, Sept. 7.)

NERVOUS AND MENTAL DISEASE OBSERVED IN COLORADO.

Dr. J. T. Eskridge read a paper upon this subject before ''The Amer. Climatological Society.'' He has not been able to decide whether there is any difference in the amount of mental work a person can do in the high altitude of Colorado and the lower altitude of other states. In regard to insomnia he concludes that for the majority of persons, especially for consumptives, sleep is more easily obtained, more continuous and more refreshing in Colorado than in the eastern states. Some persons, especially hysterical people, sleep poorly in Colorado, as also do those with active hyperæmia of the brain. The doctor concludes that persons of an inherent nervous temperament (not those who are nervous from mal-nutrition, which the climate may, and does in many instances cure) are made seriously ill by a long residence in Colorado. Chorea occurs in Colorado with about the same frequency as elsewhere, and is quite as amenable to treatment. Neuralgia is less prevalent than in low malarial districts.

INSANITY IN ITALY.

Andria Verga *Archivio Italiano per le Malatie Nervose*, XXVII, III and IV, concludes a continued article which has been running through several parts of that journal, with the following considerations:

1. Insanity has increased in Italy absolutely as well as in proportion to the increase of the general population, as shown in the following table:

Censes of	1874	1877	1880	1883	1888
Males	6'476	8,010	9,000	10,291	11,895
Females	5,734	7,163	8,471	9,365	10,529
Total	12,110	15,183	17,471	19,656	22,424
Proportion per 100,000 in- habitants	51.00	54.17	61.25	67.75	71.01

We must not, however, forget that the progressive increase of insanity may be in part only apparent depending on the larger conceptions in regard to the disorder; the loosened family ties, on account of which the parties inconvenienced by the patients' acts more readily consign them to the asylum, in part also to the more ample and healthful public care that has come to be the rule in Italy, and above all to the constantly increasing number of those under public care. In fact there were in 1874 forty-three asylums and hospitals for the insane in Italy, fifty-seven in 1877, sixty-two in 1880, seventy-two in 1883, and in 1888 the number had mounted to eighty-two.

2. Insanity in the female sex progressively increases, but always in less proportion than the men, as is clearly shown in the above table. This less proportion seems remarkable, considering the more emotional nature of females, and the perils of the delicate functions of maternity. But I reflect that the greater the predisposing causes in the weaker sex, so much the greater are then occasional causes in the stronger one due to the struggles of life, and violence as a cause of insanity is much more dangerous and frequent than in the females.

3. Insanity in Italy predominates absolutely, as well as relatively, to the general population between the ages of 41 and 60 years; as regards the civil status it is more frequent among celibates; as regards religious faith, amongst Hebrews; and as regards education, amongst those not altogether illiterate.

4. The regions of Italy in which insanity is most frequent are Emilia, the Marches, and Liguria; regions where it is least prevalent are Sardinia, Sicily, and Naples.

5. Many insane are received in a state of relapse, others are chronic and incurable by nature from the beginning, others become such by delay and neglect, and the increasingly crowded condition of the asylums illustrates the comparative rarity of complete and permanent recovery. It is advisable to have separate establishments for the treatment of curable cases (true pyschative clinics, and the chronic and incurable insane) simple asylums.

6. As regards the various forms of insanity, three points appear to be constant; first, the old classic forms (mania, monomania, melancholia, and dementia) prevail over the forms signaled in modern psychiatry; second, that the expansive forms (mania, monomania) prevail, though not to a very marked degree, over the depressive forms (melancholia, hypochondria); third, that the frenasthenia, or congenital insanity, (imbecility, idiocy, cretinism) occur in much less proportion than the acquired forms, or mental alienation properly so called, being in fact only one eleventh as frequent.

7. Some forms of insanity are constantly more frequent in the males (phrenasthenia, moral insanity, epileptic, alcoholic, and paralytic insanities); others in females (mania confurore, pellagrous insanity). Some forms prevail in certain regions, like the pellagrous insanity, unknown in Sicily, Sardinia, and almost unknown in Naples, frequent in Piedmont, Emilia, Lombardy, and most of all in Venetia; hysterical and puerperal insanity most frequent in Venetia, while cretinism infests Lombardy, epileptic insanity Tuscany, and alcoholic insanity Ligurier.

SUNSTROKE AND INSANITY.

Dr. T. B. Hyslop read an article before the psychologial section of "The Brit. Med. Assn." on this subject. He said that he accepted the classification of Morade, namely, *coup de soleil*, due to heat of the sun directly; and *coup de chaleur*, due to heat and other influences acting indirectly. Very little attention was given to the relative value of climatic and atmospheric influences, but the bodily causes—such as fatigue, bodily habits, excesses, either alcoholic, dietetic, or sexual, syphilis, etc.—were discussed at some length. These factors were stated to have an influence specially upon the general vigor of the constitution, and, by rendering a person more or less susceptible to heat, so far predisposed him to suffer from it. Out of the fifty-five cases the author had collected of insanity following sunstroke, in eight there was a history of malaria, in five of syphilis, and in seven of alcoholism. Of the forms of sunstroke, the asphyxial and hyperpyrexial appeared to lead to more important and dangerous sequelæ than the syncopal. In infancy sunstroke was given as a cause of accidental idiocy or imbecility, and six cases were recorded in which the amount of injury to the mental powers was evidenced by some imbecility or weak-mindedness and not by the characteristics of the lower grades of idiocy. It was pointed out that epilepsy was one of the most common of the sequelæ, and occurred in various degrees of severity, from slight epileptiform convulsions to the severest forms of that disease. In childhood the mental defects and convulsions were held to be collateral phenomena, both dependent upon a common cause, and suggestive of an acquired psychosis in a large measure different from the progressive deterioration of ordinary epilepsy. Insanity following sunstroke was much like that due to traumatism. The attack of sunstroke seemed to produce an acquired predisposition to insanity. Of 1,947 admissions to Bethlem, 49 (or 2.6 per cent.) were attributed to sunstroke, and an analysis of these proved the symptoms to be of great complexity. In many cases the symptoms so

closely resembled general paralysis that they were mistaken
for that disease.   The symptoms arising from locomotor ataxy,
various paralyses, epilepsy, senile dementia, etc., sometimes
complicated the task of diagnosis; but the greatest difficulty
was experienced with such affections as general paralysis,
syphilitic disease of the brain and membranes, alcoholic and
paralytic insanity.   After a brief account of the prevailing
theories as to the pathology of the affection, the paper concluded with a description of the *post-mortem* appearances found
in five cases.—(*Brit. Med. Jour.*, Aug. 23.)

PSYCHICAL RESULTS OF GYNÆCOLOGICAL
OPERATIONS.

Dr. I. S. Stone read an article before "The Amer. Med.
Assn.," on this subject, the main part of which is made up of
statements from various authorities, and which are contra-
dictory.   He quotes the well known statement of Dr. Keith
which is as follows: "Even the fact that in my cases of hys-
terectomy the removal of the uterus and ovaries was sooner
or later followed by insanity in 10% of the whole number, is
enough for me to condemn any operation that removes these
organs."   Upon which Dr. A. Martin remarks: "I have been
truly astonished to read Dr. Keith's statement.   Among my
own cases, out of whom over 1,000 have survived laparotomy,
not one patient has been insane where there had not been
very marked symptoms of mental trouble before."   Dr.
Blumer, of the Utica Insane Hospital, says that of the few
cases that have been admitted there after operation, the
patient had been generally unstable normally, and predisposed
to insanity by heredity.   In conclusion the author says:
Few facts in medicine are so well established as, that scores
of women are being rescued from insanity and some of the
protean forms of hysteria, by the gynæcologist.   The fear that
insanity may result from the operation should never deter the
surgeon from its performance; for if in doubt about it, there

must need be positive reason to suspect a mental trouble, which, according to the almost universal experience of the profession, is caused by the disease in question, and is not to be prevented by the delay of the treatment.—(*Jour. Amer. Med. Assn.*, Aug. 30.)

Dr. Tevat in "Paris Medical" points to the connection between *constipation and insanity*. We know that toxic matters exist in the intestines and when these poisonous products are retained in the system the nervous phenomena are always aggravated in insanity. He shows that the two curves of the mental condition and constipation in insanity are parallel and therefore insists on a careful regulation of the intestines with proper internal antipepsis. This is of great importance in many other nervous conditions.—(*Bulletin gen. de Therapeutique*, Aug. 25, '90.)

INFLUENCE OF THE MENSTRUAL WAVE ON PSYCHICAL AFFECTIONS.

Dr. Schüle reports three cases of alternating psychoses in women, which show in a marked manner, the influence of the menstrual, as well as of the intermenstrual period.

In Case I. menstruation was observed once, but the succeeding periods at which it was due were marked by sudden changes from exaltation to depression, or vice versa. The middle of the intermenstrual period was also frequently so marked, although sometimes the change lasted only a day.

In Case II. the patient twice had a lucid day, corresponding to the 14th day of the intermenstruum, in the midst of a continued stuporous condition, with an alarming relapse into stupor during convalescence, which also marked the middle of an intermenstrual period.

Case III. showed similar changes at the menstrual and intermenstrual period.

Goodman, Jacobi and others have shown that the physical life of woman moves in a wave, the summit of which corresponds to the time immediately preceding menstruation and the valley to the intermenstrual period. Although exact physiological observations could not be made upon the author's patients, he believes that a physiological connection between the menstrual wave and the observed alternations in the psychoses is demonstrated. It only remains to demonstrate the intermediate factors.—(*Allgemeine Zeitschrift f. Psychiatrie,* Band 47.)   G. J. K.

BLOOD AND URINE OF THE INSANE.

Dr. Johnson Smyth investigated cases of melancholia, epilepsy, paretic dementia and secondary dementia. In regard to the blood he has established the following as an average in healthy human beings: hæmoglobin 93%, red blood corpuscles 5,106,000 per cubic millimetre, specific gravity 1,056. In ten cases of melancholia he found deficient hæmoglobin, that is, diminisned red blood corpuscles with specific gravity above the average of health, that is to say, in melancholia the blood plasm is abnormally dense. It was nearly as dense in epilepsy, and became denser as the epileptic seizures occurred. It is dense in secondary dementia and in general paralysis of the insane. In epilepsy there were some exceptions to the rule named. His conclusions are, that in insane patients there is a deficiency in hæmoglobin, greatest in secondary dementia; that the blood is most dense in this condition, thus resembling senility. He finds that the renal secretion is greatest in general paralysis, less in secondary dementia, sometimes diminished in melancholia. Total solids greatest in general paralysis, less in secondary dementia, small variation in other diseases. In general paralysis, epilepsy and melancholia, the amount of urea is about equal. Uric acid increased in the excretions of the insane, and slight excess of phosphoric acid in epilepsy.—(*Jour. Mental Sci.,* Oct!)

BRIGHT'S DISEASE AND INSANITY.

This was the address on mental disorders read ,before the "Medical Society of the State of Pennsylvania " by Dr. Alice Bennett, of the Norristown Insane Hospital. Insanity is a symptom, and not always in its beginnings is it a symptom of disease of the organ whose perverted action it is, but of faulty conditions external to it. By Bright's disease we no longer understand the morbid processes affecting the kidneys only, still at some stage in this morbid process lesions of the kidney do certainly exist. Their function is interfered with, and certain waste materials being retained in the body toxic effects result. The doctor's experience leads her to the opinion that affections of the kidney are common among the insane, and uræmic poisoning is a frequent cause of insanity. While the mental manifestations may vary, the most constant symptom is that of mental pain which may range from simple depression through all degrees and varieties of persecution, self condemnation, and apprehension, up to a condition characterized by a frenzy of fear with marked motor excitement and physical apprehension. The motor centres are especially liable to affection, shown by restlessness, great activity, convulsions, or convulsive twitchings, and sometimes choreic movements, or cataleptic states. Some brains are predisposed by inheritance, or otherwise, to an easy overthrow of mental balance. In a case of this kind insanity may result from anything that increases the burden of the kidneys, diminishing their working force, and interfering with their excretions. Such causes are: improper diet, long continued constipation, sudden exposure to cold, pregnancy, or any unusual interference with the circulation, overwork of body or mind, and especially worry, and anything that lessens power of resistance. The author has noticed that in cases of Bright's disease that sudden invasion of melancholia is a constant feature, and sense of impending danger and overwhelming fear of some threatening calamity dominates the individual. The author publishes the details of a number of cases from the department

for women in the Norristown Hospital. The first group consisted of those that were rapidly fatal, twelve days to three
months; the second, were less rapidly fatal, three months to
ten years; the third, were those terminating in rapid recovery;
the fourth, those recovering after many months; fifth, those
improved and nearly stationary for years; sixth, includes
those cases running a very slow, tedious course; the seventh
group contained the histories of cases illustrating the transformation of melancholia into secondary paranoia, with delusions of personal grandeur; eighth, treats of those of puerperal origin; ninth group gives the history of two cases, also
puerperal, complicated with chorea; the tenth, relates to
cases complicated with epileptiform convulsions.

HEPATIC CIRRHOSIS AND INSANITY.

Dr. G. B. Verga, *Archivio Italiana*, XXVII, III, IV, discusses the occurrence of cirrhosis of the liver of the insane.
He found that in 1,150 patients received in the asylum at
Mombella during seventeen years that there were 121 cases
of cirrhosis. Of these cases, 12 were pellagrous insanity,
and 2 alcoholic cases. Of the total number only 4 were
females. He offers as an explanation of the rarity of hepatic
cirrhosis in alcoholic insanity the supposition that in persons
predisposed to mental disease and subjected to the injurious
effects of alcohol, the nervous system being the point of least
resistance, suffers first, and before the other organs, such as
the liver, the kidneys, and the heart, which are involved so
frequently in ordinary cases of chronic alcoholic poisoning,
can be so materially affected. The fact that pellagrous
insanity is the form in which this seems to be the most frequent, speaks in favor of affinity between the two morbid
conditions and their analogous origin, both being the result
of the introduction of a poison into the system.

MELANCHOLIA MATRONALIS.

This condition, occurring during the fourth and fifth decades of woman's life, cannot be classed as climacteric or post climacteric. It is clinically and biologically characteristic and represents "the peculiar reaction of the female brain to the struggle for existence," analogous to the progressive paralysis of the male. It furnishes the figures to fill the hiatus left in statistics by the less frequent occurence of paresis in the female. Heredity plays a less important part in its etiology than the conditions of life. Stuporous or cataleptic symptoms are very rare, disorders of sensation predominating. There is great tendency to relapse or exacerbation. Course and prognosis correspond to those of melancholia. In several cases with a fatal issue, poliomyelitic changes in the gray substance of the cervical cord were found, small hemorrhages were frequent, hyperæmia was never absent. Degeneration of the posterior columns was found in several cases.—(Dr. HALLEROORDEN, *Berl. Klin. Wochensch*, 1890.)   G. J. K.

———

IMPERATIVE IDEAS OUTSIDE INSANE DELUSIONS.

In the discussion of this subject before the psychological section of "The Brit. Med. Assn." Dr. Hack Tuke said he did not think that certain imperative thoughts and words, with which some people were haunted without being insane, but of which, on the contrary, they were perfectly conscious, were sufficiently recognized by English alienists.

The case of Dr. Johnson was described in detail, although with him the imperative idea only led to an innocent motor act which was not of a distressing character. The second case reported was that of a law student whose mental trouble had reference to negatives. He worried himself continually in his reading as to the proper place the negative should occupy in a sentence, and this imperative idea, or obsession, seriously interfered with his studies. In the third case a lady

could perform very few actions in life without first counting a
certain number of times. Her life was rendered miserable by
it. In a fourth case, related by Dr. Tuke, a lady evinced
great disgust whenever she met with particular words in her
reading. It was possible to trace this idiosyncrasy to her
aversion to a gentleman whose name contained the same letters. She also developed the dread of contamination, to
which French alienists give the name of *folie du toucher*. Dr.
Tuke pointed out that there was in some people an irresistible
tendency to touch, and in others not to touch, certain articles
of furniture or certain dresses. In conclusion, Dr. Tuke gave
the following summary:—1. Imperative ideas frequently
occurred in persons with marked insane inheritance, but this
was not a necessary factor. 2. The prognosis was not as a
rule favorable, but although the particular obsession might
not be got rid of, the patient might pass for being sane in the
society in which he moved. 3. Common to all these cases
was the bondage under which the person lay to pursue a certain trivial, or disagreeable line of thought, along witn sanity
in other respects. 4. That such mental conditions could
generally be traced to an emotional basis. 5. That the treatment was chiefly moral, and that occupation of the mind was
very important along with the avoidance of directly combating
the imperative idea.—Professor Gairdner thought that it was
impossible to escape the conclusion that eccentricities of the
kind referred to were on the border-line of insanity. They
were insane habits, and if an imperative idea or eccentric
habit interfered greatly with the manifest duties and responsi-
bilities of an individual, and was not checked by the greater
and more prevailing good sense of the individual, the question
of official or technical insanity would arise.—Dr. James Stewart said such cases as those mentioned by Dr. Tuke differed
from those who suffered from insanity, in that they recognised
their abnormal condition themselves, while the insane would
seldom, if ever, allow that they had anything amiss with
them.—Dr. Urquhart quoted a case where an imperative idea

resulted in the patient being sentenced to twelve months' hard labor. He regretted that what might be called official insanity was so far removed from medical insanity,—(*Brit. Med. Jour.*, Aug. 23.)

_____

OCULAR SYMPTOMS IN PARETIC DEMENTIA.

Dr. Oliver read an article on this subject before the "Amer. Ophthalmological Soc.," the following being among his conclusions:

1. The oculo-motor symptoms of the third stage of general paralysis of the insane—which consist in varying, though marked, degrees of loss and enfeeblement of iris response to light stimulus, accommodative effort, and converging power; lessening of ciliary muscle tone and action; weakening and inefficiency of the extra-ocnlar muscle motion—all show paretic and paralytic disturbances connected with the oculo-motor apparatus itself, all of greater amount and of more serious consequences than those seen in the same apparatus during the second stage of the disease.

2. The sensory changes of the third stage of general paralysis of the insane, which, though similar to those found in the second stage of the disorder, are so pronounced as to show a semi atrophic condition of the optic nerve head, and a marked reduction in amount of both optic nerve and retinal circulations, with consequent lessening of centric and excentric vision for both form and color, all indicate a degenerate condition of the sensory portion of the ocular apparatus, with impairment of sensory nerve action.

_____   ,

PATHOLOGICAL ANATOMY OF DEMENTIA PARALYTICA.

Without going into details in regard to gross changes, or changes in the membranes, Mendel discusses the finer changes. The basis substance shows an increase of nuclei and cells, the latter being of different sizes and shapes, and of varying behavior to reagents, show various origins. The spider cells,

normally only found in the external layers of the cortex, are
more numerous and found in the deeper layers. These
changes are, however, also found in other diseases. If the
disease is of long duration a sclerosis develops. The larger
vessels are normal or atheromatous; the capillaries show an
increase of the nuclei in their thickened walls, with hyaloid
degeneration. The ganglion cells always show sclerosis or
atrophy, which, however, are not characteristic. The atrophy
of nerve fibrils, found by Tuczek, is not confined to the cortex, nor is it characteristic. Focal lesions of various kinds
may occur, as well as a great variety of changes in the cord.
A specific pathological state has not been described as yet.
Examination of fresh cases would seem to point to the vessel
changes as the primary factor, the degeneration of the nervous elements being secondary.

In the discussion Tuczek opposed the view that dementia
paralytica was a diffuse process. Changes in the anterior
regions of the brain, especially the motor regions, are reasonably constant and are the primary factor.—(*Berlin Klin.
Wochensch.*, 1890.)   G. J. K.

NICOTINE PSYCHOSES.

The primary effect of tobacco upon the system is excitation,
the secondary, depression. The entire muscular system is
affected, also the heart and vaso-motor system. According
to Schroff, the disease called *nicotinosis mentalis* is a primary
insanity following a regular course and with distinctive symptoms. The chief characteristics are a feeling of extreme debility
and impotence with early hallucinations, delusions, and
suicidal tendencies. The patient feels ill, restless, is easily
excited, sleepless, indifferent to usual occupations, and more
or less depressed. Following these are hallucinations of
sight, hearing, and of general sensation. Much palpitation
and pain in heart. Later in the disease his delusions become
more exaggerated. He has visions of angels, heaven, and
hell; is excitable and boisterous. These paroxysms of excite-

ment are periodical, lasting two weeks or more, with intervals of quiet. In the beginning of the disease the patient is gloomy and restless, later, and as it becomes chronic, he is more quiet and feeble minded. Patients sometimes recover in the early stages, but never in the later ones. (DR. SCHROFF in *Wiener Medizin. Presse*, No. 33, 1890.)   I. L.

NICOTINE PSYCHOSES.

Dr. Kjellberg, Upsala, Sweden, has investigated the effects of tobacco in inveterate smokers and finds that they are distinctly injurious. *Nicotinosis mentalis* is a primary disease which belongs to the group of mental intoxications. It has a stage of prodroma of about three months, characterized by general malaise, uneasiness, insomnia, depression, often of a religious tendency.

After the disease is established there are three stages:

1st. Hallucinations, fixed ideas with tendency to suicide, depression, attacks of fear and outbursts of anger. The patient talks little, but logically; the nutrition is not impaired.

2nd. Exaltation, hallucinations of a pleasant type; after two or three weeks depression again followed by a slight maniacal condition.

3rd. The intervals are shorter, the patient very restless; the intellect and memory impaired. The patient observes what is going on, but is taciturn and indifferent.

Prognosis good before third stage, but after that, poor. Treatment mainly abstinence from the use of tobacco, with nutritious diet.—(*Hygiea*, Sept., 1890.)

Peculiar Form of Progressive Dementia affecting three members of a family. Prof. E. A. Homén, Helsingfors, Finland, publishes notes of these cases. The patients were the three oldest of eleven children. The disease developed at the respective ages of 21, 20, and 12, all being affected alike. It began with vertigo, occasional headache, and feeling of lassi-

7

tude, and soon after the mental vigor began to fail. During the first year the walk of all three was unsteady, similar to that of a drunken man, and was accompanied with pains in the lower extremities. Two years later, speech became difficult and indistinct. Cases one and three the last three years of life manifested complete dementia and could not speak. Case two now at the end of four years, can speak only a few simple sentences. After the disease had existed in this case about two years there were contractions in all the limbs. Case one remained in bed the last three years of his life, being perfectly helpless from this cause. Case three was also in bed the last three years of his life. Case two has been in bed one and one half years. Cases two and three showed tremor in arms and hands. On autopsy of the two who died, in case one, a girl of 26, the brain and medulla weighed 1,130 grammes above normal. That of case three, 1,160 grammes above normal. In case one, there was yellow softening in the lent. nucl. Case two, showed a cystic cavity in both lent. nucl., the cavities being of equal size. Endarteritis existed in the basal artery, left sylvian, and left middle cerebral.— (*Neurol. Centralblatt*, Sept. 1.)   I. L.

MOTOR HALLUCINATIONS.

In 1888 Seglas called attention (*Progres Medical*) to the relations of hallucinations to the function of speech, and claimed the existence of a special class of hallucinations that he called verbal, consisting in a morbid sensation of having spoken, though nothing was present to the auditory or visual sense. These he explained on the theory which Tamburini claims as his own, that hallucinations consist essentially in an excitation of the psycho-sensory centers of the cerebral cortex, and also on the undoubted fact that the mental image of a word consists really of three separate ideas, auditory, visual, and motor, which have their seats in their own special cortical centers, hence we may have verbal hallucinations that may be designated either as acoustic, visual, or graphic, or in case the phono-motor center is involved, motor hallucinations.

Tamburini, in a paper read before the Italian Congress of Alienists at Novara (*Archivio Italiano*, XXVII., V.), reports a case of a female who, without having any visual or auditory hallucinations or persecutory delusions, suffered from a peculiar sensation, as of the formation by the organs of speech of abusive and threatening language, which came on suddenly without any previous corresponding mental idea. The words were not pronounced; though the tongue was slightly moved, there was only the clear subjective sensation of the words pronouced. Tamburini discusses this case in relation with other similar phenomena, such as hallucinations of amputated limbs, and concludes as follows:

1. Aside from hallucinations purely sensory, there can be distinguished *motor* hallucinations, which can be specially verified in the sphere of the movements of speech, but may also occur in any part of the body capable of movement.

2. The seat of this kind of hallucination should be located in the so-called psycho-motor centers of the cerebral cortex.

3. According to the degree of irritation of these centers, we have either a simple hallucination of movement, or its transformation into irresistible impulse, without related convulsion.

4. Taking account, however, of the physiological data that make us admit the mixed sensori-motor nature of all the cortical centers, it is clear that sensory images, properly so-called, as well as motor ones, enter into every hallucination.

SUBJECTIVE DELUSIONS IN MENTAL DISEASE.

Dr. Draper, of the Vermont Insane Hospital, writes on this subject. Among the most common subjective delusions of the insane are the impressions of being operated upon by electrical influences. These are doubtless at first sensory hallucinations, and later become delusions through attempts at explanation. Delusions relating to the stomach are sometimes the result of previous excesses in ¡eating or drinking, which finally react upon the nervous system. In cases of hysterical insanity the sensation of being pierced through the body is not uncommon, and is probably due to visceral neuralgia.—(*Amer. Jour. of Insanity*, Oct.)

ANALGESIA AND INSANITY.

This is the title of an article by Dr. Keniston, assistant at the Middletown, Conn., Insane Hospital. The doctor's experience is, that analgesia is found in a large proportion of cases of mental disorder. It may be local or general, circumscribed or diffuse, unilateral or bilateral. The most familiar instance of analgesia among the insane occurs in phthisis, which often runs its course without pain, cough, or dyspnœa. In epileptic insanity, and especially in epileptic dementia, it may be pronounced. Among the insane, cases of pleurisy and pneumonia may run their course without pain, and the same is true of acute peritonitis, also fractures and dislocations, malignant diseases, abscesses, boils, and injuries. In general paresis, analgesia is very common in the last stages. The presence of this condition cannot be predicted in any case, but must be sought for, and it is well to look for it in every case.—(*Amer. Jour. of Insanity*, Oct.)

OBLIGATIONS OF THE MEDICAL PROFESSION TO THE INSANE.

Dr. Everts, of Cincinnati, has an article upon this subject in the Oct. number of *The Journal of Insanity*. A special obligation of the medical profession is, that physicians should have some knowledge of insanity as a symptom of disease which is amenable to medical treatment. It is pointed out that the general practitioner, rather than the specialist, comes in contact with the disease in the earlier stages, and therefore it is very important that he should be able to recognize and treat this stage of the disorder.   A physician from his daily observation of his fellowmen under circumstances when they are unconsciously stripped of much natural disguise—by which we habitually obscure ourselves in ordinary intercourse —is more capable than non-professional men of detecting mental disorder.   The learning of a physician should be of the broadest kind, and should include the natural history and historic development of the race.   Man should be studied zoologically and ethnologically through all the stages of his development, pre-historic, ancient, and mediæval, in relation to all manner of environments, racial and otherwise.   Thus he may be qualified to estimate every man's condition and characteristics, socially, physically, morally, and mentally.

PROPER DISPOSITION OF THE CRIMINAL INSANE.

Dr. Archibald Church writes on this subject in the *Medico-Legal Journal* for September.   He summarizes his views as follows: 1. That in the proper disposition of the criminal insane, criminality alone should be the criterion of classification.   2. That the criminal insane should be cared for in separate institutions.   3. That insane criminals committing capital offences should be sequestered during the period of their natural lives.   4. That insane criminals committing lesser offences should be committed for periods equal to terms of imprisonment for their crimes made and provided, and as much longer as their insanity persists.   5. That criminal insane may be liberated upon regaining their reason by the pardon of the governor, with the consent and recommendation of an advisory board.

THERAPEUTICS. ·

HYPNAL.

M. Gley (*Bulletin Gen. de Therapeutique*, Sept. '90) has experimented with this substance and finds its physiological properties to be those of chloral. There are two combinations: monochloral of antipyrin, which has 47 parts chloral to 53 of antipyrin, and bichloral of antipyrin, which has 66 parts chloral and 34 antipyrin. Injected into the veins of a dog the two combinations had the same toxic "coefficient:" 1 gramme for every kilogramme of the weight of the animal. Theoretically it appears that the monochloral ought to be less toxic than the other; but the experiments prove the reverse.

Hypnal.—This substance is the subject of an article by Mme. Dr. Frankel in the "*Bulletin Gen. de Therapeutique*" for the 30th of September, '90. She arrives at the following conclusions:

1.   Hypnal, or monochloral of antipyrin, has a well defined chemical combination, and is less soluble than chloral or antipyrin. It is easily dissolved in a feebly alkaline menstruum; consequently hypnal is dissolved in the intestines and the blood.

2.   Hypnal has hardly any taste, and no odor, which makes it superior to chloral, particularly for children.

3.   The properties of its component parts (chloral and antipyrin) are manifest in its action; thus it is both hypnotic and analgesic.

4.   The hypnotic properties of chloral are increased by the admixture of antipyrin.

5.   Hypnal produces hypnotic effects in smaller doses than those of chloral.

6.   Hypnal is serviceable in insomnia due to pain.

CHLORALAMID.

The author's conclusions are:

1. Chloralamid in doses of one to three gms. is a safe and efficient hypnotic in chronic insanity and epilepsy. All the patients were females.

2. Its action is admirable in states of excitement as well as in nervous insomnia.

3. In some cases it acts as a sedative in doses of one to two gms. given at noon.

4. Its anodyne powers are slight.

5. Tolerance is frequently established.

6. It is best to give it shortly before bed-time.

7. Its action is slower and just as sure as that of chloral; it is safer and the resulting sleep is easier and more natural.

Amylene Hydrate was given to twelve epileptic females, all old cases.

1. The remedy in daily doses of two to five tablespoonsful of a ten per cent. aqueous solution, reduced the number of paroxysms, often to a considerable degree.

2. No diminution of effect was noticed on giving it for three to five months.

3. Dangerous sequelæ were not noticed.

4. Its favorable action is not restricted to nocturnal epilepsy.

5. The state of the mind has no influence on its antiepileptic action.

In a footnote the author states that further trials upon 35 epileptic males (all old cases) seemed to be followed by an *increase* of the epileptic paroxysms.—(DR. NAEKE, *Allgem. Zeitschrift f. Pyschiatrie*, Band 47.)   G. J. K.

————

Dr. W. Hale White (*British Med. Jour.*) says: "I always prescribe it with spirit; 20 grains will dissolve in 1 drachm of rectified spirit in fifteen minutes, and water may be added to this solution without re-precipitating the drug. A good way of giving it is to tell the patient to dissolve it in a little brandy, add water to his liking, and drink it shortly before going to bed.'' The *Med. Summary*, in speaking of the administration of chloralamid gives the following formula:

Chloralamid..................4 drachms.

Spts. vini gallici..............2 ounces.

Curacao.....................2 ounces.

M. A tablespoonful (30 grains chloralamid) in water, and repeated in four hours if necessary.

ETHER INJECTIONS FOR NEURALGIA.

Dr. Kumps recommends hypodermic injections of ether for neuralgic or rheumatic pains. He does not, on account of its great effusion, inject pure ether, but mixed with an equal part of alcohol or Hoffman's anodyne. He makes an injection of one gramme of this mixture as near the seat of the pain as possible. This produces pain of short duration; the swelling which follows quickly subsides under gentle massage. The relief is often immediate, and sometimes an anæsthesia of the skin remains several hours. Ether has not the objections of morphine, but has a stimulating effect and rapidly soothes gastric malaise with headaches.—(*Bulletin Gen. de Therapeutique*, Aug. 8, '90.

SULFONAL.

Dr. Vorster reports very favorable results from the steady use of sulfonal.

As the primary action of this chemical is upon the motor sphere, its use would be indicated in states of motor hyperactivity. Given in quantities of two to four grammes in divided doses, the result is quietude during the day and rest at night. It is not necessary to produce the sedative effect at once, but gradually increase the dose from .5 gm. four times daily until a favorable effect is produced, or somnolence and a staggering gait show saturation. There is no danger of forming a sulfonal habit. The drug may be discontinued suddenly without bad effect. One patient took 580 gms., about 1¼ lbs., in 9½ months, with no bad effect on withdrawal.

Slightly depressing effects on the circulation were found in four cases, a rubeolus eruption twice, diarrhœa in seven cases. The remedy was used in 56 cases, of both sexes, presenting a great variety in the intensity of the motor symptoms. In some the result of the treatment was prominent, in others the excitement again increased in spite of continued doses. None failed to respond in some degree. Six epileptics showed a diminution of the number of attacks while under the influence of sulfonal, although the stupor was increased in two.— (*Allgem. Zeitschrift f. Psychiatrie*, Band 47.)   G. J. K.

LOCAL THERAPY AND PSYCHOSES.

O. M. Forty years old. For the last five years was treated by a number of rhinologists for chronic catarrhs of the nose and throat, with a subjective sensation of great forter. At last he created a scene in a waiting room, where he thought the people made remarks about his offensive odor. Prof. Praenkel found his nose and throat normal and sent him to the author, who found him suffering from paranoia with hallucinations of smell. He was emaciated, nervous, and avoided people, who, he imagined, talked about the stench he spread around him.

Under sedatives and hygienic measures he soon became convinced that the bad odors were imaginary and was discharged in 7 weeks recovered.—DR. EDEL BERICHT, *Allgem. Zeitschrift f Psychiatrie*, Band 47.)   G. J. K.

TREATMENT OF THE INSANE BY REST IN BED.

Dr. C. Neisser speaks very enthusiastically of the results of prolonged rest in bed in recent cases of insanity. He uses it in all cases of maniacal excitement and melancholic agitation, as well as in paranoia and paresis.   He makes an exception with young women with pronounced sexual delusions, on account of the associations.   He believes it acts in a great measure by suggesting the idea of physical disability, this idea being usually associated with a prolonged stay in bed.   Isolation and mechanical restraint are said to be almost totally unnecessary.   Absolute cleanliness and daily alcohol rubbings are necessary to prevent bedsores.   The method is not as applicable in private as in public institutions, as it requires a large staff of attendants, especially where each patient occupies a special room.   Baths are valuable adjuvants.   Paretics receive baths of from 65° to 76° F.; all others, warm or hot baths, of various duration according to indications.   The douche is not used as it may degenerate into a punishment. The time spent in bed may extend to ten or more weeks, the change to an active life being gradual.— *Berlin Klin. Wochensch.*, 1890.)   G. J. K.

The Med. Standard for October, publishes the decision of Chief Justice Thayer, of Oregon, in a recent mal-practice suit, of which the following is a brief summary.   He holds that the practice of the courts in allowing juries to determine these cases is unjust to the medical profession because it encourages the institutions of suits against physicians.   The surgeon cannot always be expected to achieve the success he desires, for the reason that causes for which he cannot be responsible may interfere.   The average juror is unfitted to decide such cases and the trial court should never allow them to be submitted to the jury unless the plaintiff has shown by proof that the defendant is guilty of the charge.   People who devote their lives to relieving the suffering of mankind should be encouraged, and the way suits for mal-practice are often conducted has the opposite effect, and is apt to leave the performance of many important duties to reckless and irresponsible empirics.

MEDICO-LEGAL.

WILLIAM MANLEY—A MEDICO-LEGAL CASE.

This is a history of a case of homicide by an insane man, published in the October number of the *Journal of Insanity*, by Dr. J. B. Andrews of the Buffalo State Hospital. It is said that a younger brother and a maternal uncle of Manley's had been insane. He was eccentric and of an inventive turn of mind. His wife states that during their engagement he requested that she should not speak to any person after they were married. Within two months after the marriage he repeated this request, and also that she should not speak to her brother nor receive visits from her friends. He became suspicious of her; would not allow her to go to the store for provisions, or to go out walking, and accused her of infidelity. He locked her in the house and threatened her life with a revolver. From the time of their marriage in 1877, to the time of the homicide in 1887, his wife left him a number of times on account of his cruelties. He would sometimes walk the floor during the greater part of the night, accusing his wife of infidelity, said that men were concealed in the cellar, and that he saw them escaping through the windows, and heard them around the house at night. If she lighted a lamp at night he said it was a signal; if she dropped the broom in sweeping this was intended to convey information by rapping. He was arrested a number of times for his violence and insanity, but was released. The delusions of suspicion which at first referred to his wife, finally included every one he knew, and even strangers. Had the delusion that preachers advised his assassination from the pulpit, that hotel keepers had his food poisoned, that people talked about him on the street and threatened his life. In the early part of '87 he wrote the following to his lawyer: "I have the pleasure to state that the backbone is broken, of one of the foulest, most gigantic, well organized and diabolical conspiracies to murder in the annals of crime. In it there were the keenest criminal intellects, both men and women, that the civilized globe could produce; in it there were the hidden and deadly instruments of science; in it there were the most craven, vindictive and insidious elements of three nationalities, combined with inflamed bigotry and pulpit fanatacism; in it there were the Knights of Labor and all secret societies; in it there was the shrewd libertine and consummate hypocrite, the author of it all, who had the police, the majority of them in Boston, the whole of them in Lawrence and Lynn, and about one-third of them in New York, in a passive form, about the same in Brooklyn, at the bidding of this insidious and cowardly monster; in it the most desperate and sin-polluted creatures available from many corners of the earth, whose only object was sordid gain; but through the all-pervading spirit of God, my own delineating power of the human character, a great force of will power, self-control, forbearance, a steady nerve and a pair of 'American bulldogs,' I have defeated, confused ·and bewildered the enemy all along the whole line, never losing one battle, from self-control and trust in God." Later in the same year he shot and killed an officer, who attempted to arrest him. The physicians who examined him decided that he was insane. Among the various delusions that he had entertained for years, was one that he had been elected president of the United States, and been kept from the position by a conspiracy. In conversation he was coherent, quiet, free from violence in conduct or action, but was so completely absorbed by his delusive ideas, as to refuse all occupation or employment. He is now confined in the Insane Hospital at Buffalo.

RESPONSIBILITY IN HYSTERIA.

In a rather lengthy paper in the *Revista Sperimental*, XVI, III, L. Bianchi reviews the subject of responsibility in hysteria and reports several cases of interest. He considers that irresponsibility should be recognized as existing in those conditions in which the hysterical psychosis runs through the phase of mental disorder in its various phases,—in which the change of character, the abnormal manner of perception and conception of the external world, the profound changes in the affective life, the intercalated attacks of ecstacy, lethargy, somnolence, etc., all indicate profound change in the whole psychic life of the individual. Another hysterical condition in which absolute irresponsibility may be imputed, is that of the cases of so-called double consciousness such as have been reported by Azam. Another class of cases has not received as much attention from medico-legal writers. These are hysterical individuals who suffer from decided somatic symptoms (such as paralysis, contracture, hemianæsthesia, etc.) and who also during the intervals between their occasional hysterical convulsive attacks, present a character with complex psychic symptoms, that is decidedly abnormal and peculiar, and which we can consider as one degree of that condition which in its more complete and classic form we recognize as double consciousness. The author reports one of the cases of this class that came under his observation in whom with the improvement, under treatment, the individual became almost another person from what she had been before.

A fourth order of cases are those who present from time to time the symptoms of grave hysteria (convulsions, catalepsy, hemiplegia, etc.), but who, in the intervals, show no disturbance whatever, but attend to their business. These cases can not be judged alike; each requires an individual study to determine the degree or kind of responsibility.

HYPNOTISM.

IS HYPNOTISM HUMBUG?

"I find it difficult to discuss the matter with patience. That the phenomena of hypnotism exist, and that they command the earnest investigation of scientific men, there is no doubt. But I do not believe the time has come to use the method as a remedial agent. We don't know enough about it. I am sorry to say I cannot draw a sharp line between some of the phenomena of hypnotism and the ideas of the spiritualists and old-fashioned mesmerists. The charlatans who use the system to humbug the public are frauds who should be suppressed. Did you ever see a collection of photographs of those people? The rogues' gallery would be a collection of gentlemen compared with them. And yet they undoubtedly have a mysterious power over certain persons. Animal magnetism won't account for it. I saw the other day the daughter of a New York physician, a delicate, beautiful young girl, completely hypnotized by one of the ugliest, most repulsive-looking Russians you ever saw."—E. C. Spitzka, M. D.

"The first thing that strikes me in connection with hypnotism is the confidence with which it is asserted that it has been proved beyond dispute to be so successful that it cannot drop. But I am old enough to remember that this was said in the time of mesmerism. Practically, mesmerism fell into desuetude fifteen years ago. Except in distant corners such a thing is scarcely heard of. Now, from some researches which have been conducted at Nancy, and stimulated by the opposition of the Salpêtrière school, we have the subject once more brought before us, and we are told of the advent of a great and important practical truth. Therefore we are told that hypnotism has established itself for all good. I have no hesitation in prophesying that before twenty-five years have passed it will be in the same position that it was twenty-five years ago."—Sir Andrew Clark (*Med. Record*, Oct.)

———

In a paper read before the first medical congress of Cuba, Dr. Malberti explains his method of treatment of the insane by hypnotism. He is of the opinion that in acute insanity, where symptoms of cerebral irritation are not well marked, hypnotism may be used with hope of permanent results.

When, however, the insanity has existed for a long time, or when there are evidences of organic brain lesion, there is no hope of benefit from this remedy.—(*Revista de Med. y Chirugia Practicas*, Aug. 27, 1890.)

SUGGESTIVE THERAPEUTICS IN MENTAL DISEASE.

Seppilli, in a paper read before the congress of Italian Ailienists (*Archivio Italiano*, XXVII, V.), finds that suggestive therapy is hardly available as a general means of treatment in insanity, since the insane are not easily brought under the influence of hypnotism, but that it may be of advantage in certain forms, more especially epileptics and hysterical cases, and that the best results are obtained in these and in dipsomania. Hypnotic suggestion should be employed only on patients who voluntarily submit to become hypnotized, and care should be taken to watch for any injurious effects. Therapeutic suggestion, given in the normal state, is of decided advantage in mental diseases, and when methodically used to combat the morbid conditions, is especially useful in simple melancholia, alcoholism, and the lighter forms of insanity. In chronic paranoia it does not seem to give any favorable results.

NEWS AND COMMENTS.

Since vivisection became a favorite mode of investigation much has been written about the pain inflicted upon animals by experimenters. The earlier investigators were doubtless rather careless of the amount of suffering they caused, but we believe that at present physicians are careful to avoid suffering as far as possible. There are good reasons for believing also, that animals are much less sensitive to pain than man, so that an injury to an animal will cause less suffering than if inflicted on a human being.

Dr. Collier, of London, has called attention to the ease with which worms, snails, and insects, survive injuries that would be fatal to higher organisms. Most of these, as well as other animals of similar organization, have no brain, but simply ganglia scattered along the length of the body. As is well known, some of these can be cut in two and yet live and even develop a head on the cut end. A cockchafer partly disemboweled will continue to eat sugar, and a beetle with a pin sticking through its body will devour insects with apparent relish.

With regard to higher animals there is abundant evidence that sensitiveness to pain is much less acute than in human beings; indeed this insensitiveness would be a natural result of the survival of the fittest within certain limits. Lions and tigers in captivity will gnaw off their tails and feet and so mutilate their bodies that death may result. Any one who has seen animals injured can testify to their survival of injuries that would prove fatal to a human being. Mr. Rowell has related the case of a horse that a few moments after having its leg broken by a carriage wheel, hobbled to the road side and began feeding with apparent unconcern. Any one who has experimented upon the lower animals can testify to their rapid recovery from serious operations, which would not be the case if they suffered much.

In the human race the sensitive nervous temperament cannot endure pain that others will easily bear.   Spencer says
the Zulus are apparently indifferent to pain that a civilized
man could not endure.   The Figi chief mentioned by this
author, who cut off his cousin's arm and compelled the victim
to sit by while he cooked and ate the limb, had to deal with a
man who was superior to a case of shock.   Dr. Collier says
when shoes were introduced into New Zealand, that when a
native experienced difficulty in getting a new pair on he would
chop off one or more toes and force the bleeding member into
the shoe without apparently feeling any 'pain.   Fright, both
in animals and man, has a power to abolish the feeling of pain.
Dr. Livingstone stated that when attacked and savagely
mutilated by a lion, he felt no pain, either mental or physical.
Is not this a sort of hypnotism by which nature dulls the sense
of pain and endeavors to divest violent death of suffering?

————

The investigations of Mr. Humphrey on skull thinning in
old age, and those of Dr. Dwight on certain peculiarities of
suture closing, summarized in our previous issue, are of more
interest when considered in connection with other facts of
brain and skull formation.   The conditions mentioned can
hardly be considered pathological, they are exhibitions of the
normal retrogression of tissue, and are part of the cycle of
vital change.   It has been found that certain skull types are
racial in character, and that there are certain racial peculiarities of skull development.   Broca has found the skull to be
the least variable of any part, and this, consequently, serves as a
means of racial identification.   The correlations of structure
are also interesting.   Among Africans the head and orbits are
long, and the hair flat.   Among Mongolians the head, orbits,
and hair are round, whereas the European type is intermediate, the head, hair, and orbits being oval.   Among Caucasians
the anterior fontanelle and sutures ossify last, while in Africans they ossify first.   The result is, as Bateman has specified,

8

that in the latter the frontal region ceases to grow first, while in Caucasians the closure is from behind forwards, and the frontal brain has a better opportunity to grow.   This later closure of frontal sutures among higher races has been considered evidence of the location of the psychic faculties in that region.   Broca concluded from the examination of the heads of house surgeons and porters in Biceitre, that intellectual work increases the brain capacity chiefly in the frontal region.

Virchow considers that idiocy is sometimes due to closure of sutures checking brain growth.   It is probable that this and other varieties of skull deformity are elements in mental deficiency, and it is also probable that arrest of skull growth may result from brain deficiency primarily.   There is probably some relation between the size of the brain and mental capacity, as examination of the brains of eminent men shows.   Yet this relation is so little understood that its value is of small account.   Napoleon and Webster had large brains, while Gambetta and the historian Grote had small brains. After all, quality is more important than quantity.   The largest brain ever examined was that of a mechanic who could neither read nor write.   Evidence seems to show that the brain power of the race is increasing, if the cranium is an index of mental capacity.   The brain capacity of the European is 94 cubic centimeters, 91 for the Eskimo, 85 for the Negro, 82 for the Australian, and 77 for the Bushmen.   Broca found that the skulls of those who lived in the 12th century, when compared with those of the 19th, showed an excess of capacity of 35 centimeters in favor of the latter.   Though there is a relation between the cortical gray matter and the mental power, yet the relation is imperfectly understood.   It has been urged that inasmuch as in some idiot brains the proportion of gray matter has been found to be large, that therefore gray matter is no index of capacity.   Many who become idiotic, however, are born with normally constituted brains, some perhaps, with all the equipment of talent, but accidents sub-

sequent to birth have checked brain growth, and have left as
an unused mass the highly organized gray matter which originally contained the promise of a better fate.

A fragment of the Michigan peninsula, the little island of
Mackinac, stands sentinel in the narrow strait that unites lakes
Huron and Michigan. How long it has looked out upon
these sister seas the wise men sayeth not, but it was old when
the now extinct lakes of the great western basin were deep
and large, and when the tropical forests on their shores were
inhabited by lions, tigers, and monkeys. This remnant of
land, so attractive to the tourist, is interesting because of its
political and natural history, and also from an accident that
occurred there it has an interest for medical men.

Mackinac has been the scene of stirring events in historic
times, having been fought for successively by France,
England, and the United States. Famous men have lived on
it, and a vice president was born there. In 1822, a young
Canadian by the name of Alexis St. Martin, while handling a
gun accidentally discharged the piece with the result that
part of his stomach was torn away. Dr. Beaumont, the surgeon of the fort, which still stands, made upon St. Martin the
observations now mentioned in every work on physiology.

While visiting there recently the writer had several inter-
views with Blind Tom, the musical prodigy. Tom was one
of a large family of negro children all of whom were normal
with the one exception. He was blind from birth, but a few
years ago an operation in London, so says his agent, gave
him partial sight. Tom is an idiot with many of the characteristics of his class. He is a voracious eater, inclined to
bolt his food, and is growing corpulent and indolent. He has
the habit of chewing his tongue and of rocking his body back
and front so common with idiots. When alone he talks to
himself constantly; part of the time declaiming in a rambling
way or talking to a friend whom he imagines to be present,
asking and answering questions to suit himself. At times he

quarrels with his imaginary companion and grows boisterous. He has no idea of the value of money, does not know his age, or his weight, or where his home is. He has no affection for his own family, no attachments, no fondness for anything but food and music. He has a wonderful recollection for dates and events in his personal history, illustrating the peculiar mechanical memory common to many defectives. Beginning with his first appearance on the stage, he can give the date and place of nearly all of his performances, even those of twenty years ago, together with trivial incidents that a normal mind would not retain a day. When younger, Tom was irritable and at times violent, but years and corpulence have tamed his temper and he is now more tractable.

His playing is wonderful as being the exhibition of a faculty which, with the exception of his memory, is a solitary possession, but when measured by any proper test it is devoid of the true spirit of music, being purely mechanical. That indefinable something that moves and charms the listener is wanting because the inspiration that intellect alone in the player can give, is lacking. His singing is better, and though his voice lacks culture it is naturally good and seems to express feeling.

His head in the convolutional region is small, cortical substance is wanting. All his talent is purely imitative. A gentleman played a very difficult piece on the piano, the first half of which was reproduced by Tom with fair accuracy. He is very suspicious and has an imbecile dislike to answering questions. It is said that he has a special aversion to chil dren, and the crying of a baby, or any evidence of suffering, seems to give him great pleasure.

The late Dr. James R. Wood, of New York, had at the time of his death the names of one hundred thousand people on his books, who had been treated by him without charge. This is certainly a large number, though perhaps not much larger proportionally of uncharged work, than other physicians do, but it illustrates the kindly spirit that characterizes the labors of the profession. There is no better missionary than the physician and none whose work in this regard is less appreciated. Few people realize the large amount of charitable work done by physicians, and this not wholly by young men, with a limited business, but by many whose only motive is that of duty, and who can illy afford the time from a remunerative practice. The constant seeking for knowledge that may be applied in the prevention of disease, the inventions for relieving human suffering that are given freely to the world, the quiet charity in the way of free medical service, all testify to the large element of philanthropy that there is in the hard and grinding duties of the physician's life.

Dr. H. M. Lyman has been appointed professor of the Principles and Practice of Medicine in Rush Medical College. He retains the professorship of Diseases of the Nervous System, which he has held for many years.

Dr. E. H. Van Deusen, of Kalamazoo, has donated to that city $50,000 for a public library building. One condition of the gift is, that a room shall be furnished for the use of "The Kalamazoo Academy of Medicine."

NEW BOOKS.

PRINCIPLES OF SURGERY.

By N. Senn, M. D., Ph. D., Milwaukee, Wis., Professor
Principles of Surgery and Surgical Pathology in
Rush Medical College, etc.

This is a great work and portrays in a striking manner the
progress of the department of medicine that is doing more
than any other for the relief of human suffering. The
achievements of modern surgery are akin to the marvelous,
and Dr. Senn has set forth the principles of the science, with
a completeness that seems to leave nothing further to be said
until new discoveries are made.

The work is systematic and compact, without a fact omitted,
or a sentence too much, and it not only makes instructive, but
fascinating reading. A conspicuous merit of Senn's work is
his method; his persistent and tireless search through original
investigations for additions to knowledge, and the practical
character of his discoveries. This combination of the discoverer and the practical man gives a special value to all his
work, and is one of the secrets of his fame.

No physician, in any line of practice, can afford to be with
out "Senn's Principles of Surgery."

"Twelve Lectures on the Structure of the Central Nervous
System for Physicians and Students," by Dr. Ludwig Edinger, Frankfort-on-the-Main. Second revised edition with 133
illustrations. Translated by Willis Hall Vittum, M. D.,
St. Paul, Minnesota. Edited by C. Eugene Riggs, A. M.,
M. D., Professor of Mental and Nervous Diseases, University
of Minnesota; member of the American Neurological Association.

There is no longer any excuse for a physician being ignorant of the anatomy of the nervous system, for with the existing works on this subject, any one can know all that the wisest
knows.

These twelve lectures give a clear and condensed statement
of the anatomy of the central nervous system, and with the
numerous and excellent illustrations that the book contains, it
furnishes a work of great value.

The style of the author is happy. Anatomical descriptions
are apt to be dry and tedious, so that those who are not
experts resort to them only in case of necessity. Dr. Edinger
has the rare merit of making them interesting reading, and
it is a genuine pleasure to follow him through his lectures.
This clearness of expression is doubtless partly due to the
translation, which is done in a faultless style. Dr. Vittum
and Prof. Riggs are to be congratulated upon their work.
The book is published by F. A. Davis.

———

"The Physician's Visiting List" (Lindsay & Blakiston's)
for 1891. Fortieth year of its publication. Philadelphia: P.
Blakiston, Son & Co., (successors to Lindsay & Blakiston)
1012 Walnut street.

This is a visiting list with a good deal of useful information
added. It contains an almanac for 1891 and 1892; table of
signs; Marshall Hall's ready method in asphyxia; poisons
and antidotes; metric system; dose table; a list of new remedies; aids in the diagnosis and treatment of the more common
ocular affections; diagram showing eruption of milk teeth;
posological table; disinfectants, (condensed from the conclusions of the Committee on Disinfectants of the American
Public Health Association); notes on the examination of urine;
incompatibility; table for calculating period of utero-gestation;
Sylvester's method for procuring artificial respiration; trans-
portation of injured persons; visiting list. For sale by all
booksellers and druggists.

FRENCH.

Morbid Affection, a pathological and psycological study— E. Laurent.

Epilepsy and Epileptic Insanity—Jul. Christian.

Researches Upon Mental Diseases—Baillarger.

Insanity in Paris—a statistical, clinical, and medico-legal study—Paul Garnier.

GERMAN.

Disturbances of Written and Vocal Speech. — Berkhan (adv).

Anatomical Researches Regarding the Roots of the Spinal Nerves in Man.—Siemerling (adv).

Paralysis Agitans.—Heimann (adv).

Morbid Phenomena of the Sexual Impulse.—Tarnowsky (adv).

Congenital Deafness, Contrib. to the Etiology and Pathogenesis of Deaf-Mutism.—Myguid (adv).

Swedish Movement Cure.—Prof. T. J. Hartelius. Translated into German by Jürgeusen and Preller (adv).

Manual of Massage, from the Swedish of E. Kleen, by Schultz (adv).

Psychopathia Sexualis, with special reference to Morbid Sexual Impulses.—Von Krafft-Ebing (adv).

Co-operation of Parts of the Brain.—Meynert (adv).

Dystrophia Muscularis Progressive—by W. Erb.

New edition of Clinical Lectures, edited by E. V. Bergmann, W. Erb; and Fr. Winckel, 1890.

Treatise on Psychiatry on a Clinical Basis.—Von Krafft-Ebing (adv).

Outlines of Systematic Craniometry.—Von Török (adv).

Special Diagnosis of the Psychoses.—Koch.

AMERICAN.

Familiar Forms of Nervous Disease.—M. Allen Starr. "Diseases of the Nervous System," is the title of a work soon to appear by Dr. Landon Carter Gray of New York.

MISCELLANY.

Dr. J. W. Good states in *The Lancet* that the servants of the Hudson's Bay Company sometimes live on an exclusive diet of meat, and sometimes solely on fish. They use no vegetables of any kind. Those living on fish alone, enjoy slightly better health than those who live on meat. Many of these people have lived on this diet for twenty or thirty years in perfect health. Mr. Wood states that a druggist at Winnepeg has lived on fish alone for four years.

REED & CARNRICK'S FOODS.—The recent destruction of one of Messrs. Reed & Carnrick's factories by fire will not, we learn from the *Dietetic Gazette*, prevent the firm from filling orders pending the completion of the new building, as their stock on hand in New York is large.

According to a recent article on longevity published in Norway, the average duration of life for men is 48.33 years, and 51.30 for women.

Stanley's recent Emin expedition was equipped entirely with Fairchild's Digestive Ferments in preference to any others, and in the recent attack of Gastritis from which Mr. Stanley suffered, he was entirely sustained upon foods previously digested with Fairchild's Extractum Pancreatis.—*Ex.*

The Civil, Military, and Naval Departments of the British government are supplied with the Fairchild Digestive products, and the Fairchild preparations for the predigestion of milk, etc., are especially preferred in India.—*Ex.*

SUCCESSOR OF THE LATE PROFESSOR WESTPHAL.—Professor Grashey, of Munich, has been appointed to succeed the late Professor Westphal in the Chair of Mental Therapeutics in Berlin.

The University of Basle now receives women as students, although it formerly refused to admit them.

Dr. T. D. Crothers, of Hartford, states that the Georgia Bromide Lithia Water is an excellent remedy for Neuralgia and Rheumatism, and predicts that these Lithia Springs will be a great resort for Neurotics.

Dr. Talcott, superintendent of the Middletown Insane Hospital, says that they give Bovinine to all exhausted and emaciated patients in that institution, and find that it acts in a rapid and satisfactory manner.

H. Mooers & Co.'s method of heating buildings, advertised elsewhere in the REVIEW, is in our opinion the best that has yet been devised.

No. 3. MARCH, 1891. Vol. I.

THE REVIEW

OF

INSANITY ᴬᴺᴰ NERVOUS DISEASE

A QUARTERLY COMPENDIUM OF THE CURRENT LITERATURE OF
NEUROLOGY AND PSYCHIATRY.

EDITED BY

JAMES H. McBRIDE, M. D.,
Superintendent Milwaukee Sanitarium for Nervous Disease.

ASSOCIATE EDITORS:

LANDON CARTER GRAY, M. D.   C. K. MILLS, M. D.
NEW YORK CITY.   PHILADELPHIA.

C. EUGENE RIGGS, M. D.   W. A. JONES, M. D.
ST. PAUL, MINN.   MINNEAPOLIS, MINN.

D. R. BROWER, M. D.   H. M. SYMAN, M. D.
CHICAGO.   CHICAGO.

H. M. BANNISTER, M. D., KANKAKEE, ILL.

$2.00 PER YEAR.   SINGLE NUMBERS, 50c.

SWAIN & TATE,
BOOK AND JOB PRINTERS,
MILWAUKEE.

INDEX.

REVIEW

OF

INSANITY AND NERVOUS DISEASE,

ORIGINAL ARTICLE.

INHERITED OPIUM HABIT.

By James G. Kiernan, M. D., Chicago.

(Fellow of the Chicago Academy of Medicine.)

The inheritance of the opium habit seems at first an isolated phenomenon, but zoologists have pointed out that pigeons whose ancestors were fed on poppies become intractable to opium.   Dr. Murrell some years ago stated that the same was true of persons descendant from Bedfordshire (Eng.) natives, whose ancestors had used infusions of poppies as a prophylactic against malaria.   The infantile mortality among these people was not stated, but nervous diseases were relatively prevalent in these districts.   Morel has shown that narcotic habits in the ancestors produced descendants in whom the normal checks on excessive nervous action were removed so that paranoiacs, periodical lunatics, epileptics, hysterics, congenital criminals, congenital paupers, or otherwise degenerate beings, resulted.   This influence is most strongly exerted when the maternal ancestor was the one affected, for to her is committed the development of the ovum prior to conception and of the child subsequently.   If either is interfered with by a habit, a being defective in some respects is the result.

The direct inheritance of the opium habit has very important consequences from every sociological standpoint and its possibility has been shown by Dr. Levenstein who found by experiments on pregnant dogs and rabbits that the use of opium during pregnancy produced either abortion or still-births, or rapidly dying offspring. Only of late has much attention been attracted to it. Probably the first to call attention to the subject was an American physician, Dr. Frank Heman Hubbard* who in a work lacking an index and replete with typographical errors, displays a keen, albeit neurologically untrained, insight into the phenomena of the habit.

His cases are as follows:

"*Case I.* A secundo-para while pregnant with her third child was given morphine to secure sleep in consequence of an agonizing neuralgia. When labor occurred the amount of morphine used was ten grains per diem. The third day after delivery her milk was suppressed to a great extent, in contrast with the previous habit of the patient, who was usually galactorrheac at that time. The babe, bright and natural at birth, when placed on the bottle became prostrate, cried, vomited incessantly, and discharged a glairy mucus with meconium. The muscles twitched continuously, and the babe died six days later from marasmus and insomnia. The mother increased the drug up to half a dram daily doses, which she was taking when two years later a girl was born. The milk was suppressed. The babe, bright and vigorous, took the bottle readily, but soon displayed the same ominous symptoms as the lost child. The mother suspected the cause and secretly gave it two drops of laudanum. The effect was magical; all the ominous symptoms disappeared. The mother increased the dose until when the child was seven years old she was taking half an ounce daily doses of laudanum. She was in excellent physical health, bright, but backward at school.

* The Opium Habit and Alcoholism. 1881.

Case II was that of a primi-para who had acquired the habit from an attempt to cure insomnia due to domestic anxiety four months previous to the birth of the child. Her milk was suppressed and the child at first bright and vigorous, displayed the phenomena already described.

Case III was that of a woman who increased her dose during gestation to three ounces daily of laudanum. Her milk was suppressed. The babe born in ordinary health took the bottle well for twelve hours, when the characteristic symptoms of opium suppression set in. Dr. Hubbard attempted to administer ten drops paregoric and three tr. ginger in a little milk and water, by passing it well into the mouth with a dropper; a large proportion being spasmodically expelled. Three drops of laudanum was given by inunction over the stomach. Three drams of paregoric were found to be the necessary daily dose, which was given morning, noon and night in milk. No change was made during the first ten days in order to allow the child to gain strength. Then the daily dose was reduced one drop until eighty drops were withdrawn when a rest of ten days was or-- dered; the process of reduction was then begun and a rest of ten days ordered, when only a dram was given daily. Then the same procedure was adopted until thirty drops only were given. After the usual rest a drop was withdrawn every other day. The babe thrived well and no crisis was noticeable.

Dr. Carson* later cited a case where the child of an opium-eating mother, whose milk was suppressed, displayed symptoms of opium marasmus, which were relieved by a grandmother who suspected the cause.

Dr. Amabile† pointed out about the same time that children or the opium-smoking Aryan wives and paramours of Chinamen were apt to die from this opium-marasmus unless given paregoric.

In 1887 Dr. Erlenmeyer in his classical work on the morphine habit called attention to the subject, but in

*Alienist and Neurologist, 1885.
†Western Druggist, 1886.

evident ignorance of the cases of Hubbard, Carson, and
Amabile.

Dr. Frank B. Earle, of Chicago, in a paper* read before the Chicago Medical Society, Dec. 1887, cited the
following case:

Case II.   Oct. 10, 1887, I delivered a 32-year-old
woman of a healthy ten-pound-girl. The mother's extreme solicitude about the child led to the discovery
that her three preceding infants had died within a few
days after birth. Concerning these deaths the following history was obtained: The first child was born
healthy, but pined away in two and a half days. The
same phenomena were present in the case of the second
child which died three days after birth. The third child
died under the same circumstances in four days after
delivery. In regard to my own case, matters progressed
favorably until the end of the third day. At this visit
I was informed that the child did not sleep and seemed
troubled with colic. I noticed that the child seemed
particularly sensitive to motion, looked pale and pinched
and singularly prostrate. The milk had not appeared
and I instructed them what to feed the child, believing
that it needed nourishment. Twelve hours later I was
recalled in great haste, and arrived at the house in a
few moments to find the child pulseless and cyanotic,
and in five minutes it was dead. Here was a history of
four apparently healthy children, born at full term, all
of whom had died before they were four days old. Upon
what basis could this be explained? I could find no
taint of constitutional disease, and no trace of syphilis
or other constitutional disease in the father. He is a
healthy, active, robust man, whose entire personal, as
well as family history is good; his brothers and sisters
have healthy grown children. The patient was a widow
with two children when he married her, and inasmuch
as the children by a former husband were living and
healthy, he placed the entire responsibiltty of these
deaths on himself. A druggist of the vicinity a few
days later informed me that the mother was a victim of
the opium habit, and since her residence in his vicinity,
a period of about five months, she purchased from

*Medical Standard, Jan. 1888.

twelve to fourteen grains of morphine sulphate daily.
Another druggist had for three and a half years previous
to her change of residence, supplied her with the same
amount. This covers a period during which the last
three children were born. The husband lived in ignorance of this habit, so that for facts pertaining to the
duration of the habit previous to this marriage and to
the amount consumed, I was dependent on the uncertain
statements of the woman herself, and on the definite
statements of the family physician, Dr. Wadsworth.
According to her own story she began at the age of
seventeen taking morphine on a physician's prescrip-
tion. Not knowing the nature of the medicine, but
finding that she felt better when taking it, she repeated
it. She was married at eighteen and the two children
before mentioned was the result of this union. She
was attended in these two labors by the physician who
first prescribed morphine, and knowing her habit he
directed that the children have soothing syrup, which
was done, and these two children survived. Two years
after the birth of this second child she became a widow,
but was again married at the age of twenty-five and has
borne four children by her present husband. In regard
to the quantity taken she informed me that since this
last marriage she has consumed from five to eight grains
daily, but that she was taking more at the time of the
birth of the children by her first husband. The physician at whose door the patient places the responsibility
of her habit states that his first acquaintance with this
lady dates back 18 or 19 years, at which time she was
called to see her mother who was an opium-eater, and
who had taken an excessive dose of tincture of opium.
He denies in toto her story of having prescribed morphine. He is positive that she was not taking the drug
at the time he attended her in her labor, and that he
did not order the babes soothing syrup. He also states
that these children were perfectly healthy and that
they were raised without the slightest difficulty. The
contrast between these statements is very apparent;
but, inasmuch as the physician is a learned reputable
man, and the woman belongs to that class of inebriates
whose every instinct is deception, the truth of the mat-

ter is easily settled. There can be no doubt about the
time she contracted the habit. According to her own
statements, she is at present consuming from five to
eight grains daily. Two reliable pharmacists state that
the quantity amounts to twelve or fourteen grains daily.
Granting that she consumes ten grains daily, what
effect would this amount have on the fœtus in
utero? Would it in any way affect its nutrition or per-
vert the functions of its organs so that it would be incapable of sustaining life when dependent solely upon
itself; or would the sudden withdrawal of a prolonged
stimulant account for the death? The child when born
was perfectly robust. "If to one of mature organism
the removal of the daily quantity of morphine produces
shock and even death, how much more should we expect to have the same condition manifest in a baby
three days old! It will be readily admitted by those
familiar with the subject that the morphine habit is
gaining ground with startling rapidity; that married
ladies and widows form the largest class who are slaves
to the habit. What, then, shall we do with pregnant
women who are victims of this degrading vice? and
what treatment can be resorted to to prevent the mortality of children where the practice cannot be abridged?
I am not prepared to answer either of the foregoing
questions. Whether or not premature birth would be
the result of an attempt to reform the pregnant woman
is as yet undetermined. In regard to the treatment of
the child where we know the habit exists, but cannot
prevent the use of morphine by the mother, the only
logical way is to administer some substitute or appropriate doses of some of the most palpable and least
objectionable preparations of opium in diminishing doses
until such time as the drug can be safely taken away
altogether." In the discussion Dr. Haven stated that
he attended the woman during one of her confinements,
and the child, deprived of the mother's milk, manifested
the phenomena described by Dr. Earle.

My own experience is limited to three cases.

Case I. was that of an opium-eating IV-para whose
daily dose amounted to eight ounces of laudanum.
She was safely delivered of an apparently healthy boy.

Her milk was almost suppressed and the child given the bottle. Twelve hours thereafter marasmus followed by convulsions set in. I was called in consultation and learned from the mother her unsuspected opium habit. Paregoric was administered to the child which checked all the symptoms. The mother stated that all her previous children had suffered similarly and had been relieved the same way. Her habit had been contracted from the use of laudanum to relieve dysmenorrhœa.

Case II. was that of a II-para who had lost her two previous babes from an unaccountable marasmus and consulted me privately as to the probable influence of morphine given hypodermically, which had been used by her ever since its administration by a Hahnemaniac for the relief of menorralgia. I tried the gradual reduction treatment advocated by Hubbard. Quinine, strychnine, and cannabis indica being gradually substituted by admixture until the last dose was withdrawn, three weeks after the child's birth. Delivery was painless, the child thrived well and no opium-marasmus resulted after the complete withdrawal of the drug from the mother who, for once had an excellent supply of milk. A year after the mother relapsed during an attack of menorralgia, for which a hypodermic of morphine was injudiciously given.

Case III. was that of a III-para who had acquired the habit from a laudanum-using mother who had come from a Bedford district in England, where a rough species of laudanum was used as a prophylactic against malaria. All the children had acquired the same habit, but could abandon it, like the grandmother, at will. The present victim had sustained much domestic worry, and in consequence of insomnia had increased her daily dram-dose of laudanum to half a pint by the eighth month of pregnancy. She had an attack of acute confusional insanity wihch lasted forty-six hours after delivery. No breast milk was secreted. The well-developed child was given the bottle, but in four hours after became marasmic and had frequent convulsions. The nurse, irritated by its whining, surreptitiously gave it a teaspoonful of paregoric, which checked all its symptoms as by magic. The paregoric was reduced, as advocated by Hubbard, when the mother's history was learned on her recovery from insanity.

All the questions which might be raised anent this subject have been raised by Dr. Earle and answered incidentally in a discussion of the treatment. The subject is not without close relation to the subject of "Chemical vaccines," now being discussed in France and elsewhere.

TRANSLATORS.

ITALIAN.

H. M. BANNISTER, M. L., Assistant Physician Eastern Illinois Hospital for Insane, Kankakee.

GERMAN.

G. J. KAUMHEIMER, M. D., Milwaukee.   I. LANGE, M. D., Chicago.

RUSSIAN.

T. KACZOROUSKI-PORAY, Chicago.

FRENCH.

C. FRITHIOF LARSON, M. D., Chicago.   J. G. KIERNAN, M. D., Chicago.

SCANDINAVIAN.

M. NELSON VOLDENG, M. D., Ass't. Physician Ia. State Hospital for Insane.

SPANISH.

HORACE M. BROWN, M. D., Milwaukee.

NEUROLOGICAL.

ANATOMY AND PHYSIOLOGY.

CONNECTION BETWEEN THE PIA MATER AND THE SPINAL CORD (by M. Asplund, Stockholm, Sweden).—According to the authors who first were engaged more especially with the above question, all neuroglia fibres occupying the periferic layer of the cord and their prolongations in the interior were only invasions of the connective tissue of the pia mater. Following, again, the opposite opinion at present most generally accepted, and in the literature chiefly represented by Boll, Key, and Retzins, His, etc., the pia mater is only in an intimate contact with the cord without any interchange of their histologically different tissues.

To Dr. Fromman belongs the honor of being the first to show the existence of extremely fine fibres from the cortex of the neuroglia of the cord to the pia mater in whose substance they gradually disappear after a more or less extended course.

Gierke also describes very thin fibres in the periferic layer of the white matter transversely passing the epi-spinal space to the inner layer of the pia (intima pia), but he has not seen them in the outer layer. Dr. Asplund has found these fibres extend from the cord to the pia, even into the outer layer. He has found them all around the cord and in its whole extension.

These fibres are found both in adults and children, both in normal and pathological conditions, at least he has demonstrated their presence in locomotor ataxia, lateral amyotrophic sclerosis, and descending sclerosis.

The article is beautifully illustrated by one large chromo-lithograph and two small lithographs showing the fibres extending from the cord into the pia.—*Nordiskt Med. Arkiv.*, Part I, Vol. XXII.   C. F. L.

EXPERIMENTS ON THE THYROID GLAND.—G. Vassale (*Rivista Sperimentale*, XVI, IV), publishes the results of his investigations on the effect of intravenous injections of the juice of the thyroid gland in dogs. It is well known that these animals do not stand thyroidectomy well and some curious results have been obtained by Schiff, Eiselsberg, Carle, and others, in extirpating and transplanting this gland in dogs. Vassale incited by these conceived the idea of experimenting with injections of thyroid juice into the veins, following somewhat after the method of Brown-Sèquard, with the extract of certain other glands. He operated on nine dogs of various sizes and weight, and on six of them performed complete thyroidectomy, and then injected the thyroid extract diluted in sterilized water and filtered, into the crural vein. Of these six three were living in health three months after the operation, making a survival of fifty per cent. as against sixteen and a half in Albertone's and Tizzonis' experiments, and less than two per cent. in those of Schiff, after this operation. In one other dog thus operated upon the injection was delayed until after the symptoms of cachexia strumipreva had become manifest, and then, as an experiment, the testicular juice of a healthy rabbit was injected with all due precautions, but without any beneficial effect. Nine days after the thyroidectomy, the extract from the thyroid of an ox (that of a dog not being available at the time) was injected with very manifest improvement in the animal's symptoms, though he finally succumbed by suffocation from excessive mucus in the bronchial tubes in a convulsive attack. In still another dog operated upon, the cachexia was allowed to develop and the injection practiced when the animal was almost in *articulo mortis*, with the result of a perfect cure. The injection was also made in a healthy dog, not operated upon, with only temporary depressive effects, uncertainty of gait, dilatation of pupils, etc.

Vassale does not conclude from his experiments that the intra-venous injection of the aqueous extract of the thyroid gland confers lasting immunity from the cachexia, but thinks that it may be safely affirmed that it rapidly relieves the severe and threatening symptoms.   His dogs that thus survived thyroidectomy generally increased in body weight.

The author promises to continue his researches.   His paper is an interesting one and also suggestive in a practical therapeutic point of view.

FUNCTIONS OF THE POSTERIOR CEREBRAL CONVOLUTIONS OF THE HIGHER ANIMALS.—Dr. Celerecki's investigations confirm the researches of Bechterew and others.   His own experiments in pathological observation have led him to believe that the occipital lobe is the immediate cortical visual centre, and that the parietal and frontal lobes are indirectly concerned in vision.   He holds that there are independent motor centres for movements of the eyes in the occipital region.—Kowalewskij's *Archives of Psychi. and Neurol.*, Vol. 16, No. 2.

THE THERMIC SENSATIONS.—The following are the conclusions of a memoir by Dr. Eugenio Tanzi (*Rivista Sper. di Freniatria*, XVI., IV., 1890), on the thermic sense in spinal affections:

1.   The pathology of syringomyelitis shows that thermic impressions arriving in the spinal cord delay there, as it appears, in the gray substance longer than do the tactile impressions, before entering the posterior columns.

2.   It is not certain whether thermic impressions become such from the fact of their longer sojourn in the posterior cornua, or whether their nature as such is not determined before their entrance there.   In other words, does a given impression become a thermic one because of its detention in the posterior horns, or is this detention because of its thermic character?   Whichever of these is true, the fact of their detention remains.

Nevertheless we must admit a few cases of inversion of the thermic sensibility, not to be explained in this second hypoth-

esis; thus, an impression of cold acting as such on a point of
sensibility is perceived as warm, or vice versa, which signifies
that the subjective determination takes place in the spinal
cord.   The thermic excitation will therefore adopt in the cord
its route (anatomical) or its method (velocity, etc.) of propo-
gation as cold or warmth respectively.

3.   The theory of the doubling of the thermic sense into
two senses, anatomically and physiologically distinct, rather
than one, though new, deserves much consideration, and is
supported by pathology.

4.   The few cases of syringomyelitis in which there has
been observed a zone of thermic hyperæsthesia, do not speak
against the many others of anæsthesia, since they are explained
in an obvious manner as the effects of a state of irritation
preceding the phases of neoplasms or softening.

5.   Whether this thermic dualism manifests itself in the
pre-spinal tract as far as the terminal expansion of the special
sensory fibres, or whether the distinction between cold and
heat is only determined in the posterior horns through their
action, it is very probable from reasons given under "2" that
the warm and cold impressions take their own course in the
grey substance, following separate tracts in different sections
of the posterior horns before entering into the sensory column.

ON THE IMMEDIATE RESULTS OF THE DESTRUCTION OF PARTS
OF THE BRAIN IN NEW BORN ANIMALS AND ON THE DEVELOP-
MENT OF THEIR CEREBRAL FUNCTIONS.—In an article with the
above title, Prof. Bechterew calls attention to the fact that
extirpation of certain parts of the brain in newborn puppies
and kittens is not followed by any immediate result, if the
operation is done before the parts extirpated are supplied with
myelin-sheaths.   The movements of newborn animals of this
class are altogether automatic and reflex, and the cortical centers are not excitable until the fibres running from them are
provided with a sheath.   This would indicate that the development of the cortical cells goes on simultaneously with that

of the fibres.   Newborn whelps are blind, deaf, and devoid of the sense of taste or smell.   At birth, only the posterior roots and the trigeminurs are provided with myelinic fibres. Of the sensory cerebral nerves the glosso-pharyngeal nerve and acoustic nerves develop first, then the optic, the fibres of the olfactory lobe remaining amyelinic until the end of the third week.—*Neurolog. Centralbl.*, No. 21, 1890.   G. J. K.

ON THE CONNECTION OF THE NASAL LYMPH CHANNELS WITH THE SUBARACHNOID SPACE.—Dr. Theodore S. Flatau claims to have demonstrated that in the rabbit and cat it is possible to inject the lymphatics of the nose from the subarachnoid space.   The injection follows the olfactory nerve fibrils, in perineural channels.   Key and Retzius had already made experiments on this point, although they were not free from error.   The author thinks that a considerable number of "reflex nasal neuroses," so-called, may be due to a derangement of the lymphatic circulation in these channels.   Further and more complete descriptions are promised.—*Deutsche Med. Wochenschr.*, No. 44, 1890.

G. J. K.

THE BLOOD SUPPLY OF THE MEDULLA AND ITS CENTERS.— Prof. Adamkiewitz, states this as follows, *Arteria sulci*, supplies the pyramids, their decussation, the fillet, the raphe, inter-olivary layer, nucleus arciformis, inner accessory olive, olivary body and especially the hypoglossal nucleus.

Arteria fissuræ supplies the posterior columns, nucleus gracilis, and especially the nucleus of the accessorius.

Special branches supply the nucleus gracilis and cuneatus, substantia gelatinosa, nuclei of the lateral columns, superior and inferior olive, and the acoustic nucleus.   These arteries run through the upper part of the restiform body.   The nuclei of the glossopharyngeus, vagus, and hypoglossus also receive special branches.—*Wiener Med. Presse*, No. 4.

G. J. K.

STRIÆ ACOUSTICÆ AND LOWER FILLET.—Von Monakow, as the result of experimental destruction of the lower fillet in cat and dog, reaches the following conclusions. The integrity of the dorsal medulla of the right upper olive, of the left striæ acousticæ and of the elongated ganglion cells in the middle layer of the left acoustic tubercle is dependent upon the integrity of that part of the right lower fillet which he denominates "the portion of the acoustic striæ."

The normal condition of the nucleus of the right lower lemniscus, a part of the right upper olive, that part of the ventral tegmental decussation passing from right to left, as well as the dorsal medulla of the left regio subthalamica is dependent upon the integrity of certain other parts of the right lower lemniscus. He distinguishes five component parts in the lower lemniscus:

1. Portion of the acoustic striæ, occupying the dorsal part, is probably the connecting route between the primary acoustic centers and cerebrum. The fibres run through the upper olive of the same, to the acoustic tubercle of the opposite side. 2. Portion of the ventral tegmental decussation. 3. Portion of the upper olive (quite small). 4. Portion of the lateral nucleus of the fillet. 5. Portions of the short fibres—central medullary portion.—*Archiv. f. Psychiatrie*, Band 22, H. 1, P. 1. G. J. K.

PATHOLOGY AND SYMPTOMATOLOGY.

WESTPHAL'S SYMPTOM IN DIABETES.—Dr. Salomonsen (*Nedeol. Tijdschr. voor Geneesk*, No. 11, 1890) reports the case of a female, with 4 per cent. sugar in her urine. By appropriate diet, the sugar disappeared, to remain absent during the month of mixed diet. Knee jerk was absent during entire period. Tabes could be excluded. The electric irritability was decreased. In extensor commun. dig. and soleus, an S Z was much stronger than K a S Z, occurring with a current which produced no effect with K a S.   Strength was reduced. There were neuralgias, paræsthesias and diminution of cutaneous reflexes.   These symptoms were referred to a multiple neuritis, due either to pathological products of tissue change, or a pathological quantity of the normal products of metabolic change.—*Wiener Mediz. Blaetter*, No. 48, 1890.

G. J. K.

TRIGEMINAL NEURALGIA IN ACUTE IODISM.—By acute iodism the author means that peculiar condition of lachrymation, swelling of the lids, and general bodily malaise following the indigestion of a single, or but few doses of iodine and its salts.   This condition disappears even if the administration be continued, and seems to expend its force within the area supplied by the carotid.   The author relates four cases (two males and two females) which presented severe trigeminal neuralgias following within an hour after taking one to two grains of iodide of potassium.   They were accompanied by swelling of the lids, redness of conjunctiva and lachrymation and disappeared in a few hours.   This leads to the suggestion that the pain was due to the hyperæmia of the nerves, and in fact, those branches were most affected which pass through bony canals.

The author has seen very little of acute iodism since giving the iodine salts largely diluted in milk or water.—DR. S. EHRMANN, *Wiener Mediz. Blaetter*, No. 44, 1890.

G. J. K.

DISTURBANCE OF INNERVATION OF THE ŒSOPHAGUS CAUSED BY A TUMOR IN THE POSTERIOR FOSSA.—Patient aet. 40. Trouble began two and a half years back with vomiting, later, weakness, and trouble in walking, slight retinits which disappeared later.   He frequently complained of regurgitation of food with no nausea, but a sense of oppression of the chest. The stomach tube demonstrated a constriction at the level of the eighth dorsal vertebra, which was overcome by a stylet in the tube. Later on, the constriction was found at the level of the fourth or fifth dorsal vertebræ. The trouble was intermittent. Pulse up to 140, temporarily reduced by digitalis and strophanthus, which also relieved the regurgitation for a time. The region supplied by the glosso-pharyngeal nerve was normal. Paralysis of right vocal cord.

Autopsy showed a fibrous tumor, of the size of an egg, arising from the lower and posterior part of the falx cerebelli, and projecting into the foramen magnum from the right side, displacing the cerebellum, slightly compressing the medulla below the pyramidal decussation and stretching the spinal accessory and first cervical nerves.—DR. J. NEUMANN, *Mendels Centralblatt*, No. 19, 1890.　　　　　　　　　G. J. K.

PERIPHERAL NEURITIS OCCURRING IN THE COURSE OF DEMENTIA PARALYTICA.—While the occurrence of peripheral neuritic changes in tabes is well established, their presence in dementia paralytica has received but slight attention.

The author describes two cases in which a peroneal paralysis of peripheral origin occurred, for which no other exciting cause could be found. One of these cases is of especial interest, in that the disease developed in an imbecile aged 24, born of an insane mother. The autopsy showed the changes usually found in dementia paralytica.—PROF. PICK, *Berlin Klin. Wochenschr.*, No. 47.　　　　　　　　　G. J. K.

BERI-BERI ON THE ISLAND OF HOKKAIDO, JAPAN.—Hokkaido is the most northerly of the larger islands of Japan. The climate and vegetation of its southern and central portions resemble those of Germany, while further north, they correspond more nearly to Scandinavia.

Although the pernicious forms of the disease are rather rare in Japan, epidemics occasionally occur in the northern part of the island with a mortality of 30 per cent. This is especially the case where large numbers are crowded together under unsanitary conditions. Certain districts seem to be exempt from its ravages. The Ainos, the aboriginal inhabitant, the Caucasians, and children under 10, seem to be immune to its ravages. Among 10,000 dispensary patients treated in 1889, 485 were found to be suffering from this disease, 385 being males. The greatest number of cases occurred in July (140) and August (109), and between the ages of 15 and 30 (302, including a few between 10 and 15 years old). During the same period, 57 males and 20 females were admitted to hospital with this trouble, of whom 11 men and 2 women died; 9 of the men died between the 5th and 9th day and 2 on the 32d day. Eleven of the 13 deaths occurred between the ages of 20 and 40 years. No exact etiological factor can be determined. An œdematous form, a dry or paralytic form, and an acute pernicious form may be clinically distinguished, although the transitions are numerous.

The therapeutic results are discouraging. Laxatives give relief, calomel often seems to act favorably as a diuretic — DR. F. GRIMM, *Deutsche Medicin Wochenschr.*, No. 43, 1890.

G. J. K.

———

ON A CASE OF TABETIC ATAXIA WITH SEEMINGLY INTACT SENSATION.—Dr. Goldscheider, at a meeting of the Gesellsch. d. Charite-Aerzte, demonstrated a case of ataxia of two years' standing, with Westphal's and Romberg's symptoms. On superficial examination, sensation to passive motion and tactile impression seemed normal. On using the author's instrument for the determination of sensitiveness to passive motion (see this Journal, No. 1), it was found that the angle of motion was several times the normal one. The use of other instruments of precision demonstrated that tactile impressions plainly felt by the normal skin were unnoticed.

The author then reviews the various theories of ataxia, and from an analysis of published cases inclines strongly to the view of Leyden (diminished sensation) as against Erb, Kumpf, and others—disturbed centrifugal conduction.—*Berlin Klin. Wochenschr.*, No. 46, 1890.

G. J. K.

ETIOLOGY OF TABES.—Prof. M. Bernhardt reports a case of pronounced locomotor ataxia, in which the only etiological factor discoverable was prolonged work with the sewing machine (15 to 20 hours per day for five years). O. Guelliot (*Union Med.*, Nos. 244, 1,882) has reported two similar cases. J. Hoffman (*Arch. f. Psych.*, Bd. XIX., 1888) has also reported a case in which tabetic symptoms were produced by severe and prolonged concussion of the entire body, in a man of 47, otherwise healthy.—*Neurolog. Centralbl.*, No. 23, 1890.

G. J. K.

———

ARTHRO-OSTEOPATHIE HYPERTROPHIANTE PNEUMIQUE.—Under this name Pierre Marie, the discoverer of acromegaly, has placed a number of cases which have been reported by Erb and others as acromegaly. This disease differs from true acromegaly in that there is no involvement of the nose or lower jaw, which, in the true disease often assumes the "sled runner" form. Marie also claims that true acromegaly always begins in youth and is accompanied by lordosis of the cervical or upper dorsal spine. Osteo-arthropathic pneumique usually results from chronic disease of the lungs. In the discussion Ewald said that, according to Marie, no case of true acromegaly had been reported in Germany.—DR. C. GERHARD, "A Case of Acromegaly," *Berlin Klin. Wochenschr.*, No. 52, 1890.

G. J. K.

———

DYSTROPHIA MUSCULARIS PROGRESSIVA.—Under this title Erb (*Sammlung Klin. Vortray*, by Bergmann, Erb and Winkel, No. 2, 1890) describes the various forms and shows their similarity. The latter is shown in the chronic development upon a family disposition; in the gradual development of a widely distributed atrophy of numerous muscles, combined with hypertrophy of others; in the peculiar correspondence of the atrophy or hypertrophy in different cases; the localization in the trunk, shoulder, lumbar region and proximal parts of the limbs, and the disturbances of locomotion and position produced thereby. The reaction of degeneration, fibrillary tremor, and disorders of sensation and of the organs of special sense are absent. The tendon reflexes disappear slowly. The correspondence in all essential points is so great that the differences may be referred to accidental and subordinate factors. Transition forms between the types have been observed, as well as obscure and undeterminable ones. The anatomical changes are the same in all the varieties. Erb regards the trouble as a trophoneurosis. The most important factor in etiology is heredity (60 per cent.). The course is frequently extremely slow. Recovery has not been observed, although improvement has been noted. The most potent therapeutic agents are electricity, massage, and gymnastics. As principal varieties Erb distinguishes the hypertrophic, the juvenile, and the infantile forms.—Review in *Deutsche Medic. Wochenschr.*, No. 48, 1890. G. J. K.

Dr. Hertel (*Charite Annalen*, XV., 1890, p. 120) reports a case of left homonymous hemianopsia, with paralysis of the left facial and hypoglossal nerves, and motor and sensory paralysis of the left side of the body, due to chronic plumbism. Complete recovery in three months.—*Neurolog. Centralbl.*, No. 23, 1890. G. J. K.

NUCLEAR OPHTHALMOPLEGIA AND ITS COMPLICATIONS.—Under this title Prof. Bernhard describes cases from his own practice and from literature, which show bulbar complications, as well of implication of the anterior horns of the cervical cord. Sudden death is likely to occur, even in cases which seem to be improving and were never alarming. The results of autopsies as far as made are negative, microscopic examinations of the nuclei have not been reported.

The bulbar and opthalmic paralyses may alternate, and different other cranial nerves may be involved at different times.

The constitutional symptoms usually are a general malaise and weakness.—*Berlin Klin. Wochenschr.*, No. 43, 1890.

G. J. K.

RECURRENT PARALYSIS OF OCULO MOTOR.—Dr. Darksheritch describes a peculiar disease characterized by periodical recurring paralysis of the oculo motor. The chief symptoms are ptosis, loss of eye movements, derangement of accommodation. The paralysis involves all the branches of the oculo motor. The cases observed usually recovered. It is sometimes of rapid onset ; in other cases it develops slowly. Headache, nausea, and general malaise accompany this order. It usually attacks children. In one case local meningitis was observed after death.—Kowalewsky's *Archives of Pyschi. and Neurol.*, Vol. 16, No. 2.

ORBITAL NEURITIS INCLUDING ALCOHOLIC AND TOBACCO AMBLYOPIA.—Dr. Herman Knapp read a paper with this title before the New York Academy of Medicine. He said that the singular cases in which people noticed a haze, or a blur, or perfect darkness in the center of vision, known under the name central amblyopia, as well as those cases where persons lost their sight for one or two days, without exhibiting symptoms of any ocular or general disease to account for it, were now known to be due to inflammation in the orbital part of the optic nerve, a retro-bulbar neuritis. It might be idiopathic, or due to alcoholism, to the use of tobacco, to syphilis, etc. It had an acute and chronic form. The symptoms of the acute form were headache in various degrees; orbital pain, increased by movements of the eye and pressure; impairment of sight in all degrees. It might begin without any warning, advance rapidly, leading, in exceptional cases, in a day or two, to total blindness of both eyes. Further symptoms were central scotoma for color and form; general diminution of color perception; moderate redness and serious effusion of the optic disk and adjacent retina. This condition might be followed by ischæmia, or the latter might be noticed from the beginning. It was due to compression of the retinal vessels, the termination revealed one of three ophthalmoscopic pictures—one, the normal condition; two, partial atrophy of the optic nerves; three, total atrophy of the optic nerves.

In speaking of alcoholic and tobacco amblyopia, he said there was diminution in sight in both eyes, coming on gradually, yet with acute aggravation; dim lines; patients saw better in the dusk; diminished perception for color and form, varying according to ring or central scotoma. According to the stage of the disease, one saw with the ophthalmoscope a fairly normal condition, or slight congestion. It was almost exclusively in males. The prognosis was good in the first stage where no atrophy of the optic disk was present; it rarely ended in complete blindness. The atrophy in the optic nerve was usually triangular, and might remain after sight had again become normal. In treatment he insisted on total abstinence from alcohol and tobacco. Among remedies were strychnia, iodide of potash, Turkish baths, mountain or sea air.

Dr. H. D. Noyes opened the discussion, and said that orbital neuritus, especially that form known as alcohol or tobacco amblyopia, had been placed on positive ground by the investigations of Utoff. The speaker had the last three months chanced to see a number of cases of this sort, and with previous observations he had satisfied himself that the cause might be either alcohol or tobacco, and that it was not necessary for the two habits to be commingled. It was formerly supposed that alcohol, without tobacco, could not produce it; but he thought it had been demonstrated that tobacco alone was quite competent to bring it about. In tobacco poisoning he had satisfied himself that in addition to the impairment of the circulation and weakness of the heart there were also ocular symptoms, including impairment of the function of accommodation, impaired control of the ocular muscles, as well as the affection of the optic nerve. It was not at all rare in the earlier period of the tobacco trouble to have persons present themselves with very little impairment of vision so far as testing by loss was concerned, while defi-

3

ciency made itself known in blindness for color in the macula lutea.

Dr. Noyes could not place entire reliance on the appearance of the optic nerve in making a diagnosis. While he fully realized that in a considerable percentage of cases the tem- poro-inferior quadrant of the disk showed a change, neverthe- less he had seen some cases in which it was not a recognizable symptom. Moreover, it was not rare to see this pale appear- ance in persons in whom there were no toxic effects whatever. Then again, the lesion did not always present itself in the form of a circumscribed quadrant atrophy, but as a pronounced general atrophy of tissue.—*N. Y. Medical Record*, Jan.

NEUROSES OF DEVELOPMENT.—Dr. T. S. Clouston chose this subject for the Morison Lectures for 1890. The first lecture is published in the *Edinburgh Medical Journal* for Jan- uary. The period in the life of a human being preceding completed development may, from a nervous point of view, be divided into two portions. If we take merely gross bulk and weight the brain is found to have attained its maximum at about seventeen or eighteen years of age, though it has grown its full growth up to a few ounces at seven years of age. From birth to seventeen is the period of growth and develop- ment together; from that time to maturity is the period of development alone. Cells of any organic tissue have different qualities, and are subject to different diseases during these two periods of increase of bulk and the perfecting of the func- tions. While cells are multiplying in number or size so as to give the organ to which they belong its normal bulk or form, they do not energize in precisely the same way as they do after growth is completed. There is more formative power and less output of energy; more oxygen consumed; more car- bonic acid and urea produced. After normal bulk has been attained development of full function is of different duration in different tissues. The kidneys, being simple in structure, attain perfection early, while the brain, being more complex,

is longer in attaining perfection of function. There is a great
and indescriable difference between the brain of the child and
that of the man in addition to their simple structural differences. Concerning development as it relates to the brain, it
can be divided into the period before puberty, and the period
from puberty during adolescence up to maturity. The periods
of brain growth and that of non-reproduction do not absolutely
correspond, but it may be held as a law that when active cell
growth ceases in the cortex, then only does reproductive function begin. Some of the most serious of pathological facts in
brain development are certain mental disturbances in the
function of the brain, and these are intimately associated with
motor, sensory and trophic neuroses incidental to the development period. It is a physiological fact that one tissue, or one
organ, can come to full structural and functional perfection
while other organs and tissues are undeveloped. Many of the
tissues, such as the blood, are developed at birth, while the
blood vessels and the heart muscle attain complete development in early childhood. There is an unwritten physiological
chapter on the order of development and perfection of the
various organs of the body. The higher nervous tissues and
organs attain maturity and perfection last, and of these the
cells of the brain cortex and connecting paths are latest in
attaining perfection. At the age of seven the brain has
attained about ninety per cent. of its maximum weight, and
after seventeen or eighteen there is no increase at all. The
unique fact about the nerve cell is the slowness with which it
develops function after its full bulk has been attained. After
most of the nerve cells of the brain have attained their proper
shape and full size it takes them the enormous time of
eighteen or nineteen years (one-fourth of life) to attain such
functional perfection as they are to arrive at. This indicates
the height, complexity, and importance of their function. It
is a rule with few exceptions, that the tissues that are of slow
development are most influenced by hereditary evil tendencies.
Concerning heredity it may be stated that a man is as much a

part of his ancestry, and his posterity is of him, as the root and stem are parts of one tree. The philosophic view of reproduction is, that it is but one incident in a continuous protoplasmic life. This view of the continuity of life enables us to grasp the idea of heredity as being one of the ordinary vital laws like that which makes a hepatic cell produce other hepatic cells during the continuance of the life of the liver. Concerning hereditary defects and looking to the influence of nerve over nutrition, it seems reasonable to attribute these early formative failures and the malformation of body and limbs, in some degree to deficient trophic innervation. Of the neuroses incidental to the period of life between birth and seven years of age the lecturer includes rickets, night terrors, and febrile delirium of children, and convulsions of teething. Rickets is a trophic neurosis in which the nervous failure concentrates itself in the bones, though many other tissues are illy nourished. Convulsions during the first dentition are related to rickets, they mark a quality of nervous unstability, and are serious in their import. The different mental faculties should have a direct mental relationship to one another, so while there may be normal variations in their relative power, yet there are limits beyond which the variation may be co. - sidered pathological. When a boy of eight develops such capacity as to be able to solve amazing mathematical problems we may conclude that there is something pathological in this manifestation of mental power. Or if a child of five attempts to kill itself on the death of its mother; or, if there is, apparently, mature intelligence at fifteen years, with little or no sense of right and wrong, we should conclude that there is some defect in the mental organization. Child prodigies are generally pathological. The premature development of any faculty in a child is dangerous to health.—*Edinburgh Med. Jour. and Brit. Med. Jour.*

ASTASIA-ABASIA ACCOMPANYING GRAVES DISEASE.—Under the title of Astasia-abasia Blocq has described an inability to stand or walk, with intact sensation, co-ordination and motor power of the lower extremities. Charcot has distinguished a paralytic, a "forme trepidante," and an ataxic form with a subdivision of the latter termed choreiform. There is great difference in regard to causation, some authors regarding it as hysterical in all cases, others as an independent symptom-group of spinal or psychic origin.

Prof. Eulenberg reports a case occurring in a girl of 18, chlorotic and the subject of exopthalmic goitre. After these troubles had improved under dietetic, electrical, and mechanical treatment, she developed an absolute inability to walk, although examination in bed showed absolutely no nervous defect. She recovered promptly under two applications of the faradic brush.—*Neurolog. Centralbl.*, Nov. 23, 1890.

G. J. K.

On the Electro-Physiology and Electro-Pathology of the Reflexes in Connection with a Case of Transverse Myelitis.—Under this title Dr. Græripner describes a case of transverse myelitis with peculiar reflexes originating in the irritation, both mechanical and electrical, of certain nerves. A lady, æt. 45, sick about six years under symptoms pointing to a transverse myelitis of the lumbar cord. Active motion below the knee was very slight in the left leg—somewhat better in the right—passive movements good.   .

The needle is not felt, neither are considerable differences of temperature on the right leg. Sensation reduced on left leg. Patellar reflexes normal. On irritating the right sole, the foot is slightly raised, and the 3d, 4th and 5th toes are extended and abducted. If the irritation is somewhat stronger, a clonus is developed. The reflex lasts two to three seconds longer than the irritation. An identical reflex was developed by irritating the skin in the area of the sural nerve. These reflexes do not occur on the left side on mechanical irritation. Electric Examin. Farradic. Indifferent electrode on sternum. Using an electrode of 20 sq. cm. area in the area of the sural nerve, the reflex as described above was obtained with a separation of the coils, of 50 mm., together with a diffuse vibration and undulation of the muscles. If the irritation was continued for a minute, a diffuse tremor and jerking of the *other* leg occurred.

Galvanic. Peroneal nerve—Right side KSZ at 4 M. A., left side 3.5 M. A. ASZ right side 4.5 M. A., left side 4 M. A. An electrode of 20 sq. mm. area applied in the area of the sural nerve, produced the described reflex on KS with 0.75 M. A. On continued irritation with K the tremor of the other leg occurred. K. O. gave no reaction; an S a slight one; An. continued, none. An O gave a stronger reaction of four to five seconds' duration, two to three seconds after opening. The formula can be written: Electrode 20 sq. cm. + 0.75 M. A.; KaSR; KaO; An'D; An OR=An OT.

With currents of 1.5 M. A. convulsive movements prevent examination. The author deduces from this that the cells presiding over both tibial nerves were abnormally excitable for the transference of reflexes, while their centripetal conduction was abolished. The insensibility to temperature suggested to the author the probability of syringomyelia.

As An S had the least intense effect, this pole was used in the treatment, both to the spine and the areas furnishing the reflex. With care in avoiding An. O, the patient soon improved so that 2 M. A. with an electrode of 40 sq. mm. produced no reflex.

This case proves that the specific polar irritation produced in sensory nerves produces specific effects on that part of the central nervous system controlling the reflexes, and that the central nervous system reacts differently to different galvanic stimuli peripherally applied, although in the normal state this reaction is not shown on account of the slight irritability. —*Berlin Klin. Wochenschr.*, No. 46, 1890.   G. J. K.

THE ACTION OF ALCOHOL ON THE SYSTEM OF CHILDREN.— Prof. Demme, in the 27th report of the Jenner Hospital for children in Berne, discusses this question in connection with the clinical data presented.

Of 114 children brought to the hospital in twelve years with defective development of mind or of the power of speech, or with imbecility, 62 or 54.3 per cent. had one or both parents, and frequently grandparents also, who were drunkards. Of 47 cases of chronic hydrocephalus internus treated in 28 years 38 showed a psychopathic ancestry. In 17 cases the father, in 4 cases the mother, and 2 cases both parents were drunkards. In 6 of these cases, previous generations gave a similar history. In 61 cases of chorea minor, intemperance of parents is noted 19 times. In 5 of these cases, the children had been dosed with liquor, the disease following a debauch. Of 98 cases of epilepsy, 7 followed acute alcoholic intoxication and recovered without medication, after prolonged abstinence. In 29 of these cases, one or both parents, and in 17, grandparents also were drunkards. Nine fathers and three mothers are stated to be epileptic. He then compares ten families of notorious drunkards with ten temperate families in similar circumstances with the following results:

TEN FAMILIES WITH FIFTY-SEVEN CHILDREN, ONE OR MORE GENER-
ATIONS INTEMPERATE.

Died of inanition and convulsions in first 6 mos.
    of life, 43.8 per cent........... 25
Idiotic....................................... 6
Retarded growth............................. 5
Epileptic 5
Congenital defects, as club foot, hare lip, etc.... . 5
Idiotic as a result of chorea.................. 1
Normal (17.5 per cent.)..................... ..10
                                              ——
                                              57

TEN FAMILIES WITH SIXTY-ONE CHILDREN, TEMPERATE ANCESTRY.

Died of inanition during first 6 mos. of life....... 3
Died of bowel complaints without convulsions.... 2
Chorea minor................................. 2
Slow mental development....... 2
Congenital defects, as hare lip (1), spina bifida (1) 2
Normal (81.9 per cent.)..................... ..50
                                              ——
                                              61

These families were selected for number of children only. He relates the case of an infant æt. 2½ months, whose convulsions were proven to be directly due to the milk of a whisky-drinking mother, also a case of diabetes insipidus in a child aged 8, due to the use of brandy. The child passed from 3 to 4½ liters of urine per day (6½ to 10 pints).—*Wiener Med. Blaetter*, No. 3–5, 1891.

BROWN-SEQUARD'S PARALYSIS.—Among the clinical histories taken in the department for mental and nervous diseases at the Copenhagen general hospital, Dr. A. E. Klær points out an unusual case of Brown-Sèquard's paralysis, if we are to follow Brown-Sèquard's own description of the disease.

Patient had received a stab wound in the middle of the back and a little to the left of the median line about 12 years ago. In the first place the author, instead of finding cutaneous hyperæsthesia together with motor paralysis on the side corresponding to the lesion, could discover nothing wrong with sensibility on this side. He thinks that in connection with this absence of hyperæsthesia, the length of time elapsed since the traumatism was inflicted, is to be considered, but does not attempt any explanation.

In the second place the author found muscular sensibility also intact on the left side; the patient could very accurately designate the position in which the limb was placed by an assistant. The absence of this one of the most prominent symptoms in the disease is to be accounted for in the regeneration of the nerve tracts for muscular sensibility. These tracts unlike the other sensory tracts, according to the views of Brown-Sèquard, do not decussate in the cord, but pass down on the same side.—*Hospitals-Tidende*, Sept. 17, 1890.

M. N. V.

A PECULIAR FORM OF SENSORY PARALYSIS.—Esner has severed the superior laryngeal nerve in the horse and found an immediate paralysis of the corresponding half of the larynx, which lasted for weeks, although this nerve is not the motor nerve for any laryngeal muscle.   The microscopic examination of the muscles showed changes similar to those described by Erb in dystrophia muscularis progressiva.   This degeneration Esner interprets as an atrophy from disuse, while he regards the paralysis as the severest degree of ataxia.—*Men. Mediz. Presse*, No. 45, 1890.   G. J. K.

SENSORY DISTURBANCES IN FOCAL LESIONS OF THE BRAIN.— Dr. L. Darkschewitsch reports a case of right brachial monoplegia, with great reduction of sensation.   Autopsy showed a cavity due to an absorbed tubercle, about the size of a hazelnut, in the white substance immediately under the cortex of the left posterior central convolution.—*Neurolog. Centralbl.*, No. 23, 1890.   G. J. K.

A CASE OF GENERAL CUTANEOUS AND SENSORY ANÆSTHESIA. —A soldier, 22 years old, during convalescence from a light typhoid, was seized with trembling, great lassitude, convulsions, and a cutaneous anæsthesia beginning in the limbs, but soon extending over the whole body.   There was also sensory anæsthesia.   Neither needles, cautery, ice-water, nor electric stimulation were perceived by the skin or the muscles. Mucous membranes of nose, eye, mouth, œsophagus, larynx, rectum, and urethra were totally anæsthetic.   In consequence there was neither hunger nor thirst, nor any desire to defæcate or urinate.   Taste and smell were totally abolished; sight and hearing normal; strength diminished; fatigue and muscular sense absent.   Walking and standing alone with open eyes only possible with assistance.   Walk *not ataxic*.   Pupillary reaction, swallowing and lacrymation normal, as were patellar, abdominal and cremaster reflexes.   Plantar, conjunctival cough and nasal reflexes absent.   The mental condition was also abnormal.   He became very emotional, had frequent dreamy spells, did not know where he was.   On closing the eyes and plugging the ears, the patient immediately fell into

a deep sleep, from which he could only be awakened by loud
noises or bright lights.   Voluntary motion was only possible
under the continuous control of the eyes.—Dr. M. Heyne,
Deutsch. Arch. f. Klin. Med., XLVII., p. 75 *Neurol. Centralbl.*,
No. 23, 1890.

Prof. Ziemmsen (same Journal, p. 89) gives the further history of the case.   For nine months there was no change.
After that there was slow improvement, although mentally the
recovery was not complete.   Ziemmsen adds a second case,
that of a woman, æt. 58, who had borne 16 children.   After
headache and paræsthesia lasting for days, she suddenly complained of vertigo, weakness, and trembling of the right arm.
In the course of eight weeks, during which headache, vomiting and high pulse continued, she developed, first on the right,
then on the left side, intention tremor, great increase of tendon reflexes, and complete cutaneous anæsthesia.   Loss of
strength and of muscular sense, and progressive loss of memory; anosmia and agensia on the right side.   On closing eyes
and ears, sleep lasting one minute occurred at once; death in
14 weeks. Microscopic and gross examination gave no results.
Prof. Ziemmsen believes the trouble to be functional and
allied to the psychoses, especially to melancholia.—*Neurolog.
Centralbl.*, No. 23, 1890.   G. J. K.

Isolated Peripheral Paralysis of the Supra-Scapular
Nerve.—Dr. Benzler reports a case of this rare trouble, only
four other cases having been published.   There was no ascer-
tainable etiology.   There existed atrophy and paralysis of
the supra- and infra-spinatus muscles, with partial reaction of
degeneration.   The case is of interest as showing the function
of the supraspinatus muscle, proving Duchennes' statement,
that this muscle lifts the arm and draws it forward and out-
ward.—*Deutsche Medic. Wochenschr.*, 1891, No. 3.   G. J. K.

Landry's Paralysis.—Prof. Klebs, in an address before the
Berlin Society of Internal Medicine, states that he has found
in a case of this disease, hyaline thrombi in the capillaries

surrounding the central canal of the cord as well as miliary hæmorrhages around the vessels. There was also a fibrillary exudate with small round cells around the large ganglion cells. The paralysis occurred in the course of an unrecognized tuberculous pericarditis, and Klebs believes that the products of the bacillus tuberculosis are responsible for the changes found in the cord.—*Deutsche Medic. Wochenschr.*, 1891, No. 3.   G. J. K.

ETIOLOGY OF TETANUS.—The following abstracts of recent memoirs on the pathogenesis of tetanus are taken from an extended review of the recent literature of the subject by P. Wagner in *Schmidt's Jahrbucher*, Nos. 10 and 11, 1890. After noticing the researches of Carle and Rattone, Nicolaier, Rosenbach, Brunner and Peiper, and others, he gives extended extracts from other papers, the more important of which are the following:

Dr. S. Kitasato, *Ztschr. f. Hyg.*, VII., 1889, sums up in these conclusions:

1. Tetanus is an infectious disease due to a specific bacillus.

2. The cause of human tetanus and of that from inoculation is one species of bacillus, which is identical with that first described by Nicolaier, later confirmed by Rosenbach and others.

3. This bacillus is present in the pus of the tetanic wounds of men and animals subjected to experiments. It frequently develops spores in the pus, but if investigated early enough, it often appears as sporeless rods.

4. This bacillus from the secretions of tetanus patients, and animals in which it has been artificially produced, and also that from pure cultures, produces tetanus in animals.

5. The varying views held previously in regard to the bacteria, are undoubtedly to be explained by the circumstance that the disease has been examined in different stages, and the earlier the death occurs the more rarely do the bacilli form

their spores in the pus. The bacilli themselves are always present and a spore forming bacillus can be cultivated from sporeless tetanic secretions.

Dr. Kitasato describes the bacillus of tetanics as an anærobe bacterium that grows only with absence of air. It thrives under hydrogen, but not in carbonic acid. It can be cultivated in the common peptone containing weak alkaline, agar, and also in gelatine. This last deliquesces with evolution of gas, which is not the case with agar. They give to the cultures a very repulsive smell and they may be carried on in continued cultures, without losing their virulence, as is the case with some other bacilli. The colonies on gelatine plates have much the general appearance of hay bacillus, but the deliquescence of the gelatine is much slower. They thrive best at a temperature of 36 to 38 C. and their spores, which are round and rather thicker than the rods, are attached at the ends, producing the bristle appearance as described by Nicolaier. They are very resistant to heat and chemical action.

They possess a peculiar, but very slow motion, and spore containing bacilli are motionless. They take readily the colors from the usual aniline reagents.

II. A memoir by G. Tizzoni, J. Cattani and E. Baquis (*Beitr. z. Path. Anat.*, VII., 1890), is the subject of an extended analysis. The authors, basing their researches on material furnished by three fatal cases of tetanus, sum up their conclusions as follows:

1. In tetanus there are found two species of bacilli, with end spores both pathogenic.

2. Each of these bacilli has special morphological, biological and cultural peculiarities.

3. By inoculation one of these bacilli produces typical tetanus, the other gives rise to a disorder also fatal, but in which the tetanic phenomena are very mild and restricted.

4. In certain nutritious substances (bouillon, agar) the bacillus of acute tetanus undergoes a diminution of its pathogenic property and causes in animals a chronic disorder,

accompanied with slight tetanic symptoms or none at all, therefore may the two clinical forms of tetanus stand in relation with the two above mentioned species of bacilli as also with the weakening of the power of the bacillus with round end spores.

5. The bacillus with round end spore in ordinary cultures requires the absence of air, but in rabbits' blood it thrives even though air is admitted. The authors found that the ablation of the original point of inoculation did not hinder the fatal result, which speaks for the existence of a secondary infection and indicates therapeutic uselessness of amputation.

III. Th. Kitt (*Centra Bl. f. Bakteriol*, VII., 10, 1890) has, through a series of very thorough experiments, shown that infectious traumatic tetanus which is produced by Nicolaier's bacillus, also occurs from the bacillus found in the human subject and in the earth. He also found that pure cultures could be produced without the application of heat, and that the bacilli preserved their virulence from four to sixteen months.

IV. Ricochon (*Gaz. hebd*, XXXV., 35–36, 1888) discusses the etiology of tetanus and concludes it is a true infectious disease that may be transmitted through the atmosphere as well as by contact. He agrees in general with Verneuil as to the equine origin of tetanus, but he admits the possibility that the infection may be autocthonous in the soil of certain regions. Vallas (*La Prov. Med.*, I., 1889) also holds to this view and as against the theory of the exclusive equine origin of the disorder. There are regions, for example, in central Africa, where tetanus is very frequent in man, and horses are very rare or unknown.

V. Chantemesse and Widal (*Bull. Med.*, 74, 1889) have been able to produce tetanus in rabbits by inoculation with dust from crevices in the neighborhood of beds on which patients with tetanus have lain. Dust from the walls, hangings, mattress, etc., produced no results as was the case also with inoculation from the inner organs of tetanic animals.

Repeated inoculations caused an increase of the time of incubation until by the fourth or fifth remove they became ineffective. They produced pure cultures of Nicolaier's bacillus, but could not with these effectively inoculate all species of animals. It is noteworthy that they found that dust containing tetanus germs became inocuous after it had been exposed three days to the light under a bell glass.

VI. Babes and Puscarin (*Centr. Bl. f. Bakteriol*, VIII., 3, 1890), have experimented with Kitasato's pure cultures, and compared them with others taken from animals suffering from tetanus, and have come to similar conclusions as regards the virulence of the bacilli. They were unable to produce tetanus with emulsions from the inner organs or from the blood, but found that the brain emulsions produced fatal results in animals without tetanic symptoms.

VII. Gotti (*Memorie della R. Acad. della Scienze dell Institutodi Bologna*, IV., 9, 1889) concludes a paper on some experimental investigations regarding the etiology of tetanus with the following:

1. That the experimental tetanus produced in animals by infected earth is clinically identical with that observed after wounds in the human subject.

2. That in the majority of fatal cases of experimental tetanus, the characteristic micro-organism described by Nicolaier and Rosenbach are found in quantities at the point of inoculation.

3. That in experimental tetanus propagated through a series of animals, the alterations at the point of inoculation decrease with diminution or disappearance of the Nicolaier micro-organisms and with this it becomes more difficult to produce further transmission of the disorder.

4. That finally the bacteriological investigation of the fluids and organs of animals that have succumbed to experimental tetanus, does not reveal either the Nicolaier bacillus or any other lethal micro-organism, thus supporting the opinion of Rosenbach based on Brieger's researches, that the

symptoms of tetanus are to be attributed to a toxic substance developing itself in the localties of the body where the tetanic secretion is produced.

VIII. Morisani (Napoli, 1888) (*Cbl. f. Chir.*, XVI., 4, 1889), sums up his conclusions of an experimental study of the causes of tetanus as follows:

1. Purulent necrotic products from human subjects of tetanus, in which the presence of tetanus bacillus was demonstrated, produced by inoculation tetanus in the lower animals; and at the point of inoculation there is produced an inflammatory process in the secretion of which the above bacillus is, with others, to be found.

2. Inoculations of these secretions from animal to animal produce tetanus with the inflammatory process at the point of inoculation.

3. These inflammatory products, when kept in closed vessels till the disappearance of the bacilli, are still able to produce tetanus, though with a longer period of incubation.

4. In these cases, at the point of inoculation there occurs a simple necrotic process which does not prevent the healing of the wound and by the usual methods the micro-organisms are not to be detected.

5. The headed tetanus bacillus which flourishes with the needle formed tetanus bacillus, causes after isolation, through cultures and inoculation in animals, simple gangrene at the point of inoculation, in the products of which it appears to be so altered as to seem identical with tetanus bacillus.

IX. From a number of interesting researches carried on in the laboratory of Prof. G. Gormani which are abstracted at some length by Wagner we extract only the following:

1. E. Pariatti has been able to analyze the cultures obtained from tetanus patients so far as to isolate two species of bacteria evidently identical with those observed by Kitasato. The one with oval spores was recognized by him as identical with the *Clastridium fœtidum Liborius* and was not tetanogenic, the other was the Nicolaier bacillus, and was

exceedingly virulent. Using these two bacilli he found that he could produce tetanus in dogs, though with difficulty and generally only in the immediate neighborhood of the inocula-tion. Dogs that had survived one experiment of this kind had an immunity against further inoculations.

2. G. Sormani, in a number of experiments on the effect of various agents upon the tetanus virus, found that in one case at least, iodoform applied to the surface of the wound rendered the tetanus secretions innocuous. He found also that amongst other agents, all of which had more or less effect, that nitrate of silver in one per cent. solution destroyed the tetanus spores in one minute, and was in all respects the most powerful. He also made some experiments on the pos-sibility of conveyance of the tetanus infections through inhalation with negative results.

These results were published in the *Reforma Med.*, V. and VI., 1889, 1890.

X. R. Schwartz, *Reforma Med.*, VI., 1890, carried on in Tizzoni's laboratory in Bologna, investigations on the vitality of tetanus germs in sterilized and unsterilized water. In the former their virulence was unaffected, but in the latter it became gradually weakened through the increase of the usual bacterial forms, but when these were eliminated it regained its former strength. In unsterilized sea water peculiar chemical conditions seemed to affect its virulence somewhat. In all cases the toxic power returned when favor-able conditions were restored.

XI. Brieger and Frænkel, *Berlin Klin. Wochensch.*, XXVII., 11 and 12, 1890, publish important researches on the chemical composition of the tetanus poison, and have isolated a toxal-bumin produced by the tetanus bacillus, which in guinea pigs caused after four days, paralysis, spasms and death. It seemed to differ from tetanotoxin in that it was less virulent. Kitasato and Weyl have isolated an apparently identical product.

XII. Tizzoni and Cattani, *Cbl. f. Bakteriol*, VIII., 3, 1890, have investigated the tetanus poison. They found the tetanus bacillus culture in gelatine retained its virulence fully, while meat broth cultures soon lost their dangerous properties. If the culture was filtered through a Chamberland's filter so as to be perfectly free from mucus, the gelatine preparation continued toxic while the other was inocuous.

The animals showed the first tetanic symptoms about ten or twelve hours after the injection and they succumbed, generally, in from twenty-four to thirty-six hours. The effective injections were made under the skin in the sciatic nerves, into the circulation, or under the dura mata. Very large quantities were injected into the stomach without any effect. The authors have endeavored to isolate the toxic principles from the cultures and obtained finally a golden yellow intensely toxic crystalline substance, but do not identify this with the tetanotoxin of other authors. They believe it to be an immediate derivation of albuminous substances, but maintain strongly that it belongs to the class of soluble ferments.

XIII.   Dr. Kund Faber (*Berlin Klin. Wochensch.*, XXVII., 31, 1890), obtained by filtration from a very virulent tetanus culture, a filtrate perfectly free from bacteria, which by injection produced a perfect experimental tetanus in animals. The filtrate was strongly alkaline, but equally poisonous after having been rendered acid with dilute tartaric acid. It completely lost, however, its poisonous properties by five minutes' heating to a temperature of 65° (R?). It was effective not only through the tissues, but by intravenus injection. In this last case the effect was quicker and more severe and began at once with trismus and general tetanus.

Dr. Faber considers that the local cramp, which is without exception one of the earliest symptoms in animals, is not so infrequent in the human species and that it has some importance in the production of the complete disorder. Thus he explains the favorable effects sometimes observed from nerve stretching or nerve section by the cutting off of the local irritation.

4

XIV. D. J. Fontan (*Gays. Hebd.*, XXXVI, 25 and 26, 1889) concludes a paper in which he publishes some eighteen cases of tetanus with the following propositions:

1. Tetanus occus very frequently in hot regions, especially in India, West Africa, and Madagascar.

2. It is more commonly of traumatic than of spontaneous origin, and effects children by preference, and apparently also the colored races.

3. Its connection with certain telluric influences is at least doubtful.

4. It occurs usually in little epidemics; sporadic cases are rare.

5. Its contagiousness in the same house, hospital ward, etc., is established by numerous instances.

6. In most of the regions in which tetanus is frequent, there is a great abundance of animals that are liable to tetanic infection.

7. In hot countries the frequency of human tetanus corresponds to that of equine tetanus.

8. The human individuals who suffer from tetanus are usually those that have certain definite relation to the animals.

9. Examples of transmission of tetanus from horses to man have not been found with certainty, but very probably occur.

10. There are no facts which contradict the theory of the equine origin of tetanus.

XV. C. Hægler (*Beitr. z. Klin. Chir.*, V., 1, 1889) publishes a fatal case of tetanus in which he took from the fresh corpse small pieces of the subcutaneous cellular tissue, of an arterial clot, and of the medulla oblongata, in all of which the Nicolaier–Rosenbach bacillus was found. Inoculation with this material caused fatal tetanus in mice, guinea pigs, and rabbits. The blood taken from the dorsal veins of the guinea pigs *in extremis* also showed the presence of these bacilli and by inoculation produced tetanus.

XVI. Dr. C. Brennen (*Deutsche Ztschr. f Chir.* XXX., 6, 1890) has investigated the symptoms that had been described by Rose as characteristic of tetanus localized in region of the cranial nerves. This author mentions as characteristic in these cases the following:

1. Convulsions of the muscles of the throat and glottis and (2) paralysis of the facialis on the side of the injury. As regards the first of these he finds it constantly occurring in cephalic tetanus in man, as well as that due to injuries of other parts of the body. Paralysis of the facial, which seems to be a very common symptom on the side of the lesion when the face is the seat of the original injury in man, does not occur in experimental tetanus in the lower animals; on the contrary there is in them invariably facial spasm. From his analysis of these results and his study of the recorded cases in man, Brennen concludes that in many cases the facial paralysis reported is the result of faulty observation or else an accidental complication not essentially belonging to the disorder.

XVII. V. Gautier (*Revue. Med. de la Suisse Rom.* IX., 12, 1889) has collected seventy-four cases of puerperal tetanus of which thirty-six occurred after abortion, and thirty-eight after normal, or almost normal labor. No influence could be attributed in these cases to season, age, number of previous labors, or time of the abortion. The most of the subjects were from the lower classes. In most of the cases there was some manual interference in the labor or abortion. Fifteen times there was partial or complete removal of the retained placenta, thirteen times tamponing, and seven times severer operations. An equine origin could be demonstrated in no case. The infection occurred in the great majority during the first two weeks, and sixty-four of the seventy-four were fatal. The symptoms, pathological findings, showed nothing special. The same was the case as regards treatment.

In a more recent memoir *Zeitsch. f. Hyg.*, VII., 3, 1890 (And by Nowak in *Schmidt's Pub.*), Kitasata and Weyl publish the results of chemical investigations of pure tetanus cultures. We find five different substances in them, tetanin, tetanotoxin, iodol, phenol, and butyric acid, and express their conviction that the tetanus bacillus must produce a poison of much greater virulence than that of tetanin. A future publication on this point is promised. H. M. BANNISTER.

THE PRESENT STATUS OF OUR KNOWLEDGE OF TETANY.— Dr. Herman Schlesinger attempts an etiological classification. That group called by Jaksch, acute relapsing tetany, he denominates epidemic tetany. In a second group he gathers all cases, of whatever etiology, whose characteristic is the tetany and calls these pseudotetany, or tetanoid conditions. These include the toxic forms (chloroform, ergot, mucin, pellagra), the tetany of infants (worms, rickets, or intestinal catarrhs), those accompanying infections (malaria, typhus, scarlatina), those of cerebral origin, and those following extirpation of the thyroid. The latter may be differentiated clinically, by the etiological factor, its severe course, its transition into epilepsy, the beginning of the spasms in the calves of the legs, severe catarrhs, and the discovery of mucin in the tissues. Schlesinger would class this last form as mucin poisoning. He has found the "facial nerve reflex" in other diseases.—*Allgem. Wien. Med. Ztg.*, No. 30, 32, 1890. *Neurol. Centralbl.*, No. 21, 1890. G. J. K.

ON THE SENSORY AND AUDITORY NERVES AND RESISTANCE OF THE SKIN IN TETANY.—Dr. F. Choostek concludes from investigations in Kahler's clinic, that tetany is accompanied by a heightened mechanical and electric irritability of the sensory nerves. During the attack this irritability is at its height in all the nerves, and fluctuates with the other symptoms, disappearing at first in the head. The increased galvanic irritability is shown by the slight current required and the shortness of the interval between the local and radiating sensations. In six of seven cases the acoustic nerve responded to galvanization with the complete formula, the nerve being irritable in only 15 per cent. of normal subjects. The resistance of the skin is normal.—*Wien. Mediz. Presse*, No. 42, 1890, 1671.

G. J. K.

APOPLEXY DUE TO ECHINOCOCCUS-EMBOLUS OF CEREBRAL ARTERITIS.—Girl, æt. 12, was seized at school with violent hemicrania and vomiting. The next morning she was found comatose, with reflexes totally abolished. The left side was colder than the right. Before death, which occurred in a few hours, numerous erythematous spots appeared over the body.

Autopsy, confined to head, showed a hæmorrhage into right thalamus, with rupture into the ventricle. The fourth ventricle was filled with bloody fluid, extensive diffuse hæmorrhage into arachnoid. The left sylvian and deep cerebral arteries and basilar artery were found plugged by collapsed echinococcus cysts. No similar case has been reported.— DR. DÆHNHARDT, *Mendel's Centralblatt*, No. 19, 1800.

G. J. K.

ON A PECULIAR FORM OF DISEASE OF THE CENTRAL NERVOUS SYSTEM IN THREE MEMBERS OF A FAMILY.—The symptoms consisted mainly in muscular weakness and insufficiency, so that all movements were either excessive or defective. Standing and walking were difficult and only possible with solid support; walk not ataxic; reflexes normal; no Romberg symptom; sensation subjectively and objectively normal. Speech was explosive and shouting, as the patients were unable to gauge the proper amount of muscular effort needed. The play of the facial muscles was also excessive. There was pronounced ataxic nystagmus. Two of the patients had optic atrophy to a moderate degree. A certain irritability of character and suspiciousness was pronounced in all cases.

The subjects of this disease were three brothers aged respectively 49, 46, and 40 years. In the oldest, whose trouble was intermediate in severity to the other two, the trouble began in the 14th year. The second brother, a sailor, who was least affected, remained well until the age of 30, when he suffered shipwreck, soon after which the trouble developed. In the youngest brother, who was most affected, the trouble was noticed in the 10th year. This brother died of pneumonia. Autopsy showed a central nervous system normal to the naked eye in everything but size, although the normal proportion between different parts was present. Cerebrum weighed 1,020 gm (normal 1,160), rest of brain 120 gm (normal 160, 170). The entire brain and cord resembled that of a boy of 10. The cerebellum was reduced from ¼ to ⅓ in all its measurements. A section of the medulla at the abducens nuclei measured 1.7 cm. by 1.5 cm. (normal 2.7 cm. broad, by 2 cm. high). The cord showed the same reduction in size. (Normal measurements in parenthesis.)

Cervical enlargement 11 x 6 mm. (15 x 9).

Middle dorsal cord 8 x 6 mm. (11 x 8).

Lumbar enlargement 9 x 7 mm. (11 x 8.5).

Microscopic examinations showed absolutely normal tissue, except in the spinal nerves, where there was a great preponderance of the fine over the coarse nerve fibres. The muscles and peripheral nerves were normal.

The clinical features of this disease resemble those of cerebellar atrophy.

The father of these three patients was healthy, as was one sister, whose two children were also normal. Of the ten descendants of another defective sister, five showed inequality of the face, tremors, etc. A third defective sister had no children. Of the twelve descendants of an uncle, who himself had cleft palate, two had cleft palate, one was born blind. Of the seven descendants of a normal paternal aunt, one had a trouble resembling that described, but to a slight degree only.—DR. M. NONNE, *Arch. f. Psych.*, Bd. XXII, h. 2.

G. J. K.

A CASE OF APHASIA AND PSYCHICAL DEAFNESS, WITH AUTOPSY.—O. Heubner reports the case of a man aet. 64, who after an apoplectic attack, showed the following symptoms. He had lost voluntary speech and the understanding of written and vocal speech, but retained the power of repeating spoken words, of writing spontaneously and on dictation, and of reading, but without any conception of the meaning of the words. He also had a symbolism. As the sound picture of a word was understood and the motor conception of a word was good, motor and sensory aphasia were absent. Heubner explained this combination of symptoms by a lesion of the cortex at a point where the different conceptions stored in the brain are united to form the idea of the word. This spot H. placed in the marrow of the left first temporal convolution. Autopsy showed a focus of yellow softening, mainly occupying the cor⁻tex, at the point where the lower parietal lobe merges into the posterior branch of the supramarginal gyrus and the first temporal convolution. The area of softening crossed the lower part of the lower parietal convolution and the middle part of the supramarginal gyrus almost to the fissure of sylvius, being 27 mm. long. It also extended for a distance of 65 mm. in the sulcus between the first and second temporal convolutions. By this lesion the first temporal convolution was isolated to a considerable degree from the rest of the cortex, although it was intact otherwise. The patient was not word-deaf; he could repeat; he was not cortically, but psychically deaf. H. believes that this case proves that the center for simple cortical deafness is in the first temporal convolution and the center for psychical word-deafness near it.—Reprint, abstracted in *Deutsche Medic. Wochensh.*, No. 44, 1890.   G. J. K.

———

AURAL VERTIGO (Menière's Disease).—Dr. Mettler reported a case with comments to the Philadelphia Neurological Society. He holds that the term "Aural Vertigo," is a misleading one, inasmuch as the aural symptoms are not necessarily indicative of trouble in the ear. He discusses at length the relations of the sensory tract to the auditory nerve and the cerebellum. He concludes that any disturbance in the course of the aural tract is liable to give rise to symptoms of Menière's disease. The lesion may assume many forms, but the important point is, that it be not limited to the labyrinth. In many of the cases reported it is probably outside of it.

In the so-called Menière's disease the source of irritation may sometimes be in the course of the semi-circular canals, but very often it is elsewhere in the auditory tract. In 46 cases of necrosis of the labyrinth only twelve had vertigo. We are not justified in assigning all cases of vertigo with loss of hearing to lesions of the internal ear.—*Jour. Nervous and Ment. Dis.*, Jan.

CHRONIC HYDROCEPHALUS WITH ALMOST COMPLETE DISAPPEARANCE OF THE CEREBRUM.—Prof. Henoch reports the autopsy of a child 3½ months, in which only minute parts of the cerebrum remained, in a highly atrophic state, above the corpora quadrigemina. The peripheral portions of the nerves of special sense were atrophied. Nursing and all functions were performed normally. This case supports Soltmann's view, that the movements of the new born are purely reflex.—(*Charite Annalen*, 1890), *Neurolog. Centralbl.*, No. 24, 1890.

G. J. K.

PRIMARY ACUTE ENCEPHALITIS.—Strümpell (*Deutsch. Archiv. f. Klin. Med.*, XLVII, H. 1 and 2) describes this condition, which must be differentiated from the group of cerebral palsies of childhood. Children from 1½ to 4 years old, previously healthy, become suddenly sick with fever, vomiting, and headache. In from a few hours to several days violent cerebral symptoms, such as convulsions and coma occur. After some time the acute symptoms subside, leaving as a permanent lesion, a more or less extensive unilateral paralysis with frequent secondary changes, such as contractures, hemiathetosis, or disturbances of growth. In a certain number of cases embolism is the pathological factor; although in the greatest number, an acute inflammatory process, ending in the production of a cicatrix, is the pathological cause. S. has dropped the term polioencephalitis.—*Neurolog. Centralbl.*, No. 23, 1890.   G. J. K.

PATHOLOGY OF PACHYMENINGITIS INTER. HÆMOR.—Dr. Wigglesworth has an article in the January number of *The Jour. of Insanity* on this subject.   In the *Jour. of Ment. Sci.* for Jan., 1888, the doctor published an article in which he held that pachymeningitis is not the result of inflammation, but due to effusion of blood into the subdural spaces.   The present article is a reply to certain criticisms of this view by Dr. E. C. Seguin.   He holds that the different layers found in pachymeningitis are no evidence of inflammatory action, else the layers of fibrin in an aneurismal sac would also be evidences of inflammation.   The origin of the laminæted membrane he conceives to be this: the blood attaches itself to the entire membrane of the dura and the fibrin precipitates itself on the surface of the clot so that the membrane, which speedily forms, exhibits on section, a dark core bounded by paler lines, which appear to consist of fibrin and leucocytes.—*Jour. Insanity*, Jan.

CASE OF OVERGROWTH OF THE SKULL.—Dr. Thompson publishes an interesting case in the January number of *The Edinburgh Medical Journal*.   The patient suffered with epilepsy from infancy and died at fifty-three.   The right side of the skull was normal.   The enlargement affected the left side of the bones of the face and skull.   The condition was one of hypertrophy of the affected bones, and also of nodular and bossy thickening of their surfaces.   These thickenings were identical in structure with the nodes of the long bones found in syphilis.   There was also enlargement of the grooves, canals, and foraminæ, for nerves and blood vessels on the same side, especially those which served for the transmission of the branches of the fifth nerve.   There was enlargement of the corresponding branches of the fifth nerve, and it was also observed that the bony alterations followed with the greatest accuracy the distribution of this nerve.   This may be considered a case of tropho-neurosis.   Mr. Jonathan Hutchinson, to whom the skull was sent, considered that the specimen showed sufficient grounds for the inference that there exist certain fibres in the fifth nerve which preside over the nutrition of the bones to which they are supplied, and which, when their function is exalted, may cause overgrowth and tumor formation in their area of distribution. The entire brain weighed fifty ounces, of which the left hemisphere weighed sixteen, and the right twenty-five. The enlargement of the skull, therefore, prevented the left hemisphere from developing.

PATHOLOGY OF THE ABDOMINAL SYMPATHETIC.—Prof. Talma of Utrecht, believes that many of the symptoms attributed to hysteria, hypochondria, neurasthenia, or nervous dyspepsia are due to disordered function of the abdominal sympathetic. A great variety of pains and paræsthesias in the abdomen, as well as feelings of "goneness" and abnormal motor phenomena on the part of the viscera are due to abnormal innervation. Excessive secretion of very acid gastric juice, or of watery urine, may be due to this cause. He considers spasm of the stomach, due to this cause, of great importance in the etiology of gastric ulcer, the spasm producing anæmic areas which are acted upon by the gastric juice.—*Deutsche Medic. Wochenschr.*, No. 50, 1890. G. J. K.

CHANGES IN THE BRAIN IN DROWNING.—Dr. B. Salemi Pace (*Il Pisani*, XV., 1890) has made experimental researches on the condition of the brain in drowning and finds microscopically no alterations of the brain substance worthy of note. The vascular alterations are inconstant and uncertain, thus agreeing with the majority of medico-legal writers on this subject. Microscopically, however, he found a constant hyperæmia of the smaller vessels, with occasionally ruptures o the more minute.

The constancy of this symptom is a medico-legal point of ome consequence, if it can be found also constant in the human subject in cases of drowning, accidental or otherwise. Dr. Salemi-Pace experimented with guinea pigs.

Vasomotor Disturbances of the Skin in Traumatic Neurosis.—Although most systematic writers mention the fact that such disturbances may occur, especial attention has not been paid to them. Dr. H. Kriege divides them into three groups. The first, includes the abnormal flushing often observed and hyperidrosis, and is due to an abnormal excitability of the vaso dilator centers. The second group comprises the transient urticarial phenomena, but without the subjective sensations of the true eruption. A slight blow, or similar irritation, may cause the formation of wheals 1 cm. high, lasting from 15 min. to several hours. This he refers to an abnormality of the vaso dilator centres combined with defects of structure in the vessel wall. The third group comprises the localized cyanosis, which may attain such a degree as to merit the name of local asphyxia (Raynaud's disease). This is probably due to vaso-motor spasm.—*Arch. f. Psych.*, Bd. 22.

G. J. K.

THERAPEUTICS.

Treatment of Tabes.—Prof. Strümpell believes tabes to be a post syphilitic degeneration　In estimating the value of any therapeutic method, and in fixing a prognosis, we must have a clear idea of the course of the disease when untreated. The refinement in diagnosis of tabes has reached such a degree that we see a great number of rudimentary and abortive forms, which present certain symptoms only, otherwise remaining well. Especially are the symptoms due to irritation prone to disappear. A second factor, is the fact that especially in tabes do we find severe functional symptoms to appear, and to slowly disappear uninfluenced by treatment. A third factor is the fact that hygienic measures exert a great influence upon the general welfare of the patient. While S. admits the value of baths, electricity, suspension, etc., in the improvement of symptoms, he claims they certainly cannot restore degenerated fibres. The specific treatment of tabes cannot be addressed to the tabetic process, but is always indicated for its influence on accompanying gummous affections of the nervous system, and in removing the general syphilitic dyscrasia. A number of cases have been published which tend to show that the symptoms of tabes may be produced by gummous affections of the cord, and S. believes quite a number of cases to be combinations of gummous and tabetic processes. A specific treatment for tabes is still to be found.— *Wiener Mediz. Blætter*, No. 43, 1890.   G. J. K.

DIETETIC TREATMENT OF EPILEPSY.—Dr. Jno. Ferguson in a lecture on this subject says: "The following remarks on diet are applicable to idiopathic epilepsy.   In this disease we have a form of convulsive movement that involves the highest centres.   An epileptic seizure is an excessive liberation of convulsive energy.   It is a well known law that all the nitrogen compounds are particularly likely to be unstable and of an explosive nature.   The brain substance contains much nitrogen and, as such, tends to be an unstable compound.   Acting upon this principle the author restricts his epileptic patients to vegetable diet, thus reducing to the smallest amount the quantity of nitrogen entering the system.   Under this plan it is difficult for the nervous system to build up the unstable molecules so ready to fall into new combinations and precipitate a discharge of energy.— *Ther. Gazette.*, January.

TREATMENT OF SCIATICA.—In regard to this subject Dr. Pritchard says: (*American Jour. Med. Sciences*, Jan.) "The establishment of the fact that neuritis is a pathological condition in certain cases of sciatica, has led to the belief that it is the only condition ever present.   The author believes, however, that there are certain cases of this affection which should be classed as neuralgic rather than neuritic.   In the neuritic type the pain is duller and less continuous, concluding in paroxysms. Movement of limbs, especially forcible extension, gives pain; anæsthesia rapid in onset at times and in limited areas. Sometimes swellings along course of nerve and tenderness, trophic changes, paresis or paralysis of muscles and reaction of degeneration. In the neuralgic form pain is more constant, and aggravated by movement, anæsthesia rare, no swelling nor tenderness, no trophic changes, paresis, paralysis, nor reaction of degeneration. In sciatic neuritis one of the best remedies is phenacetin in doses of 7½ grains given every four or six hours. A combination of iron, quinine, and arsenic is also good. One of the most important items in the treatment is absolute rest. The patient should be kept in bed, and if necessary, a splint applied to prevent the limb from being moved. This splint should be worn until the symptoms have disappeared. Either cold or hot applications as may be agreeable to the patient may be applied, with morphine if necessary. When the acute stage has passed, gentle massage is also advised. The continuous galvanic current should also be applied, not exceeding five milliampéres in strength Nerve stretching is valueless.''

A New System of Administration of Static Electricity. —Dr. W. J. Morton read a paper with this title at the Dec. meeting of the New York Neurol. Society. His new system comprised the development by an influence machine of a rapidly interrupted and graduated current by means of a circuit-breaker introduced into a circuit with and without condensers, and in the medical application of this current without and within the human body by moistened sponge or other electrodes, just as in the case of the ordinary galvanic and faradic currents. It involved the removal of the spark, in itself more or less disagreeable and painful, and often difficult to localize especially about the face and neck, away from the patient's body, and yet retained all the physiological effects of the kinetic or current part of the circuit. The spark was no longer a direct feature of the administration; it occurred at some distant part of the necessarily closed circuit, and in modified form now became mainly a regulator for timing the discharge of the equalizing potentials. The circuit-breaker consisted of a pair of metallic ball electrodes, introduced at any point of the circuit, having a narrow air space between the balls; the circuit "made" when a small spark overcame the resistance of the intervening air, and "broke" when it failed to do so, and the current was due to rapidly successive equalizations of the differences of potential of oppositely charged condensers, whether prime conductors or with the addition of Leyden jars. The circuit-breaker served (1) to afford time, infinitely brief, to the prime conductors and condensers if used to charge; (2) to regulate the frequency of the discharge, and collaterally the frequency of the succession of transient currents so that their aggregate might be classed as a steady current; (3) to determine the strength of the current. This latter might be varied at will and with the utmost nicety from a just perceptible to a most powerful effect. The spark circuit-breaker practically represented the vibrator in the primary of an induction coil, the specific inductive capacity of the air replacing the spring and its magnetic attractability. In describing the physical properties of the franklinic interrupted current the author said that it was neither a sudden and transient form, spark, or shock, nor an ineffective continuous flow, but a succession of relatively small sudden discharges. A single spark would produce but a single contraction instantly recovered from; a continuous flow, no effect. The franklinic interrupted current produced the effect of physiological tetanus. It therefore stood distinct and by itself as capable of producing a result unattainable by either the galvanic or the faradaic current. Applied to a motor point, the franklinic interrupted current produced most vivid and persistent muscular contraction with a minimum of pain; applied farther back on the trunk of a motor nerve, it threw large groups of muscles into contraction. The contraction was peculiarly painless as compared with that of faradaic coils, and the influ-

ence was remarkably diffusive.   Accompanying a contraction
of a large group of muscles was a peculiar sensation of light-
ness and buoyancy of the member.   It was applicable to every
form of muscular paralysis, for there was no practical stimulus
to nerve and muscle except the electric, and none more energetic than this form of it.   Its effects upon the Hallerian
irritability of the muscular tissue included an effect upon the
lymphatics, and to this might doubtless be referred many clinical results of relief, as in lumbago, and all forms of muscular
rheumatism, subacute and chronic rheumatic affections of
joints, ovarian or pelvic pain, sciatica or other neuralgias.
One of the characteristics of this current was its power of
relieving pain.   Gynæcologically, this system of conveying the
current within the cavities of the body opened out a wide and
promising field of clinical results.   From a very considerable
experience the author was satisfied that this current penetrated more deeply into the human body than did the galvanic.
In conclusion the points brought forward were: (1) The gen-
eralizing of what the author had announced as an isolated fact
in 1881, that a regulated interruption in the otherwise inoper-
ative circuit of a Holtz machine would produce in another
part a current adapted to electro-therapeutic practice.   This
current was now designated the franklinic interrupted current.
It included the adaption of the parts of a Holtz machine to
produce the results.   (2) A new electrode combining this current with the various terminals.   (3) The practice of intro-
ducing franklinic electricity in current form into the interior
cavities of the human body.—*Jour. Nerv. and Ment. Dis.*, Jan.

SURGERY AND TRAUMATIC NEUROSES.

CONTRIBUTIONS TO CEREBRAL SURGERY.—A. N., æt. 20, had
otitis medea suppurativa following influenza, with violent
headaches.   The right mastoid region was painful on pressure
at two points about 3 and 4 centimeters behind the meatus.

Occasional vertigo, but no spasms or focal symptoms. As the tympanic cavity filled immediately after cleaning, it was decided that it must communicate with an abscess cavity in the region of tenderness. On operation the periosteum was found adherent to the bone. The skull was opened at the level of the auditory meatus. A cavity of the size of a cherry was found, which communicated with the cavity of the cranium. On removing the bone over this, the dura was found discolored and not pulsating, with pus oozing through it. A probe readily passed through the dura, inward and backwards for a distance of 4 cm. into a cavity in the brain substance. A drainage tube was inserted. Rapid and uninterrupted recovery ensued. The only symptom pointing to an intracranial abscess was the painful spots. The case shows that it is important to uncover the entire area of the inflamed dura in suppurative pachymeningitis externa.

Case two was that of a lad of 18, who had had otorrhœa for 14 years. There were excruciating headaches, vertigo and staggering gait. Operation showed an extensive cholesteatoma of the mastoid cells, as well as an extensive extradural abscess, which required the removal of the mastoid process, the posterior inferior angle of the parietal and of part of the occipital bone for its exposure. Recovery was rapid.

Case three was that of a boy of 15, who had received several blows on the head with a club. On admission, 48 hours after, he was comatose and restless, with some rigidity of the muscles. The left side showed some paralysis, and its temperature was somewhat reduced. Examination showed two fissures of the right parietal bone. The bone between them was removed, exposing the pulsating dura covered by a thin clot, on removing which, a tear in the dura was found. This was enlarged by a transverse incision which evacuated a considerable quantity of clear fluid. Aspiration did not discover any pus, although brain debris was aspirated. The effect of the operation was good, the sensorium becoming clearer, the rigidity disappeared, although the left hand and foot remained paralyzed and the fever high.

Later on somnolence returned.   Examination showed bulging of the brain through the wound of the dura, and a pressure necrosis due to the hard edge of this membrane. No inflammation of the visible membranes.   As acute ventricular dropsy was suspected, an aspirator needle was introduced and several tablespoonfuls of fluid removed, but without result.   Death on 7th day after receipt of injury.   Autopsy showed purulent basal meningitis, which extended to the convexity at the left frontal lobe and the cerebellum, together with considerable softening at the seat of injury.—DR. EGON HOFFMAN, *Deutsche Medic. Wochenschr.*, No. 48, 1890.   G. J. K.

At a meeting of the London Obstetric Society, Jan. 7, Dr. Playfair read a paper *On The Removal of the Uterine Appendages in Cases of Functional Neurosis.*   His conclusions were as follows: 1.   That the removal of the appendages was not a legitimate procedure in cases of functional neurosis. 2.   That when marked structural disease of the appendages coexisted with several neurotic conditions, the latter should be treated in the first instance, in the hope that operation might be avoided.   3.   That in hystero-epilepsy and hysteromania the results of operation had been so unsatisfactory that it was a procedure of very doubtful expediency, and not to be recommended.   Spencer Wells concurred in all that Dr. Pritchard said.—*Lancet.*

CEREBRAL ABSCESS FROM OTORRHŒA.—Dr. Pritchard reports (*Arch. of Otology*) two cases of cerebral abscess with operation and recovery.   The first case was that of a man twentythree years of age, who from childhood had suffered from chronic ear disease.   A few months before the operation he had several attacks of partial loss of consciousness, and later . convulsive attacks with twitchings of left side of face.   Shortly previous to the operation he was drowsy and incoherent and

5

had repeated attacks of convulsions. The skull was trephined three inches above the meatus, and one-half inch in front
of it. Offensive pus was found outside the dura. Patient
completely recovered. The second case was very similar
except that the patient had marked delirium for some days
previous to the operation. The skull was trephined one and
one-quarter inches behind the meatus, and the same distance
above the cerebral base line. An abscess was found in the
cerebral substance which was evacuated and irrigated. This
case entirely recovered. After the operation there was transient word deafness in both cases.

TUMOR OF THE CENTRUM OVALE.—Dr. L. C. Gray reported
a case at the meeting of the New York Medical Society. The
patient, a man of 38, had been brought to the speaker from
Richmond, Va., by Dr. I. H. White, two weeks after motor
paralysis had begun in one lower extremity. As there was a
large tumor in Scarpa's triangle, and as an angieoma had been
removed some time before from the popliteal region, the
speaker had been led to believe that the symptoms had been
caused by an intracranial tumor. There was motor paralysis
of one lower extremity and paralysis of the muscular sense,
and slight headache, but no mental symptoms whatever. The
patient had been advised to go home and settle his affairs,
and to have trephining done if he grew worse. Two weeks
after he had been brought to New York, he was found to have
a motor paralysis and a paralysis of the muscular sense, also
of the upper extremity on the same side as the paralysis which
had first appeared. He was gradually becoming comatose.
The headache was very much worse. There were no changes
in the optic disc, and very slight impairment of the tactile
sense or motor or sensory disturbances beyond that mentioned.
The patient had been operated upon by Dr. Wyeth, but
neither by palpation nor by exploration could any tumor be
found. Death had occurred two days after the operation and
at the autopsy the growth was discovered as above described.

The speaker said that neurologists were divided in opinion as to the exact location of the muscular sense, and he thought that this case was unique as indicating, within a period of very rapid development, the precise locality.

TRAUMATISM OF THE BRAIN.—Drs. Richard Dewey and B. L. Riese of Kankakee publish the histories of two cases in the January number of The *Journal of Insanity*. The first case was that of an insane man who, apparently, drove a nail entirely through his skull, injuring the motor region of the brain, producing aphasia and paralysis of the arm opposite the side injured. An operation was performed and local inflammation was found and the nail removed. The second was the case of an insane person who was struck by a butcher's cleaver by another insane patient with a resulting fracture of the occipital bone, and exposure of the occipital lobes of the brain to the extent of three inches by two. The wound was dressed antiseptically and healed by first intention. The patient entirely recovered. It is interesting to note that immediately after injury the patient walked some distance and was not at any time unconscious.

LOCALIZATION OF INTRA-CRANIAL LESIONS.—In discussing this subject before The New York State Med. Soc., Dr. C. K. Mills of Philadelphia, said that the so-called signal or initial symptom, while of value in localization, is sometimes misleading. The motor signal symptom has, perhaps, as often misled, as it has correctly led the surgeon. In active irritative lesions, especially in their early history, it is an aid, but it may change as the lesion progresses. The early irritative lesion will give place after a time to a destructive one, and then the point of cortical irritation shifts.

It must be remembered that in every case of either hemispasm or monospasm (whether reflex, dural, nephritic, toxic, or hysterical), the spasm really or apparently begins with an initial symptom in the limb or face. This may indicate that the beginning of the cerebral discharge occurs in the area of the cortex, which is the seat of the representation of the movement, but it would be unwise always to operate with such indications,—the true cause of the spasm being either a general condition, or a localized one in some distant peripheral area. His experience would lead him to rely most upon the signal, or initial symptoms, in tumor and fracture, cases in which localized irritation is often extreme. In this connection it is important to remember that the centres for highly specialized movements have a high degree of irritability and hence are likely to be affected by lesions of adjoining areas. Lesions of the areas for grosser movements are more significant because the highly specialized centres are likely to discharge first. Conjugate deviation of the eyes and head has a localizing value only when it takes the form of an isolated tonico-clonic spasm, and stands alone as a manifestation. This conjugate movement is one which occurs synchronously with various lateral movements of the body, however produced, and it may be a coincidence, apparently primary, in a case where the real initial phenomenon is either in the upper or in the lower limb. If asked when conjugate deviation might be given a cortical localization, I should say, probably only when the movement takes the form of an isolated, tonico-clonic spasm, and stands almost alone as a manifestation. Some of the cases of spastic torticollis may be due to cortical irritation.

Another cause of failure in attempts to apply the principles of localization has been in giving relatively too much importance to motor localizing symptoms. Too much stress is sometimes placed upon these symptoms, particularly on more or less circumscribed spasmodic manifestations. Motor symptoms are the simplest, most readily understood, and those which have most frequently furnished a clue to the proper site for operation. Cutaneous sensory phenomena and symptoms belonging to the special senses are more likely to be overlooked, and, with one or two exceptions, are not as striking when noticed as those which are referable to the motor sphere.

An excellent rule for trephining in cases of brain abscess connected with aural affections is laid down by Barker, who goes so far as to say that nine-tenths of all the abscesses of the cerebrum are situated within a circle three-fourths of an inch in diameter, the centre of which is one and a half inches behind and above the centre of the meatus auditorius. This is, perhaps, putting it too strongly, but it is not far from the truth, and the rule bears upon the question of the mistake which is sometimes made in trephining, particularly when the operation is guided by motor symptoms.

When motor symptoms appear in such cases, they are usually the result of the extension of the abscess to a considerable distance from its place of origin, and also from its focus of greatest suppuration. If trephining is performed, it should, as a rule, be done one to three inches back of the point indicated by the motor symptoms, or else two trephinings should be performed, one from a consideration of the motor symptoms, and the other at the probable primary seat of the pus formation near the ear.

In cases of abscess—whether aural, traumatic, or from whatever cause—the possibility of distant infection causing either abscesses or purulent meningitis should not be overlooked. Occasionally in such cases either the primary or secondary abscesses are very deep seated. The abscess is usually solitary, although two or more may be present. The temporal lobe is the most frequent seat, next, the cerebellum. It is not advisable to operate for localizing symptoms in spastic or paralytic affections which are congenital, or date back to infancy. The greatest success of late years has been in trephining for endocranial hæmorrhage. Localized lesions associated with syphilis or alcoholic diseases of the brain do not justify operation. This, as a rule, applies also to tubercular lesions. Of 300 cases of brain tumor reported by Starr as occurring in persons under 19 years of age, 152 were tubercular. Eight of 20 cases of tumor of the brain reported by Osler were tubercular. Of 28 cases of my own, 7 were tuber-

cular, and in every one of these cases the disease was very
diffuse or multiple.　Dr. Mills especially condemned trephining for general paralysis of the insane.　Operations in this
disease can only be of temporary relief, if any.—*Jour. Nervous and Ment. Dis.*, Dec.

SECONDARY SUTURE OF PERONEAL NERVE.—Dr. Jno. B.
Deaver reports a case where the injury occurred during the
War of the Rebellion, an iron ball having struck the trochanter
major of the right side and penetrated to the bone.　A year
or two after the close of the war patient noticed a small lump
on the outer side of the right knee which caused pain when
struck, referred to the course of the external popliteal.　This
swelling increased until it became the size of an English walnut, and about a year ago was removed, but the ends of the
nerve were not sutured.　This operation was followed by
paralysis of the extensor muscles of the leg.　Dr. Deaver
sutured the ends of the nerve with catgut and aseptic silk.
Some months after, sensation returned, but no recovery of
motion.　It is probable that within a year or two motion will
be recovered, as there is no doubt of the union of the nerve.
Dr. Mills in discussing this case said that he had treated this
man for three or four weeks.　He made a careful electrical
examination, and found the reactions of degeneration.　There
was also active neuritis.　This was treated by mercurial and
belladonna ointment, freely applied, and the internal use of
salicylates and tonics, with rest part of the time.　Hot douches
were also employed.　This was followed by improvement.
He examined the man frequently, and on one or two occasions
thought that there were signs of return of motion in the
muscles.　His impression was that the man would eventually
get back power in these muscles.　The time that he was under
treatment was not sufficient.　Before the man returned to his
home he was provided with an apparatus similar to that used
in club-foot, the result of peroneal palsy; and with this he was
able to get along pretty well.—*Jour. Nervous and Ment. Dis.*,
December.

STRETCHING OF THE FACIAL NERVE IN CLONIC FACIAL SPASM.
—Dr. Robert Schott relates a case operated on by Prof.
Kraske. The case was of five years' standing, and all other
therapeutic measures were in vain.. The nerve was reached
by an incision 1 cm. in front of and parallel to the ramus of
the jaw. The presence of cicatricial tissue, the result of
hypodermic injections, rendered operation difficult. Four
hours after operation a complete paralysis of all the muscles
supplied by the upper and middle branches of the nerve was
noticed. The paralysis of the lower branch was not complete.
The paralysis gradually improved, until, four months after
operation, there was a complete relapse. Twenty-three other
cases have been found recorded. Five of these are valueless
on account of insufficient observation. In the nineteen cases
there were two cures; six were improved, the remaining eleven
cases being complete failures.—*Deutsche Medic. Wochenschr.*,
No. 44, 1890.   G. J. K.

STRETCHING OF SCIATIC FOR LATERAL SCLEROSIS.—Drs. W.
W. Keen and C. K. Mills reported a case in which the operation was unsuccessful. This patient, a young married woman,
when she first came for treatment, had the symptoms of an
incipient lateral sclerosis, the disease being in such an early
stage that some doubt existed as to whether we had not to
deal with an hysterical rather than an organic spasmodic
tabes. The sole symptoms were exaggerated knee-jerks,
slight ankle clonus, stiffness in the lower extremities, with a
slightly halting and stumbling gait. Here was a good opportunity, if ever, of curing spinal sclerosis, and the patient was
told that we would try to exhaust the measures known for its
treatment. She proved a persevering patient, staying with
Dr. Mills for several years, during which time she was treated
with all manner of drugs, with galvanism, massage, and the
cautery, and for several months by forcibly stretching of the
limbs without cutting down on the nerves. In 1883 the nerve
was stretched. Recovery from the operations was rapid and

perfect. After the effect of the ether passed off, she had some numbness at the plantar surface of the left foot, most at the toes, and also slight loss of sensation along the outer edge of the sole of the foot. The plantar reflexes were exaggerated, but ankle clonus, which was present before the operation, had disappeared. The knee-jerks were still marked, but less so than before the operation. Spontaneous tremors in the limbs were present at times, and she had occasionally pain in the left knee and ankle. These were the only points of interest in connection with the case after operation.

The woman remained for many months under observation. Her gait and reflexes seemed for a few weeks slightly better, but the improvement was neither marked nor permanent. The disease re-asserted itself, and steadily although very slowly got worse until she passed from observation.—*Journal Nerv. and Ment. Dis.*, Dec.

In the same journal Drs. Mills and Deaver reported a case of *Spasmodic Torticollis*, unsuccessfully treated by ligaturing of spinal accessory nerve. The nerve was exposed at·its point of exit from the sterno cleido muscle and ligatured with silver wire. Collier has reported in *The Lancet* a similar case in which the same operation resulted in complete recovery.

In the same journal Drs. Mills and Deaver report two cases of successful stretching of the sciatic nerve. In both cases complete recovery followed within a month of the operation.

TAPPING AND DRAINING THE VENTRICLES.—Dr. Mayo Robson reports cases on which he operated, and concludes with the following remarks:

1.   If there be time, a purgative to be administered the day before operation.

2.   The head to be thoroughly shaved and washed, and the site of the operation to be rubbed over with benzine.

3.   The chief convolutions, the fissure of Rolando, and the

base line to be mapped out, according to the needs of the case, with nitrate of silver.

4. A carbolic dressing to be applied, which is to remain on until the time of operation.

5. The anæsthetic, if one be required, to be the A.C.E. mixture.

6. The observance of strict antiseptic precautions.

7. In órder not to lose the landmarks after the flap has been turned back, the scalp to be perforated at the necessary point by a small drill, so as to make an easily recognizable mark on the skull which will show after the flap has been reflected.

8. If the trephine opening have to be a large one a semi-lunar flap should be reflected, but if, as I would propose in these cases of simple tapping of the ventricles, the trephine opening be only a small o..e, then a simple vertical incision through the scalp should be made about three-quarters of an inch or an inch in length.

9. If drainage have to be effected the trephine opening must be larger, and then the dura mater must be reflected by an incision passing two-thirds or three-quarters round the circumference, one-sixth of an inch from the bony margin of the opening.

10. The avoidance of puncture in the course of the main vessels and through the known important parts of the brain, if possible.

11. The needle or director to be removed and reintroduced for every fresh puncture, so as to avoid lateral movement when within the brain.

12. The punctures to be made in the course of the fibres and vertical to the surface.

13. If a drainage tube is to be left in, a fine director should be used, along which the small extremity of Professor Lister's sinus forceps can be pushed, so as to dilate the puncture, after which the drainage tube can be easily introduced by means of forceps.

Since writing this paper I have drained the lateral ventricles in a rapidly advancing hydrocephalus in an infant, but
unfortunately I did not allow sufficient length of drainage tube,
and when the brain began to shrink the tube slipped out of
the ventricle, and failed to drain properly. Dr. Fraser's
charts helped me considerably in this case also.—*Brit. Med.
Jour.*, Dec.

SURGICAL TREATMENT OF EXOPHTHALMIC GOITRE.—Case I.
Male æt 17. Required urgent tracheotomy on account of
compression of trachea. The symptoms were typical. The
middle and left lobes of the gland were removed. Seven
months afterward the right half was not noticeable and all
other symptoms had disappeared.

Case II. Male æt 47. Emaciated, pulse too rapid to be
counted, otherwise typical symptoms. Pulse became quiet
in chloroform narcosis. The right half of the struma, of the
size of a small fist, was removed with a great deal of hæmorrhage. Discharged in about ten days. Two months afterward resumed his work as a shoemaker, and has remained
well since (six months). Repeated examination showed a
gradual contraction of the visual field after the operation.—
Dr. F. Lemke, Deutsche Medic. Wochensch., No. 2., 1891.

G. J. K.

PSYCHOLOGICAL.

INSANITY IN THE COLORED RACE OF THE UNITED STATES.—
Dr. A. H. Wittmar read an article upon this subject before
The World's Medical Congress. He states that in 1860 there
were only 766 colored insane in the United States. In 1870
there were 1,822 of this class. In 1880 there were a little
over 6,000, which shows a rapid increase since emancipation.
He shows that the struggle for existence in freedom, and dis-
regard of the laws of health, are probably responsible for the
increased amount of insanity. He considers that the types
of insanity affecting the white and colored people are
essentially the same. Suicidal tendencies are unusual among

the insane colored people. The percentage of recovery among colored people is about the same as among the whites. —*Alienist and Neurologist*, January.

INSANITY IN AUSTRALASIA.—(From a letter by Dr. Ross, in the January number of *The Journal of Insanity*.) There are 10,639 insane in Australia. These figures include the native races. The recovery rate has been about 42 per cent. In New South Wales, during the past ten years, it has been found that

1 Irish in every 93 of the population was insane.

1 English in every 135 of the population was insane.

1 Scotch in every 155 of the population was insane.

1 Chinese in every 188 of the population was insane.

1 Australasian in every 579 of the population was insane.

The lunacy laws of the colony are based upon those of England, with such local changes as may be necessary. The institutions have at their heads legally qualified medical men. It is stated that their insane asylums are located as follows:

New South Wales, Gladesville, 795; Parramatta, 1,002; Callan Park, 772; Newcastle, 245 (special institution for idiots and imbeciles); Tempe, 108 (private institution, but has fifty state patients); Parramatta (criminal) 51; Victoria-Yarra Bend, 940; Kew, 1,150 (including idiots in special cottages); Ararat, 600 (criminals in special detached building); Beechworth, 550; Sunbury, 550, for chronic insane; South Australia-Adelaide, 250; Parkside, 530; Queensland-Goodna, 874 (includes a house for chronic insane at Ipswich); Toowoomba (recently opened); Tasmania, New Norfolk, 307; Cascades (convict insane); Western Australia, Fremantle, 132; New Zealand, Auckland, 389; Christchurch, 368; Seacliff, 496; Hokitika, 105; Nelson, 98; Wellington, 272; Ashburn Hall, 42 (private institution).

GENERAL PARALYSIS AT GHEEL.—To those who may not have heard of Gheel it may be well to state that there is a colony of several thousand insane people at this place in Belgium. These patients are cared for partly in private families, and partly in an institution. The chief medical director of the institution is Dr. Peeters, whose study of general paresis is based upon the admissions to Gheel from 1856 to 1885. The admissions for this term were 4,299 males, and 3,357 females, making a total of 7,656, of which 520 males and 148 females were paretic, a total of 668 paretics. There has been a gradual increase in the number of paretics admitted during that period. From 1856 to 1865 inclusive the number was 125; from 1866 to 1875 inclusive, 213; from 1876 to 1885 inclusive, 323. It will be seen that over 8 per cent. of all patients were paretics. A large majority of the paretics (270) developed the disease between 40 and 50 years of age. There were two cases between the ages of 70 and 80. The civil condition was as follows: single of both sexes, 215; married, 372; widowed, 77; unknown, 4. It is shown to be twice as frequent among men, and seven times as frequent among women living in towns as among those living in the country. Intemperance is given as the cause in more than half the cases in the men. In regard to syphilis as an etiological factor, he states that the histories of the cases are so imperfect that evidence on this subject is wanting. Only four of his cases were undoubtedly syphilitic. Female paretics live longer than male paretics.—*Bul. de la Soc. de Med. Ment. de Belgique*, March and June, 1890.

INFLUENZA–PSYCHOSES.—Dr. R. Jutrosenski, from a study of 104 cases from Prof. Jolly's clinic, reaches the following conclusions:

1. A nervous disposition is of importance in the causation of psychical disturbance by influenza.

2. These disturbances may occur at any period, although they are most frequent during convalescence.

3. The mental disturbance may assume any form, the melancholic and hypochondriac varieties predominating.

4. Sex has no influence.

5. The greatest number of cases occur between 20 and 50 years of age.

6. In the insane, influenza almost invariably causes increase of mental symptoms.

7. In cases previously healthy the prognosis is good.— *Deutsche Medic. Wochensch.*, No. 3, 1891.   G. J. K.

THE GERMS OF DELIRIUM.—This is an article by Dr. Tanzi of Genoa, translated by Dr. Joseph Workman. It is an attempt to explain the hallucinations and delusions of the insane as a reversion to normal ancestral modes of thought. In other words, the writer holds that, as a rule, the delusions of the insane are ideas that at some time in the history of the race have been normal conceptions. Concerning the delusions of persecution, he says the inclination of the paranoiac to invent persecutors in order to explain what happens to him, is well known. He thinks *everything* that happens to him to be due to some personality, like the child, or the savage, who finds a personality in every force, and attributes his own ideas to some individual. This element in savage mentality is illustrated by many examples. These morbid creations of the paranoiac are related to the fetiches of the negro, the evil spirits of the Ashantee, and the spirits, witches, goblins, dwarfs, etc., of other savages. In the middle ages all nature was demonized, and we see these normal beliefs of that time reproduced in the delusions of the insane. The author thus accounts for delusions of poisoning, religious delusions, ambitious delusions, the tendency of some insane to coin words with mysterious meaning, the prejudices regarding certain numbers, enigmas, conjurations, magic, double personality, etc. The theory of the author is, that these special supernatural beliefs were universal and normal in uncivilized man, and that later mental acquirements have dispensed with them, and that disease in attacking the higher and more recent acquirements allows these old and slumbering superstitions to reassert themselves. Insanity, therefore, if we understand the author, is a dissolution of the higher mental faculties, leaving the earlier developed and more stable faculties to be the prominent and controlling element of the mental life.—*Alienist and Neurologist*, Jan.

ILLUSIONS OF MEMORY (PSEUDO-REMINISCENCES) IN POLYNEURITIC PSYCHOSES.—The conclusions reached are:

1. Illusions of memory and delusions based thereon, are not infrequent in polyneuritic psychoses.

2. These delusions may be very stable, or extremely changeable.

3. One of the most frequent subjects of these delusions relates to death or burial.

4. These delusions may occupy the patient to such an extent as to present the picture of partial insanity.

5. The illusions of memory are usually based on the faint traces of some actual incident.

6. The "trace" of memories probably consists in the persistance of function in certain nervous elements, even if in a very slight degree.

7. The assumption is tenable, that these latent traces may be associated in definite groups.

8. The illusions of memory probably result from the unconscious association of these traces and their perception.

9. The illusions are probably caused by some defect in the association of these "memory traces," most likely by the omission of certain members necessary to normal recollection and ideation.

10. The discovery of the combination of such latent "traces" and of their influence upon ideation, would explain many interesting phenomena in normal and pathological psychology.—DR. S. S. KORSAKOW, *Allgem. Zeitsch. f. Psych.*, Bd. 47. G. J. K.

THE EVOLUTION OF DELUSIONS FROM IMPERATIVE CONCEPTIONS.—Dr. J. G. Kiernan writes on this subject in the *Alienist and Neurologist* for January. The tendency of recent literature is to regard imperative conceptions as distinct in kind from their morbid mental phenomena. Tuke states that imperative conceptions do not necessarily imply insanity. Kiernan shows, however, the intimate relation which exists between these phenomena and insane delusions, and gives illustrative cases showing the relation between imperative conceptions and delusions. Imperative conceptions are often recognized as such by paranoiacs. Hallucinatory phenomena of paranoiacs may long exist unassociated, but side by side with imperative conceptions. He endorses Ribot's view of the "ego," that it is a co-ordination and cohesion of the states of consciousness and accompanied by other states which though perhaps unconscious, are elements in the result. An imperative conception may occur in an otherwise healthy mind and never develop into delusions but disappear with improved health.

———

Dr. Robert gives an account of a boy whom he saw in the mountains near Barcelona, who manifested some rather peculiar mental symptoms. The doctor seems to consider the case a form of epilepsy, although we should be inclined to think it hysterical. There was no history of heredity, syphilis, or injury. The patient was addicted to the use of alcohol. He would suddenly quit his work and go out doors and stand and gaze at the sun for an indefinite time, or until taken away by force. If not removed, he would soon begin to scream and cry and throw himself about until he would fall into a state of collapse. Then he would return to his usual occupation. At other times the patient was normal.—*Revista de Med. y Cirugia Practicas*, Oct.

———

UNILATERAL HALLUCINATIONS.—In an inaugural dissertation Dr. Geo. Souchon supports the theory of Schuele that with an existing psychical disorder, a lesion of the peripheral sensory apparatus is pathogenetically important. If the peripheral lesion is amenable to treatment, the prognosis is good. —*Neurolog. Centralbl.*, No. 21, 1890.   G. J. K.

MASOCHISM.—Under this title Krafft-Ebing has described a peculiar form of sexual perversion. The name is derived from the German novelist Sacher-Masoch, who has treated of several subjects of it in his tales.

In the intercourse of the sexes the male is the conquering, aggressive part, and the female the resisting, or at most, passive part. The sexual act typifies the ascendency of one over the other. Among savage tribes we often find courtship to consist of abduction, with or without the infliction of blows.

Masochism is that form of sexual perversion in which the male receives a certain, at least mental gratification, by being maltreated by the female. It is the reverse of sadism, which inflicts brutalities.—From further studies in Psychopathia Sexualis; *Wiener Mediz. Blætt.*, No. 52, 1890, p. 818. G. J. K.

———

DIABETES IN INSANITY.—Dr. G. H. Savage read an article before the meeting of the London Medical Society on this subject. He said: Most diseases had a special mental aspect; thus with disorders of the digestive tract depression was common, while the mental tone of consumptives was bright and hopeful. With diabetes there was a tendency to a depressed mental tone, and beside this there were some specially interesting connections between the two disorders. He did not claim all cases of diabetes as of neurotic origin, but he believed that some were. The facts which pointed to the central nervous origin of some cases of diabetes were overwhelming. Experiments on animals, disease and injuries in man, had shown that brain disease did in some cases excite the disease. Writers on general medicine admitted that there were similarities between the causation of diabetes and that of nervous disorders; only very few asylum physicians had, however, recognized the relationship which he believed to exist. Dr. Clouston of Edinburgh, however, had described a class of diabetic melancholiacs; and Dr. Maudsley, in his Pathology of Mind, had clearly pointed out that there was a very distinct connection by origin between diabetes and neuroses.

Diabetes was to be regarded as a group of symptoms which for convenience had to be looked upon as forming a fairly definite diseased process. Among the insane he had not found sugar at all commonly in the urine. The facts upon which this paper was founded were drawn chiefly from Bethlem Hospital. He had records of forty patients in that hospital who had diabetic relations, ten of them·having diabetic parents or grand parents, fourteen having diabetic brothers or sisters, twelve, aunts or uncles, and three, cousins suffering from this disease. Besides these there were twelve insane patients who had both insane and diabetic relations, and ten patients who were both insane and diabetic. Nearly all the cases with diabetes and insanity were melancholic. The patients who had diabetic symptoms and then became insane almost all lost some, or all, of these symptoms during the period of mental disorder. There appeared to be a kind of alternation between the two disorders. He submitted cases in which acute diabetes was replaced by acute melancholia, this latter giving place again to diabetes, which once more had been replaced by temporary mental depression. In two other cases elderly men suffering from diabetes became melancholic some time before the fatal termination of the disease, and during their mental disorder no sugar appeared in the urine. In one case of general paralysis of the insane the diabetic symptoms, which had lasted two years at least, disappeared for over a year, only to return a few months before death. In conclusion, he pointed out that similar causes might give rise to either insanity or diabetes, that diabetes occurred in the same families as did insanity, and that there might be an alternation, so that insanity occurred in the one generation and diabetes in the next; again, in diabetes itself the symptoms, one or all, might be replaced for a longer or shorter time by mental symptoms.—Dr. Beevor said he had examined the urine of a large number of cases of epilepsy for sugar with negative results, and there seemed to be no especial relationship between diabetes and epilepsy.—Dr. Angel

6

Money pictured to himself in patients about to become insane a nerve perturbation traveling about from one part to another and producing, when settling down, either diabetes, general paralysis, or paroxysmal neuralgia.   It was curious that the diabetes disappeared so completely in these cases —Dr. Sidney Philips asked if diabetes were common in epileptic mania.—Dr. Hadden inquired if glycosuria were included in the paper.—Sir William Roberts said that the urine of apoplectic people was often glycosuric, and these cases should not be classed with real diabetes.—Dr. Savage, in reply, said that sugar was not common in epileptic mania.   It was very difficult to separate glycosuria from diabetes in insane cases, for directly they mixed their characters became masked.   Some of the patients certainly suffered with true diabetes, but others were glycosuric.—*Lancet*, Nov.

MONOMANIA.—Dr. H. M. Lyman of Chicago read a paper upon this at the December meeting of The Medico-Legal Society.   The author prefers the term "paranoia" to that of "monomania."   He holds that we are now in a position to restrict the use of the term to persons who are congenitally defective, and who are dominated by one or more systematized delusions, so that their reasoning faculties remain in a large measure intact, the principle difficulty being that they reason wrongly from certain data which are presented to their hampered minds.   We may say there is the genus paranoia.   Then there are species in which certain forms and delusions are most prominent; for example, there are patients dominated by delusions of an expansive character, "grand delusions;" others in which the delusions are depressive; others in which these two characters are combined; others entertain delusions of a religious or erotic nature.   The paper contained a number of illustrative cases.   In the discussion of the paper, Dr. D. R. Brower said that the great majority of paranoiacs had, in his opinion, a certain degree of responsibility.   Paranoiacs should, to a certain degree, be held responsible for crimes,

the degree of responsibility must be determined in each case.
Paranoiacs are dangerous characters, and are sooner or later
homicidal.   Concerning the diagnosis of paranoia, it is important to consider that the defectiveness applies to the physical
as well as to the mental organization.   They are defective in
the shape of the head, in the eyes, in the arms, and in the
legs.   Dr. Church said that he believed Peter the Hermit,
John Bunyan, and Swedenborg, were paranoiacs.   He called
attention to the fact that the great crimes done by insane men
are done by those who are trusted.   Many paranoiacs have
hallucinations of hearing which render them especially dangerous.   Systematized, persistent delusions with hallucinations of hearing, justify an unfavorable prognosis.

INTERCURRENT DISEASES AND SERIOUS INJURIES IN RELATION
TO RECOVERY FROM INSANITY.—Dr. W. T. Granger read a
paper on this subject at the January meeting of The New York
Neurological Society.   In 2,000 cases of insanity the author
had never seen acute articular rheumatism, and but three
cases of pneumonia, three of typhoid fever, and one of
diphtheria.   Sore throat was frequent, erysipelas not uncommon, epidemics likely to occur.   Pain has little effect upon
the mental condition of the insane; sometimes it aggravates
it; sometimes it is borne with indifference.   In the author's
experience there was not infrequently temporary improvement
in the insane when afflicted with intercurrent diseases.   Permanent improvement of the insanity following acute disease
must be very rare.—*Jour. Nerv. and Ment. Dis.*

THERAPEUTICS.

HYPER-ALIMENTATION IN ACUTE DELIRIUM.—Dr. Guicciardi,
Congress of Novara (reported in *Il Pisani*, XI., 1890).   In
eleven female subjects of acute delirium, the author had only
four deaths, and of the other seven there were five complete
cures, and two incomplete (*dementia consecativa*).   In these

cases the treatment used, aside from injections of ergotin,
was methodic, and ample alimentation by means of the feeding
tube.   This was practiced four or five times in a day, whenever the patient was in a suitable condition to receive it.   The
good results of this treatment were probably to be explained
by the fact, that death in this disorder results from exhaustion
of the nervous system, and the feeding assists in supporting
the system against the disease.   This is not at all inconsistent
with the theory, that the disorder is due to an infection.

———

ON THE SEDATIVE AND HYPNOTIC ACTION OF ATROPHIA AND
DUBOISIA.—Dr. Nic. Ostermeyer, as the result of practical
observations upon insane patients reaches the following conclusions in regard to atropia:

1.   In the insane, atropia has primarily a sedative, and secondarily a hypnotic action.   The latter is due to the paresis
of the sensory and motor nerves and the reduction of reflex
irritability which produce a predisposition to sleep.

2.   In regard to certainty and intensity of action, atropia is
far inferior to hyoscin, but has the advantage of not causing
the same degree of bodily lassitude.

3.   Repeated doses cause tolerance, with reduced or absent
action.

4.   Unpleasant symptoms, consisting of diarrhœa and
emesis of short duration, were observed once in 14 cases,
treated by the alkaloid.

5.   Atropia often acts as a sedative where morphia and
hyoscin fail.

6.   Owing to its uncertainty of action, it is not adapted to
general use.

In regard to duboisia, his conclusions are:

1.   Duboisia sulphate, is, like hyoscin, a sedative and
hypnotic of prompt and intense action in psychoses attended
by excitement, without the unpleasant sequelæ of hyoscin.

2.   Its action is frequently manifested in from 10 to 15
minutes.

3. Duboisia sulphate is principally hypnotic, sleep occurring in from 20 to 30 minutes, in most cases.

4. The efficient dose is from 2 to 3 mg., in states of great excitement; 1 to 1.5 mg. in simple agrypnia.

5. Symptoms of intoxication, or sequelæ, were not observed.

6. Tolerance is established by continued use. Increase of dose, or pause in the administration of the drug, are then indicated.

He advises the substitution of duboisia for hyoscin in cases of cardiac or vascular lesions.—*Allgem. Zeitsch. f Psych.*, Bd. 47.   G. J. K.

On the Action of Hyoscin in the Insane.—As the result of 914 doses of this drug given to 18 patients, Dr. Serger reaches the conclusion that "its value as a sedative is very problematical," and agrees with Grauck that it is not suitable for use in the insane. Its disadvantages are: Its inconstant action, variable dose, early establishment of tolerance, short duration of its action when it does act, and its unpleasant sequelæ, such as dry throat, difficulty of swallowing, hebetude and weakness, and depressing action on the heart.—*Allgem. Zeitsch. f. Psychiatrie*, Bd. 47.   G. J. K.

The Hypnotic Action of Effervescent Mixtures (Tartaric Acid and Bicarbonate of Soda).—Umberto Stefani, *Archivo Italiano per la Malatie Nervose*, Nov., 1890, publishes a long experimental investigation on the hypnotic action of effervescent mixtures, from which he concludes that it exerts a hypnotic action, perhaps inferior to that of hyoscyamus and probably less than that of sulfonal or uralium, but always a decided action, and that it can be recommended in the milder forms of insomnia, the more so since with the exception of some slight gastric phenomena, it produces no disturbances of the system, is without danger, and the sleep caused by it is in all respects like natural sleep.

CHLORALAMID.—"Dr. G. Genersich has prescribed Chloralamid in thirty-two cases, giving thirty grains at night.   This
dose was generally sufficient to induce sleep within half an
hour.   A more certain effect and a longer sleep was obtained
when forty-five or sixty grains were prescribed.   He considers
Chloralamid preferable to other hypnotics, both because it acts
more decidedly and because it is less unpleasant to take.   It
must be remembered that its effect is negative when sleeplessness is due to pain.   It is not by any means a dangerous
drug, but headache and vomiting may occur after a very large
dose.   It does not seem to affect the digestion nor the renal
functions.   The pulse generally becomes softer and more frequent."—*Med. Record.*

———

CHLORALAMID IN INSANITY.—Dr. Umpfenbach reports (*Ther.
Monatsh.*, No. 10, 1890), on the use of this drug in doses of
2 to 6 gm. in 55 cases (14 cases of mania, 13 paranoia, 3
melancholia, 8 paralytic dements, 9 cases of excitement in
dementia, and 8 of epileptic excitement).   Success was
attained in 30 cases, no result in 12 cases, and but slight
results in 13.   It acted best in paranoia, paralysis, and the
short excitement of epilepsy, least in the excited dements and
the fresh maniacs and melancholiacs.   The remedy may be
given for months and has no after effects.   It was also given
in one case each of essential tremor, hereditary chorea, and
disseminated sclerosis, the tremor and chorea being improved
by doses up to 2 gm. per day.—*Wiener Mediz. Presse*, No. 47,
1890.   G. J. K.

———

GALVANOTHERAPY OF THE BRAIN.—Herman Gessler (Wür-
temberg *Med. Correspbl.*, No. 21, 1890) recommends that the
current be directed from the forehead to the nucha.   Although
this does not admit of reaching the diseased focus with the
greatest intensity of current, it obviates the unpleasant consequences frequently observed.   He uses flexible copper elect-

rodes 6 x 15 cm. He recommends that central galvanization be tried in all cases of apoplexy, though not earlier than 3 months from the seizure.—*Wien. Mediz. Presse*, No. 43, 1890,

G. J. K.

Deleterious and Toxic Action of Sulfonal.—Dr. H. Bresslauer reports 7 cases of sulfonal poisoning (5 ending in death) among 74 cases of various troubles treated by the drug. The primary symptom was usually constipation, obstinate and prolonged, requiring frequent high injections. The action of the drug, frequently postponed on account of its insolubility, suddenly became manifested in deep stupor, vomiting, weak and rapid pulse. The urine became scanty, of a dark red to violet color, (indican), and contained epithelial and hyaline casts, as well as changed blood coloring matter.—*Wiener Med. Blætter*, 1891, No. 1 and 2.   G. J. K.

MEDICO-LEGAL.

Two Cases of Traumatic Hysteria.—Dr. Henry Hun read a paper with this title at the recent meeting of the New York State Medical Society The first case had resulted from a railway collision, and the second from a severe fall in the street. In the first instance there had been hysterical convulsions, temporary insanity, hemianæsthesia, and paralysis of motion and sensation in one leg. Both cases had presented many symptoms typical of hysteria. Seclusion in a hospital and vigorous and painful treatment had been necessary before any improvement was manifested. The first patient had been cured, and the second greatly improved. In the first case the cure had not taken place for more than a year after substantial damages had been awarded. Indeed, both patients had been awarded large damages, but the progress toward recovery had been in no way modified by this fact.   Simulation had been carefully considered and excluded. The speaker said that the question as to the amount of money which was to form just compensation for an injury received was always a difficult one to answer, and this was especially true when in regard to injury of the nervous system, whether organic or functional.   In the case of organic nervous disease —such as cerebral tumor, myelitis, neuritis, etc., which might result from an injury—these lesions were always of such slow development that several years might elapse before the symptoms were pronounced, and therefore the injury to the nervous system was not fully manifested till long after the question of damages had been settled.   The question was not less difficult in the case of those functional diseases of the nervous system resulting from injury, and the difficulty in these arose partly from the danger of deception, because many of the symptoms were easily feigned, and the patient had great temptation for such feigning of symptoms, and partly from the great obscurity of the pathology of these diseases.   These cases of functional nervous diseases, which were described by Erichsen under the name of spinal concussion or railway spine, were now generally regarded as of cerebral origin, and known by the name of traumatic neuroses.   They depended in their etiology quite as much on the fright as on the physical injury. In conclusion, it was suggested that, if any claim for damages was made on account of a functional nervous disease resulting from injury which was due to the negligence of a corporation or an individual, it would be wise if such corporation or individual were to offer to pay for the special hospital treatment of the patient in the hope of obtaining a rapid cure.   In a considerable number of cases the offer might be accepted, and in the other cases the fact that the offer had been made would put the defendant in a better light before the jury, for no expert could deny that, had a course of treatment been adopted, the patient would have stood a better chance of recovery.   Furthermore, on deciding upon the amount of compensation, it should be remembered that a considerable number of these cases were easily and rapidly cured under proper treatment, especially after the question of damages had been definitely settled.

THE COMEDY OF HYPNOTISM.—Prof. F. Fuchs of Berne, in an article with this title, expresses the opinion that most subjects are deceivers, either wittingly or unwittingly. The consideration with which a "good" subject is treated makes it an object for impostors to acquire proficiency, so as to be assured of an easy life in a hospital. He relates a number of cases which he observed in clinics, which seem to prove his point. The tone of the article indicates that the author believes the physicians themselves to be the subjects of auto-deception, although the editor, in a note, expresses the opinion that it is intended to ridicule only certain extravagant claims.—*Berlin Klin. Wochensch.*, No. 46, 1890. G. J. K.

Dr. C. F. MacDonald has an article in *The Jour. of Insanity* for January, on "Recent Legislation for the Insane in the State of New York." It is a resumé of some of the investigations into county insane asylums, which led to recent legislation. He concludes as follows: Among the less obvious, but not less valuable, effects of the new law may be noticed:

First.—The principle of State care for the insane (already adopted by the State in 1836, and given wider scope in 1865), is not only reaffirmed by the act of 1890, but, going beyond all previous legislation, carries the principle to its legitimate conclusion by committing the State to the entire support of its insane wards. That the insane are the wards of the State has long been established, but not until now has the State undertaken to fulfill the obligations of guardianship, by providing solely and entirely for the maintenance of these wards. This is perhaps the most important feature of the new law.

Second.—By districting the State, and by obliging each State hospital to receive all the insane in its district, the recent legislation destroys at once the unscientific and pernicious system of making a legal distinction between acute and chronic cases of insanity.

Third.—By accommodating the insane in small, detached cottages, the new law provides for the classification of the insane upon a medical basis, almost unlimited in its possible subdivision and extension; and also gives to the medical officers of large asylums the long desired opportunity of *individualizing* the treatment of insanity.

Fourth.—The interests of the tax payers have been guarded in the new law, as never before, by limiting the cost of building, equipment and furnishing to a fixed rate *per capita* sum of moderate and proper dimensions, namely, $550.

Finally.—The act of 1890 not only makes it obligatory upon all counties of the State, but three—and with these three it is permissive—to place all their insane under the care of the State, but makes it for the financial interest of each county to do so, thereby creating a State system of care for the insane, which contains within itself the elements of self-perpetuation and extension.   Thus the State of New York, in its wisdom, has finally and unequivocally determined that it will assume the care and control of all its insane who are unable to obtain private care.   This determination has been slow in coming, but it is believed, by those best qualified to judge, that it will stand.   The people of this commonwealth are strong in the faith that "nations are never impoverished by the munificence of their charities;" and that failure on the part of a State to make provision for all its dependent inmates is "poor economy and worse philanthropy."

CRIMINAL STATUS OF INEBRIETY.—Dr. T. L. Wright, in closing a long article on this subject says:

First.—Drunkenness appears to be true insanity.

Second.—It is an insanity augmented and intensified by several efficient causes of mental disorganization operating together, and in the same direction.

Third.—It is an insanity unbroken by lucid intervals, for there is no abatement of the toxic impression of alcohol as long as that poison remains in the system.

Fourth.—Drunkenness is not able to hide itself, or even disguise its own features—in other words, the drunken man cannot control his actions.

Fifth.—The law declares intoxication to be a crime, yet it affords and protects facilities for unlimited intoxication.

Sixth.—The law declares that criminal courses shall not excuse other crimes growing out of them; yet it provides the conditions for the establishment of criminal courses.

Seventh.—The public in its aggregate capacity knows that drunkenness is unable to control its own actions, yet by its permissive attitude respecting drinking resorts, it panders to the morbid, or possibly, vicious appetite for intoxication, thus becoming itself materially responsible for the crimes of alcohol.

Eighth.—When the law declares that *voluntary drunkenness is no excuse for crime,* the word "voluntary" becomes amenable to criticism. When the dipsomaniac, or even the vicious idler, is invited, lured, seduced and beguiled into drinking shops by the blandishments and the sensuous attraction of such places, the public, under whose auspices they are conducted, is a party to the inevitable results. The diseased or the perverted will, yielding to temptation under such circumstances, should not, in fairness and honesty, be esteemed to act with freedom; to act *voluntarily.*—*Alien and Neurol.,* Jan.

New Medico-Legal Questions Relating to Inebriety.—Dr. T. D. Crothers, in discussing the legal status of the insane, closes his article as follows: "The general problems which are presented in these medico-legal cases are:

First.—Was the person an inebriate or one who drank spirits at all times or at intervals? If this fact is established beyond question, his sanity and mental capacity may be most reasonably and naturally doubted.

Second.—What was the mental condition and circumstances of the person at the time of the commission of the act in in question? Was he sane? Was the act reasonable and just in its effects and consequences? If not, the first suspicion is strengthened, and the irresponsibility of the person must be assumed, and the legal theory must be reversed; the sanity must be proven, not the responsibility.

Third.—The medical man has only to gather the facts and have the reasonable assurance of their accuracy. From this he can point out the most reasonable conclusions which are sustained by such facts. The question is one of preponderance of evidence, which, if it points to defective consciousness of act and conduct, and inability of control, is far more likely to indicate impaired mind, or insanity, than any other condition. The limits of scientific study will not sustain and will not support assumption of boundary lines of responsibility and irresponsibility."—*Alien. and Neurol.*, Oct.

NATIONAL AND STATE CHEMISTS IN THE COURTS OF LAW.—In an article upon this subject Mr. Clark Bell suggests that congress should create public officials to be called the chemists of the nation, whose duty it should be to conduct all investigations in supposed poisoning cases. The writer asks the co-operation of the National Association of Chemists of America, in securing congressional action upon this subject. —*Medico-Legal Jour.*

NEW BOOKS.

Twelve Lectures on the Structure of the Central
Nervous System, for Physicians and Students, by Dr.
L. Edinger; Second Revised Edition with 133 Illustrations; Translated by Dr. W. H. Vittum; Edited
by Dr. C. E. Riggs; F. A. Davis, Publisher.

A long-felt want is supplied by the appearance of this book.
Cerebral anatomy is no longer the hobby of a few neurologists
who take particular delight in unravelling the mysteries of
brain-structure, but it has become a subject of the greatest
importance for every intelligent general practitioner. The
day is not far distant, if it has not already arrived, when it
shall be justly demanded of every medical man that he recognize promptly, cases of cerebral and spinal diseases calling
for immediate surgical interference; but to be able to do this
he must have some, and a sufficient knowledge of the structure of the central nervous system. Such knowledge the book
under review supplies, and it is not knowledge at second
hand that the author gives. He is well known to neurologists
as one of the most promising brain anatomists of the present
day, and his thorough grasp of the subject is shown on every
page of this book. It is a book "for physicians and students"
as the title page puts it; the latter will get all the information
they desire on any branch of this subject, and the former will,
in special paragraphs added to each chapter, find the practical
bearing of the diagnostic importance of the facts therein
discussed.

The author has shown his own good judgment by omitting
theories, and giving only well founded facts. His own
researches he has introduced carefully, as in the chapters on
the Embryology and Comparative Anatomy of the Brain, and
on the Spinal cord tracts. He claims for them only such consideration as they justly deserve. We consider the intro-

duction of such matter an excellent feature of this book. The reader feels that the author is not a mere compiler, but an independent worker, and the student furthermore gains an insight into the points and questions which need special elaboration.

The arrangement of the book differs also from the "cut and dry" arrangement of older anatomies. The author gives first a review of the methods which have led up to our present knowledge of cerebral and spinal cord structure. He then begins his subject proper with a chapter on the Embryology of the central nervous system in which the fore-brain, and its connections with the other parts of the cerebrum are treated with a clearness that no other author has been able to attain. This chapter is followed by a chapter on the general conformation of the brain, and one on the convolutions and fissures of the surface of the cerebrum. This chapter is written largely on the lines laid down by Ecker; in the next edition we have no doubt that Eberstaller's work will be considered. The theory of localization is impassionately considered, and we may quote from this part of the book to show the author's views and the excellence of the translation:

"It is evident that all motor, and many sensory-psychic functions, may start from deep-lying centers, but the higher we ascend in the scale of animal life, the more is the cortex concerned in cerebral activity, and consciousness plays a more important part. * * * In mammals all possible degrees of variation are observed. Thus is explained the fact that irritation of certain tracts of the cortex will give rise to muscular action, and yet movements may be executed after those particular parts of the cortex have been removed. In man the greater part of the surface of the hemispheres has become indispensable to the proper performance of such movements."

In the following eight chapters the minute structure of the brain and spinal cord is carefully presented. With the aid of excellent illustrations, every level of the central nervous system is made clear to the student; even the subthelamic region in Edinger's hands becomes a terra cognita. We know of no better presentation of the origin of the cranial nerves than is contained in Lectures XI. and XII., and Fig. 132, showing the origin of the cranial nerves should be freely used for class room lectures and be allowed to supersede the somewhat antiquated diagram from Erb's large treatise on Diseases of the Nervous System. The final summary of "brain tracts" is a fitting close to this capital little book.

A now famous teacher of Anatomy was in the habit of saying of Hyrtl's text book, that it was the only anatomy one could read in bed; it contains jokes and anatomy; the student is apt to remember the jokes and to forget the anatomy. Edinger presents his subject in such a fascinating way that you can read his book, if not in bed, at least in an easy chair, and you will get the gist of the whole book too; and yet it is pre-eminently a book for close study.

The translators, Dr. Riggs and Dr. Vittum, have done their work faithfully and well. The book does not read like a translation, and the reviewer after a careful perusal can say that he has found no error of any importance. The translation is worthy of the original; it needs no higher praise.

The lectures are well printed, and the illustrations from the original plates are brought out distinctly; but we do wish the publisher would hereafter use thicker paper, and by way of compliment to the æsthetic taste of the average physician, he might hereafter abandon the variegated paper on the inside of the cover, which is better fitted to the walls of an old-fashioned tea-store than to the cover of a serious medical book.

NEW BOOKS IN GERMAN.

Baer—Drunkenness and its Prevention (adv.).

Eberstaller—The Frontal Globe of the Brain (adv.).

Preyer—Hypnotism (adv.).

Wetterstrand—Hypnotism and its Applications in Practical Medicine (adv.).

Forel—The Drinking Customs in their Social and Hygienic Aspects (adv.).

Ziegler—Contributions to Pathological Anatomy and General Pathology (adv.).

Kölliker—Injuries and Surgical Diseases of the Peripheral Nerves (adv.).

Jastrowitz—Treatment of Insomnia (adv.).

J. L. A. Koch—The Psychopathic Deficiencies (adv.).

Ziehen—Handbook of Physiological Psychology (adv.).

Krafft-Ebing—Further Contributions to the Study of Psychopathia Sexualis (adv.).

Werner—Religious Insanity (Popular, not Scientific), Review.

Lombroso—The Man of Genius; translated by Fraenkel. (Review.)　Classifies genius as "a degeneration psychosis of the group of moral insanity."

Chatelain—Insanity, from the French, by O. Dornbluth. Preface by Krafft-Ebing (adv.)

Daxenberger—Gliomatosis of Cord and Syringomyelia (adv.)

Eichhorst—Manual of Special Pathology and Therapy, Vol. III.　Dis. of Nerves, Muscles, and Skin (adv.).

Freud and Rie—Clinical Studies on the Cerebral Hemiplegia of Childhood (adv.).

Macalester—Sarcoma of the Cord and its Membranes (adv.).

Singer and Münzer—Contributions to the Anatomy of the Central Nervous System (adv.).

Wilheim—Electricity and its Application in Neurasthenia and Allied States (adv.).

Dohrs—A Case of Pachymeningitis Hæmor. Intern., with Large Bilateral Hæmatoma (adv.).

Knoke—Subnormal Temperatures in the Insane (adv.).

Lehr—Vasomotor Neurasthenia and its Treatment (adv.).

THE HUMBOLDT LIBRARY.

Two recent issues of the Humboldt Publishing Co. are of special interest. One is Björnström's work on Hypnotism. This is a historical and critical review of the entire subject and deserves to be widely read. It is untechnical and as interesting to the laity as to the profession. The other volume is Mentigazzi's work on the Physiognomy of Expression. This is the most recent work on the subject and it is considered in the light of modern science. This series of publications is valuable and their cheapness (15 and 30 cts. per volume) puts them within the reach of all. The publishers have adopted the right course to popularize science.

ILLUSTRATIONS

BY

SANGER BROWN, M. D., CHICAGO.

Experiment upon the brains of living animals and clinical observation have together fairly established the location of the motor areas as represented in the accompanying charts, and the same may be said of vision and general sensory perception. But up to the present time there has been no satisfactory evidence of any kind to show, as was formerly asserted, that the angular gyrus is concerned in vision (and I have therefore not included it) and the centres indicated as presiding over taste, smell, and hearing must be regarded as highly hypothetical.

EXPLANATION OF SHADING ON PLATES: The motor areas are all related to movements on the opposite side of body, and the same is true for the area for general sensibility, marked "sensory." The occipital lobe supplies functional sensibility to half of each retina on the same side.

The hypothetical areas of vision (angular gyrus) and hearing are described as being crossed.

SANGER BROWN.

Fig. 1. *Fissures.—c. f.* Central, or Fissure of Rolando.—*s. f.* Fissure of Sylvius.—*p. c. f. s.* Superior, and *p. c. f. i.* inferior Præ-central.—*f. s.* Superior, and *f. i.* inferior frontal.—*p. t. c. f.* Post-central, practically the anterior part of *i. p. f.* the intra-parietal.—*c. m. f.* Marks about the position of the upper end of the calloso-marginal.—*p. o. f.* Parieto-occipital.— *t. r. o. f.* Transverse occipital.—*l. o. f.* Lateral occipital.—*s. t. f.* Superior temporal, or parallel.—*m. t. f.* Middle temporal.

Lobes.—F. Frontal.—*P.* Parietal.—*O.* Occipital.—*T.* Temporal or Temporo-sphenoidal.

Convolutions.—a. f. Ascending frontal, or præ-central.—*fl.* Superior,— *f2.* middle, and—*f3.* inferior frontal.—*a. p.* Ascending parietal, or post-central.—*s. p.* Superior parietal (lobule) continuous on the mesial surface with the præcuneus (*Q.* outline of Fig. 2.).—*i. p.* Inferior parietal (lobule) continuous in front with—*s. m.* the supra-marginal, and behind with—*ang.,* the angular (pli courbe).—*s. o.* Superior,— *m. o.* middle, and—*i. o.* inferior occipital. The division of this region into convolutions is inconsistant and unsatisfactory.—*s. t.* Superior,—*m. t.* middle, and—*i. t.* inferior temporal convolution.

Fig. 2. *Fissures.—s. f.* Sylvian.—*c. c. l. f.* Fissure of corpus callosum. *c. m. f.* Calloso-marginal.—*c. f.* Central, or Fissure of Rolando, its upper end.—*p. f.* Paracentral.—*s. p. f.* Subparietal.—*p. o. f.* Parieto-occipital.— *c. l. f.* Calcarine.—*i. o. f.*—Inferior occipito-temporal, or collateral.—*h. f.* Hippocampal, or dentate.

Lobes.—F. Mesial aspect of frontal.—*Q.* Quadrate, or præcuneus.— *C.* Cuneus.—*O.* Occipital.—*T.* Temporal.

Convolutions.—f. i. Superior frontal.—*p. c.* Paracentral lobule.—*g. f.* Convolution of corpus callosum, or gyrus fornicatus. That part of it lying on the upper aspect of the corpus callosum is sometimes called the gyrus cinguli. It extends from below the genu in front, to below the splenium behind, where it becomes rather suddenly narrowed to form—*i.* the isthmus, and then continues on as—*h.* the hippocampal convolution (subiculum cornu Ammonis, or uncinate convolution).—*u.* Uncus.—*g. d·* Gyrus descendens.—*o. t. m.* Middle, and—*o. t. i.* Inferior, occipito-temporal.—*f. d.* Dentate, or fascia dentata.

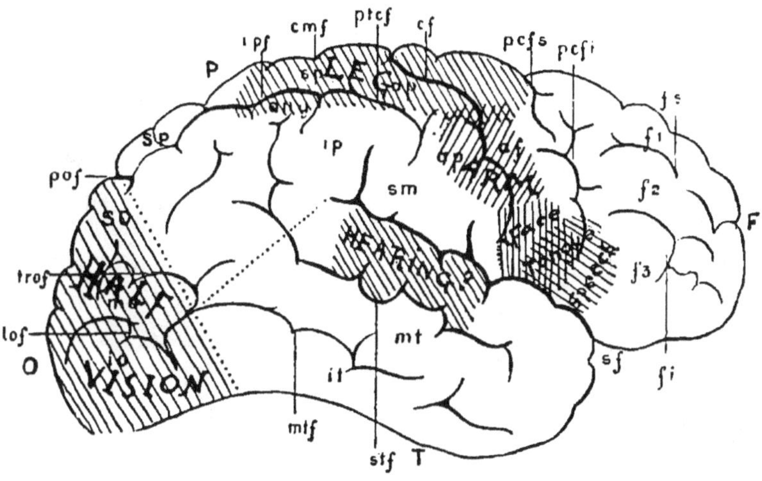

FIG. 1.

FIG. 2

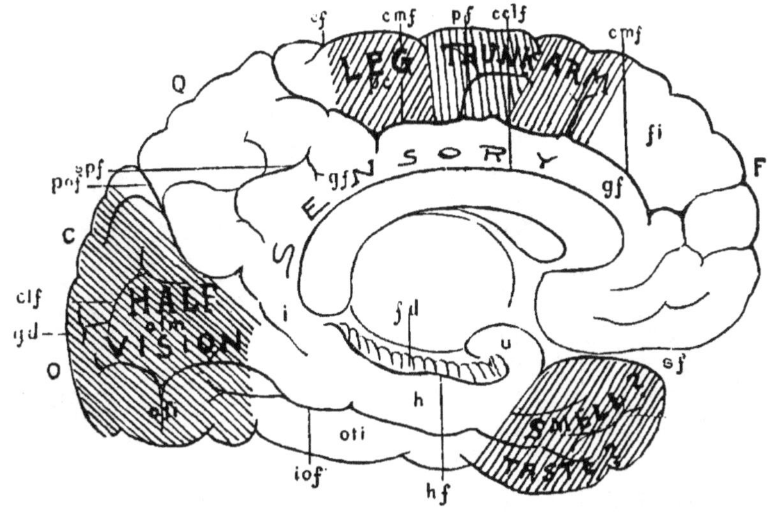

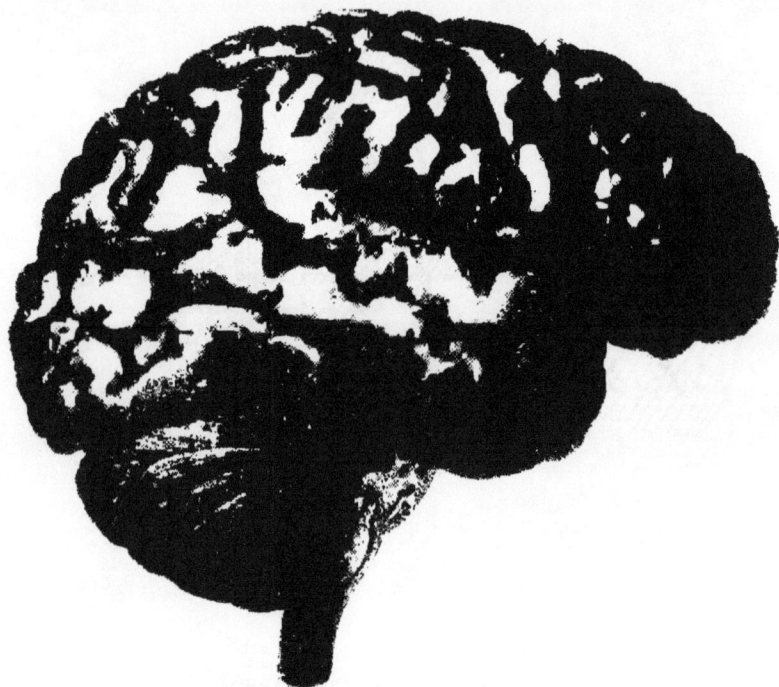

FIG. 3.

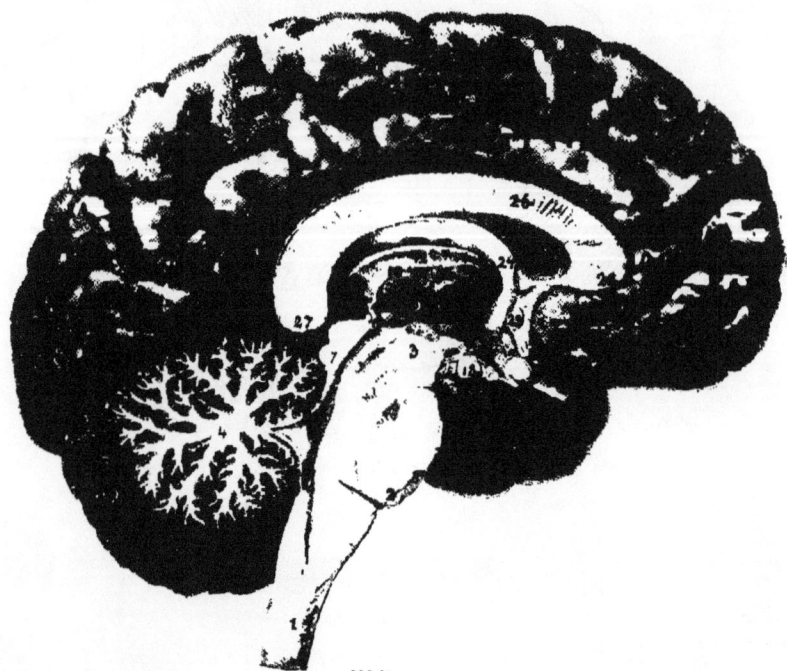

FIG. 4.

Fig. 3. *1*. Medulla oblongata.—*2*. Pons varolii.—*3*. Cerebellum.—*4*. Flocculus.—*5*. Frontal lobe.—*7*. Occipital.—*8, 9*. Fissure of Sylvius.

Fig. 4. View of internal or mesial aspect of the cortex, the corpus callosum and hind brain being cut down the middle line.—*1*. Medulla oblongata.—*2*. Pons Varolii.—*3*. Crus cerebri cut obliquely.—*4*. Cerebellum.—*5*. Acqueduct of Sylvius.—*6*. Valve of Vieussens.—*7*. Corpora quadrigemina.—*8*. Pineal gland.—*9*. Its posterior peduncle.—*10*. Its so-called anterior peduncle, with the swelling of the ganglion habenulæ close to the gland. *11*. Great transverse cerebral-fissure.—*12*. Upper, and *13*. Lower and internal surface of optic thalamus.—*14*. Soft or middle commissure.—*15*. Infundibulum.—*16*. Pituitary body.—*17*. Tuber cinereum.—*18*. Corpus mammillare.—*19*. Posterior perforated space.—*20*. Third nerve.—*21*. Optic nerve.—*22*. Anterior commissure.—*23*. Foramen of Monro.—*24*. Fornix.—*25*. Septum lucidum.—*26*. Corpus callosum.—*27*. Splenium.—*28*. Genu and rostrum.—*29*. Fissure of corpus callosum, also called the ventricle of the corpus callosum.—*30*. Convolution of corpus callosum, gyrus fornicatus, or cinguli.—*31*. Internal surface of superior frontal convolution.—*32*. Calloso marginal fissure.—*33*. Inferior occipito-temporal convolution.—*34*. Parieto-occipital fissure.

No. 4. JUNE, 1891. Vol. I.

THE REVIEW

INSANITY AND NERVOUS DISEASE

A Quarterly Compendium of the Current Literature of

Neurology and Psychiatry.

EDITED BY

JAMES H. McBRIDE, M. D.,

MILWAUKEE, WIS.

$2.00 PER YEAR. SINGLE NUMBERS. 50c.

MILWAUKEE, WIS.
Swain & Tate, Printers, 387 Broadway,
1891.

m. J.
10.17.1892

INDEX.

REVIEW

OF

INSANITY AND NERVOUS DISEASE.

ORIGINAL ARTICLE.

DANGERS OF THE BROMIDES
IN EPILEPSY.

BY HARRIET C. B. ALEXANDER, M. D., CHICAGO.

About a quarter of a century ago Dr. W. A. Hammond* called attention to a phenomenon, since frequently observed, yet pretty generally ignored. He found that the convulsions of a traumatic epileptic were replaced by furor under the use of the bromides. Nearly coincident with Dr. Hammond, a German alienist, Dr. Stark,† reported several cases. Dr. Stark was of opinion that the effect was due to the suppression of the convulsions rather than to any untoward effect of the bromides. The subject appears from the first to have received most attention in America, for some thirteen years later Dr. H. M. Bannister‡ of Kankakee called attention to the same fact. He reported three cases:

Case I. was that of a powerful, good-humored man, liable to frequent attacks of *grand mal.* He was usually mild and good humored; even the post-epileptic state was that of stupor. Bromides made him unmanageable, violent and homicidal, or querulent, irritable and suspicious.

---

*Jour. of Psychol. Med., 1868.
†Allg. Leitschrift f. Psych. B. XXXII.
‡Jour. of Nerv. and Ment. Dis., 1881.

2

The two other cases cited were similar:

Case II. was that of a semi-demented patient, who became talkative, querulent, and suspicious under the bromides.

Drs. Jewell,* Kiernan,* Moyer,* Rockwell,* Sequin,* and Spitzka* about the same time stated that they had observed similar cases. Sequin's case was as follows:

Case III. A twelve-year-old boy was the victim of *petit mal*, which took the form of "chills." When the "chills" were suppressed by the bromides the boy became turbulent and unmanageable.

Dr. Rockwell reported the following case:

Case IV. The attacks of a female victim of *grand mal* were suppressed by the bromides. In consequence she became irritable, depressed, and suspicious.

Some years subsequent to the discussion of. Dr. Bannister's paper, Dr. C. H. Hughes† of St. Louis, reported the following:

Case V. The attacks of a victim of *petit mal* were suppressed by bromides and replaced by kleptomania in consequence.

Four years ago I reported the following cases which came under my care at the Cook County Insane Hospital:‡

Case VI. The patient has a hysterical mother, a paranoiac father, and periodically dipsomaniac brother. She had frequent paroxysmal attacks of nymphomania in early puberty, which disappeared at the age of twenty to give place to epilepsy. She is furiously nymphomaniacal precedent to a convulsion. In her case, mental disturbance precedes, succeeds, and takes the place of a convulsion; in other words, she has the pre-, post-, and psychical equivalent types of epileptic insanity. In the inter-epileptic period she is good-humored and somewhat stupid. Under the use of bromides her attacks

---

*Jour. of Nerv. and Ment. Dis., 1881.
†Alienist and Neurologist, 1883.
‡Medical Standard, Vol. I.

disappear, but she becomes irritable, querulent, and suspicious. Mixed treatment (conium, arsenic, ergot, and the bromides) does not have this effect.

Case VII. In the collateral ancestry there are several epileptics. Her brother has *petit mal.* She has the pre-, post-, and equivalent types of epileptic insanity; both these and the attacks are at times replaced by imperative homicidal conceptions. Long continued furor with vivid causal erotico-religious auditory and visual hallucinations result from the bromide treatment.

Case VIII. is that of a woman who comes from a markedly neurotic ancestry on both sides. By medical advice and for therapeutic reasons, her father paid a man to marry her. She has had three still-born children and five others died in convulsions. Her insanity is of the type described in the two previous cases. She is frequently troubled by coprolaliac imperative conceptions between the demonstrably insane periods when she ties a bandage around her mouth to aid her will in restraining the coprolaliac utterance, which it does. She is usually obliging and good-tempered between her attacks. Under the bromides she becomes sullen, querulent, and is no longer able to restrain her coprolaliac tendencies.

Case IX. The patient born of a neurotic family, has an early history of paroxysmal debauchery; later, epileptic attacks took the place of these paroxysms. She is usually good-humored between attacks; the use of the bromides is followed by an irritable, suspicional, treacherous state. Mixed treatment reduces her attacks without deteriorating the mental state. The insanity in her is pre- and post-epileptic.

Case X. A ten-year-old girl, is descended from very neurotic maternal, and congenital criminal, paternal ancestors. She has the same types of mental trouble as the first three cases. The fit, as in most epileptic children, is preceded by terror, then follows a purposeless running (*epilepsia cursiva*), then a fall and then the full motor explosion. Under the use of the bromides these phenomena disappear, but are succeeded by irregular kleptomaniacal attacks and inter-epileptic querulent, suspicious, irritable, mental condition.

Case XI. A female has *petit mal.* She denies all
epilepsy. Until long after her marriage, epilepsy,
although it clearly existed, was never suspected until
she awoke her husband one night by beating his face
with a slipper while unconscious. In the inter-epileptic
period she is mild-tempered, good-humored, and suave.
She becomes first irritable and querulent during the
inter-epileptic period, then· paroxysmally, furiously
excited and has vivid auditory and visual erotic-religious
hallucinations, and is coprolaliac. Mixed treatment
has no such effects.

Since then I have observed the following cases:

Case XII. A thirty-four-year-old woman has *grand
mal* followed by a dazed condition. Under the use of
the bromides these attacks were replaced by nymphomania with decided erotic manifestations attended by
religious hallucinations and furious masturbation. The
use of ergot removed these manifestations and the
alternation of ergot with the bromides prevented them.

Case XIII. A forty-two-year-old woman had attacks
of *grand mal* at the menstrual period and *petit mal* in
the interegnum. These under the bromides were both
replaced by furor under the bromides.

Dr. L. W. Baker* reports three cases in which the
use of the bromides led to violence and decided mental
disturbance.

The *Lancet*† says editorially concerning the same
subject:

In many cases the patients become wild and maniacal
from the prolonged use of the drug, and in the asylums
this condition is well recognized under the title of
"bromomania." In former times the same class of
persons continued along about the same from year to
year and did not require to be sent away to the asylums.
This was before the bromides became the routine means
of treatment in epilepsy at all the various dispensaries
and out-patient departments. It is at those institutions
that much damage is done by the drenching of the
epileptics with the bromides. Then, too, many of these

patients will go from one dispensary to another, and thus get loaded with an amount of bromide preparations that is far in excess of the intent, knowledge, or conjecture of the respective prescribers. When, therefore, a practitioner finds that any of his epileptic patients give indications of an increase of excitability and violence, it will be well to investigate into the amount of their drug-consumption; it may be the invasion of bromomania.

The question naturally rises whether these effects be due to idiosyncrasies in regard to the bromides to the "untoward effects" of these or to the suppression of the epileptic explosions. That the bromides do produce untoward effects analogous to the stupor of epilepsy there is abundant evidence to prove.

Voisin* points out that staggering gait, mental apathy, somnolence, abulia, delusions and hallucinations result from the bromides. He has observed aphrodisia also produced from its use.

Labordet has observed sexual excitement to result from their use, and Dr. J. E. Winters‡ has observed instances of visual hallucinations due to the bromides. Dr. Kiernan§ has reported instances where aphrodisia is produced by the bromides, and analogous cases are reported by Dr. G. J. Monroe.* That these untoward effects of the bromides closely simulate the effects produced in epileptics there can be no doubt, yet the weight of authority, and indeed the weight of evidence, is in favor of the opinion that these phenomena result from the suppression of the epileptic explosions. Dr. Bannister|| says:

" The fact that in these cases the suppression of the epileptic attacks by the bromides was accompanied by cerebral excitement and outbursts of maniacal furor is

· *Bull. gen. de Therap. LXXIII.
 †Gaz. Med. de Paris, 1869.
 ‡Medical Standard, 1887.
 §Medical Standard, 1891.
 ||Medical Standard, Vol. I.

strongly suggestive that the attacks themselves are somewhat of the nature of a safety valve in some cases, and that the epilepsy is itself an alternative to acute and dangerous mania."

Dr. Spitzka has said: "To give the bromides alone is to postpone the explosion and generally intensify it. The very fact that a sudden suspension of bromide administration is followed by a severe explosion is clear evidence that the bromide acts rather like a load keeping down a safety valve. To use a rather coarse simile, I prefer to tap an overloaded and continually refilling cistern to simply putting on a water-tight cover, or perhaps combine both."

My own experience, based on therapeutic uses of ergot and other agents adopted to avoid these results, tends to support this view of Stark, Spitzka, and Bannister, and to demonstrate that the use of ergot on this "minute explosion" plan of Spitzka is an excellent therapeutic procedure.

NEUROLOGICAL.

ANATOMY AND PHYSIOLOGY.

CEREBRAL FISSURES.—G. Valenti, *Attidella Soc. Toscana di Scienze Naturali in Pisa*, Vol. XI., (abstract by Christiani in *Arch. Italiano*, XXVIII., I., Il., Jan. and March, 1891), from an examination and comparison of 160 human brains and a corresponding number of brains of the Simiæ and other mammals, offers the following conclusions which are of interest in an anthropological and psychiatrical point of view.

In the frontal lobe of man the following, when found, are simian, or degenerative peculiarities:

A predominant development of the superior frontal convolution.

The anastomosis of the superior and median frontal convolutions across the middle portion of the superior frontal sulcus.

A slight development and simplicity of the lower frontal convolution, and especially the unity of the anterior branch of the Sylvian fissure.

Smallness of the inferior præ Rolandic fissure and increased distance of its inferior extremity from the fissure of Sylvius.

The following characteristics indicate a high grade of development:

Smallness of the superior frontal convolution, and especially its being single at its root.

Integrity of the superior frontal sulcus.

The existence of a transverse sulcus in the median frontal convolution.

A single prae-central sulcus.

In the parieto-occipital and temporal lobes the following are degenerative characteristics:

A greater development of the occipital lobe than the normal.

A more oblique direction of the interparietal sulcus causing the superior parietal convolution to be narrowed posterially.

A prolongation backward and upward more than ordinarily of the external branch of the fissure of Sylvius.

A deficient development of the angular gyrus.

Simplicity of the arrangement of the tertiary sulci of the parietal and occipital lobes.

Depth and unusual length of the superior occipital fissure (corresponding to the perpendicular fissure).

The following characteristics indicate a higher development:

Multiplicity of the tertiary sulci ot the parietal lobe.

Smallness of the occipital lobe and especially under development of the superior occipital sulcus.

The existence of anastomotic convolutions between superior and inferior parietal convolutions.

Increased development of the external convolutions of passage.

ON THE PATHOLOGICAL HISTOLOGY OF THE NEUROGLIA.— Weigert has investigated this subject with the help of his new stain. He found the proliferation of the neuroglia identical in nature in locomotor ataxia, Friedreich's ataxia in multiple and amyotrophic lateral sclerosis, and in the ascending and descending secondary degenerations. W. insists that the proliferation is not due to an increase of the idioplastic properties of the neuroglia, but a diminished resistance of the surrounding tissues. This, however, presupposes that the ability to proliferate is unimpaired. If it is impaired, a scar or a cyst results, although there are numerous transitions. The energy of proliferation seems to decrease with age. In the molecular layer of the cerebellum in progressive paralysis, a great increase of the neuroglia was found, as in all other parts of the brain. In two typical cases of glioma no increase of fibres could be demonstrated. In the tumors, all trace of them was absent. The author leaves us in doubt whether this is due to a defect in method, whether the tumors were not gliomas, or whether the cells had lost the ability to form fibres. In four other cases of glioma, fibres were found. The tissue surrounding the cavity in syringomyelia is composed of proliferated neuroglia.—*Centralbl. f. allg. Path. in. Path. Anat.* 1890, Bd. I., *Neurol. Centralbl.*, No. 1, 1891.   G. J. K.

THE NEUROGLIA OF THE CENTRAL NERVOUS SYSTEM.— Weigert claims to have found a method of staining the neuroglia a dark blue, but has not finished his trials. His sections corroborate Ranvier's opinion that the fibres simply surround the cells, and that the cell protoplasm and the fibres have nothing in common. The fibres are smooth. Postmortem changes soon take place and the fibre dissolves into a granular material. It is not identical with the neurokeratin of the peripheral nerves. In these, W.'s method produces no stain as in the centres. At the point of exit, a small bundle of neuroglia accompanies the fibres for a short distance. The network is most dense on the surface. In the cord all the fibres are separated by layers of this substance. Lissauer's zone is rich in neuroglia, the substantia gelatinosa very poor. The neuroglia fibres are most profuse around the central canal and in the medulla, in the olives. In the cerebellum, they are numerous in the white layer, rare in the granular layer, Purkinge's cells being surrounded by a very fine network. The white substance of the cerebrum has a very dense network; the deep layers, rich in nerve cells, a very slight one. —*Anatom. Anzeiger*, 1890, No. 19, *Neurol. Centralbl.*, No. 1, 1891.   G. J. K.

TROPHIC CENTER OF THE SPINAL NERVES.—Dr. P. Krouthal reports a case of amyotrophic lateral sclerosis and bulbar paralysis, in which the following lesions were found in the cord: The cells of the anterior horn were completely atrophied, the pyramidal tracts very much so. The posterior and the lateral cerebellar columns were normal, the anterior lateral columns slightly atrophic. The cells of the posterior horns

and of Clarke's columns were normal, as well as the posterior
roots.   The anterior roots, totally atrophic, with the exception
of a few fibres.   The spinal peripheral nerves were absolutely
normal.   Several similar cases have been reported and quite
a number in which the nerves were found atrophied with intact
anterior roots, while the anterior horn was in some of these
cases, normal.   The author confesses his inability to give a
satisfactory explanation of the absence of even a degeneration
from disuse, but suggests that the ganglion cells in the nerve
itself took up a vicarious action.—*Neurolog. Centralbl.*, No.
5, 91.   G. J. K.

A NEW METHOD OF STAINING.—Dr. T. L. Tiehen reports
excellent results from the following method: Small pieces of
the brain or cord are immersed in a mixture of equal parts of
1 per cent. solution of sublimate and 1 per cent. solution of
auric chloride in which they remain from three weeks to five
months.   The sections can be differentiated in dilute Lugol's
solution (1:4) or Tr. iodine of similar strength, dehydrated in
alcohol and clove oil and mounted in balsam.   The advantages claimed are, that the contour of the cell, nucleus, and
nucleolus are stained leaving the body semi-transparent.—
Neurolog. Centralbl., No. 3, 91.   G. J. K.

THE RESULTS OF FATIGUE OF NERVE CELLS.—C. F. Hodge
publishes in *The American Journal of Psychology*, some interesting experiments which he conducted with the view of
determining the result of fatigue occasioned by the electrical
stimulation of cells of the spinal ganglia.   His method is to
expose a spinal nerve of the cat, and after stimulating it by
electricity until exhausted, kill the cat, and examine the
condition of the nerve cells.   He finds that as the cell is thus
worked the nucleus gradually loses its sharp outline, and at
the same time it becomes clouded and filled with darkly
stained granules.   If the stimulation is continued the nucleus
shrinks smaller and smaller and becomes so dark in osmic

acid specimens as to be hardly distinguishable from the almost black nucleolus. The process of recovery is the reverse of that of fatigue. At the end of six and one-half hours the cell protoplasm is apparently quite recovered, although in many cases the cell does not recover for twenty-four hours. He finds that the shrinkage of the nucleus is directly proportional to the length of the stimulation, and in general, inversely proportional to the period of rest.

PATHOLOGY AND SYMPTOMATOLOGY.

ACUTE ANTERIOR POLIOMYELITIS ASSOCIATED WITH MULTIPLE NEURITIS.—Dr. Gowers (*Lancet*, Mar. 21) reported a case at the March meeting of The London Clinical Society. The poliomyelitis was caused by a cold bath taken in hot weather. Before the development of the palsy which usually follows, the arms and legs became extremely tender. Careful examination showed that the sciatic and popliteal nerves in the legs, and the ulnar and median in the arms, were involved. The symptoms of neuritis soon disappeared, but those of atrophic palsy continued.

In commenting on the case, Dr. Gowers remarked on the probability that the pain and tenderness met with in other cases of poliomyelitis, and usually referred to the joints, was really seated in the nerves, which suffer mechanical disturbance at every movement of the joints. Although poliomyelitis was probably produced through the agency of a blood state, the latter was certainly often generated by exposure to cold, and effective chiefly in the hot season of the year, which in some way predispose either to the susceptibility of the individual or to the production of a toxic influence. Exposure to cold was capable also of causing the other and more common form of multiple neuritis, the parenchymatous variety,

in which the nerve fibres suffered primarily, the sheaths
scarcely or not at all, except in their nerves, and the incidence
of the effect was peripheral and more perfectly symmetrical.
A difference must exist in the blood state due to cold in the
two classes of neuritis, and it was important to note the associations of the two forms thus produced. In this connexion
the author referred to the leading features of the two forms of
neuritis, the parenchymatous and the adventitial, using the
latter term as a designation for the variety in which the connective tissue elements were primarily affected. Recent
researches were referred to which suggested that there were
two varieties of poliomyelitis, the primary affection in one
being of the nerve cells, in the other the basis substance of
the grey matter, the former being analogous to the parenchymatous, the latter to the adventitial forms of neuritis, and
reasons were given for regarding the poliomyelitis in the case
described as of the second class. The paper concluded with
an allusion to the probability that an actual organized virus
sometimes produced the adventitial form of neuritis by the
direct influence of the organisms on the connective tissue
elements, and especially on the sheath, as was seen in leprosy,
and might be the case in syphilis; while, on the other hand,
the parenchymatous form was certainly often due to a product
which the organisms produced by their growth in the body, as
one active agent, alcohol, was produced out of the body.
The question was suggested, as deserving study in this connexion, whether diphtheritic neuritis commencing during the
primary disease was not chiefly adventitial, and that which
began when the disease was over in greater degree parenchymatous, and the contrast was noted between the adventitial
neuritis of the active stage of syphilis and the parenchymatous
later form which often constituted part of the lesion of tabes.

ACUTE ASCENDING PARALYSIS.—Dr. Jos. Eichberg, of Cincinnati, published in *The N. Y. Med. Record* the history of a
case of this disease, his conclusions being as follows:

1. The term "Landry's paralysis," as being wholly unscientific, should be dropped entirely, and the name "Acute progressive motor paralysis" substituted.

2. That by this term we understand an acute infectious process, a pathological entity, whose virus expends its effects primarily upon the gray matter of the anterior horns of the cord.

3. That, as in other intoxications, the economy may be so prostrated by the poison that time is hardly given for the anatomical changes characteristic of the affection, before the patient succumbs; hence, the inability to find notable lesions in cases of very short duration.

4. That, in cases of longer duration, lesions characteristic of inflammation of the gray matter and of the cord can generally be met with, in the form of inflammatory stasis, perivascular infiltration, microscopic hemorrhages, swelling of the ganglion cells of the axis-cylinders, and the destruction of the ganglion cells; the intensity of the lesion usually being greatest at the point first involved; *i. e.*, the lumbar part of the cord. It is, however, to be borne in mind that, as in the case of diphtheria, the local manifestation is not necessarily initiated at any one part; so that we may have exceptional instances in which the process begins above and extends successively to lower levels.

5. That, secondary to changes, anatomical or molecular, in the ganglion cells of the anterior horns of gray matter in the cord, or their congeners in the medulla, lesions are found in the peripheral motor nerves, comparable to those which follow the separation of every nerve fibre from its ganglionic connection; and these lesions are to be ascribed to disturbances of nutrition of the ganglion cell itself.

THE DEGENERATIVE CHANGES OF THE GANGLION CELLS IN ACUTE MYELITIS.—As a result of very extensive experiments Dr. M. Friedmann finds that the degeneration always begins in a circumscribed part of the cell and seems to affect first the chromatic substance. The death and dissolution of nucleus and processes is secondary and due to the death of the cell. The process is therefore the reverse of a coagulation-necrosis in which the nucleus dies first. The degenerative changes are best studied in sections hardened in alcohol, and stained with magenta-dahlia or safraum.—*Neurolog. Centralblatt*, No. 1, 189.   G. J. K.

A CASE OF CEREBRAL SYPHILIS.—Dr. H. T. Pershing, of Denver, publishes a case in *The Medical News* for April, and in conclusion says:

Syphilis is indicated with sufficient probability to determine the treatment, irrespective of other signs of the disease and the patient's history, in the following cases:

1.   Sudden cerebral hemiplegia in patients under forty-five, in whom atheroma, high arterial pressure, and the causes of embolism, notably endocarditis, can be excluded.

2.   Progressive multiple cerebral palsies.

3.   Insomnia and nocturnal headache, followed either by cranial nerve palsy or cortical irritation.

4.   Sudden stupor or coma, without other assignable cause. Somnolence, resembling that of alcoholic intoxication, with pain in the head and aimless, automatic actions.

5.   Paretic dementia, in which syphilis cannot be excluded, especially with prodromal nocturnal headache, insomnia, or somnolence and early epilepsy.

MORVAN'S DISEASE.—Dr. Archibald Church, reports a case of this disease in *The Chicago Medical Record* with remarks. The case was that of a man who at twenty-five years of age developed spinal curvature. About the same time had a felon on the right ring finger. Four years later, after exposure to cold, left hand, wrist, and arm became greatly swollen and painful. The fingers and wrist became deformed in consequence, and joints partially ankylosed. Three years ago had a felon, which was absolutely painless, on the right index finger, and, as a consequence, the end of the finger came off. Within the same year the middle finger of the right hand went through the same process. There was no loss of sensibility in the hands. The prick of a pin, however, was not felt as a painful sensation. There was also inability to distinguish heat and cold. His gait was shambling and difficult to describe. The analgesia in the hands, the distrophy of the wrist joint and vessels, indicate Morvan's disease. The retention of tactual sense with the marked blunting of painful and thermic impressions would indicate syringomyelia. This is, therefore, probably a mixed case.

ON "MORVAN'S DISEASE" (SO-CALLED).—Prof. M. Bernhard, of Berlin, in the critical discussion of a case of Morvan's disease, points out the great similarity between this disease and syringomyelia, shown in this and many other published cases. He advances the opinion that Morvan's disease, or "pareso-analgesie" is merely a variety of syringomyelia, the differences being due, probably, to the extent of the gliomatosis of the cord.—*Deutsche Medicin Wochenschr.*, No. 8, 1891,

G. J. K.

OCULAR SIGNS AS AIDS TO DIAGNOSIS.—Dr. Belt of Washington, has an article upon this subject in *The Medical News* of May 2. The following is the part that relates to ocular signs as aids to diagnosis in nervous affections:

Nettleship says: "Reflex iridoplegia (the Argyll-Robertson pupil) is one of the most valuable of the early signs of locomotor ataxy." Inequality of the pupils seems to be very frequently a precursor of insanity. Aside from dilatation caused by paralysis of the third pair of nerves and the local influence of mydriatics, according to Meyer and other authorities, the pupils are found dilated in hysteria, in hypochondria, in the later stages of meningitis, in hydrocephalus, etc. Sudden dilatation of the pupils during the administration of chloroform is a very important danger signal. Contraction of

the pupils, according to Swanzy, is found in the early stages
at least of all inflammatory affections of the brain and its
meninges, in tobacco amblyopia, at the beginning of an
hysterical or of an epileptic attack, in the early stages of intra-
cranial tumors situated at the origin of the third nerve, or in
its course. In cerebral apoplexy the pupil is at first contracted, according to Berthold, who points out that this contraction is a diagnostic sign between apoplexy and embolism,
in which latter the pupil is unaltered. In lesions above the
dorsal vertebræ myosis occurs. Some authorities regard
myosis as one of the earliest signs of tabes, while others do
not. In acute mania the pupil is usually much dilated, and
when this mydriasis is changed for myosis, approaching
general paralysis may be prognosticated.

Optic neuritis, loss of sight in portions of the field, and
paralysis of certain ocular muscles, frequently give us
such aids as to enable us to diagnosticate with comparative
certainty not only the presence, but often the location of
cerebral lesions. According to Gowers, neuritis is present at
some period in at least four-fifths of the cases of tumor of the
brain. He says the value of optic neuritis as an indication of
the existence of an intra-cranial tumor is very great, and that
it may be the only unequivocal sign of the intra-cranial
disease. Nettleship says: "Although pointing very strongly
to organic disease within the skull, neuritis is not of itself
either a localizing or differentiating symptom." In conjunc-
tion, however, with other symptoms that are frequently
present it is of much value in localization.

Neuritis limited only to one eye, generally indicates disease
of the orbit. Loss of half the field of vision (hemiopia) is
one of the most important aids in the localization of cerebral
lesions. When binocular it indicates disease at or behind the
optic chiasma. In the majority of cases the hemiopia is
homonymous, that is, the right or left lateral half of each
field is lost. Loss of the right half of each field points to
disease of the left optic tract, or of some part of the left occip-

ital lobe, or in the fibres connecting these parts. Temporal hemiopia points to disease at the anterior part of the chiasma. When lateral hemiopia co-exists with hemiplegia the loss of sight is on the paralyzed side. Swanzy says: "We may conclude that the hemiopia depends upon occipital lesion, if it be unaccompanied by hemiplegia, motor aphasia, or paralysis of cerebral nerves;" also, that contraction of the pupil in a case of hemiopia, when the light is thrown on the blind half of the retina, indicates that the lesion causing blindness is back of the corpora quadrigemina. The absence of this reaction indicates that this lesion involves the corpora quadrigemina, or the optic tract. Sometimes there is incomplete hemiopia, that is, only one quadrant of the field of vision is affected. Hun reports a case of this kind, in which the left lower quadrant in each field was blind, and where the autopsy showed a lesion strictly limited to the lower half of the right cuneus. Remember the fact that loss of a certain half of the *field* means loss of function of the opposite half of the *retina*.

Paralysis of the ocular muscles also aids in locating cerebral disease. Complete paralysis of the third nerve, without any other paralysis, is almost always due to lesion at the base of the brain, on the same side. Hemiplegia of one side, coming on simultaneously with paralysis of the third nerve of the opposite side, is a common sign of disease of the crus cerebri. When paralysis of the third nerve occurs with hemiplegia of the opposite side of the body and other cerebral symptoms, it is usually due to pressure on the nerve where it runs beneath the cerebral peduncle. According to Nothnagel, this localization is still more certain when paralysis of the facial and hypoglossal nerves exist on the same side as the hemiplegia.

Paralysis of the sixth nerve, according to Swanzy, simultaneous in its onset with hemiplegia of the opposite side of the body, indicates a lesion of the pons, usually a hæmorrhage on the side corresponding to the paralyzed nerve. By itself paralysis of the abducens is of little value in the localization of cerebral disease from the fact that it is so often of peripheral origin.

8

A Case of Bilateral Acute Opthalmoplegia following the Use of Spoiled Meat.—Dr. G. Gutman reports a case occurring in a healthy man æt. 30.   There was complete paralysis of all external eye muscles, the internal ones being normal.   This state lasted about two weeks.   As Cohn and Grœnouw have described similar, though not identical cases, and no other etiological factor could be found, G. believes the trouble to be due to the action of a toxine on the nuclei of the involved nerves.—*Berlin Klin. Wochensch.*, No. 8, 1891.

G. J. K.

On Typical Abnormalities of the Visual Field in Functional Diseases of the Nervous System.—Dr. H. Wilbrand (*Jahrb. d. Hamburg, Staats, Ka.*, 1890), classifies these as follows:

I.   Fatigue contractions.   These belong in a certain degree to the physiological peculiarities of the eye, but may occur in functional nervous disturbances, and lead the inexperienced observer into great error.

II.   The uniform, concentric contractions (temporary and permanent forms).   In this class belong the disturbances in hysteria, hystero-epilepsy, epilepsy, delirium tremens, neurasthenia, traumatic neuroses, and acute melancholia.

III.   Unilateral hysterical amaurosis, and bilateral transient amaurosis, the latter being often a part of an epileptic aura.

IV.   Transient reduction of central vision, without demonstrable scotoma and with normal visual field.

V.   Those cases in which a contraction of the visual field is added to symptoms due to focal cerebral lesions.   Its importance is in these very difficult to estimate, e. g., when it accompanies a heminanopsia.   The concentric contraction must be estimated as a functional trouble accompanying the focal lesion.

VI.   Intoxication amblyopia, always in both eyes.   The form of the visual field varies with the acuteness or chronicity of the atrack.   It is only in the early stages that the trouble is functional.

VII.   Fugitive scotoma, due probably to spasm of the cerebral arteries, and accompanied by transient complete or incomplete homonymous hemianopic defects.   Concentric contractions are probably due to involvement of the arteries of
chiasma and optic nerves.

As an important point in differential diagnosis, the author
points out that even with a functional concentric contraction
of high degree, orientation in space is not impeded, while it
is rendered quite difficult by small defects due to focal lesions,
even with normal visual acuity.—*Neurolog. Centralbl.*, No. 1,
1891.   G. J. K.

_____

AGORAPHOBIA AND CONTRACTION OF THE VISUAL FIELD.—
Dr. A. Nieden reports a case of agoraphobia in a man, æt. 34,
who was otherwise healthy.   Later in the course of the
trouble, claustrophobia (fear of closed spaces) became associated with it.   The patient also noticed indistinct vision for
objects not centrally situated.   Examination showed a visual
field, in both eyes, of barely one-third the extent of the normal.   Under appropriate psychical and medicinal treatment
(bromides) the visual fields expanded and the morbid fears
disappeared   to   a   considerable   extent.—*Deutsche   Medic.
Wochensch.*, No. 13, 1891.   G. J. K.

_____

ON A PECULIAR FORM OF NYSTAGMUS.—Dr. C. S. Freund
reports the following case which presented a form of nystagmus hitherto undescribed.   A healthy young recruit of 18,
after severe exercise awoke with pain behind the eyes, diminution of vision, and a strong tremor of both eyes.   Both eyes
were in a state of extremely rapid horizontal vibration, which
was accompanied by sparkling and by apparent motion of
external objects.   He soon found that the tremor became very
much less on strong fixation of an object, and disappeared
altogether upon monocular vision, as in closing either eye.
It reappeared, however, at once, on opening both eyes.   The

author also found the tremor to diminish as soon as the eyes were brought to a lateral terminal position. The pains behind the eyes disappeared upon the application of sinapisms to the nape of the neck daily for a fortnight. Examination showed the eyes to be normal, with the exception of hypermetropia and a small central scotoma for red in the right eye. Glasses of +3 D. were given. An intermittent clonic blepharospasm of the upper lid was noticed on lightly closing the eyes. The nystagmus, which had persisted for over six months, disappeared almost completely upon the application of a galvanic current of 4 M. A. to the head for five minutes, and was cured with one week's treatment. A short relapse was also amenable to treatment. In searching for a cause for this trouble the author found the following additional symptoms. The pulse showed decided acceleration, some arhythmia and occasional intermission. A small goitre was detected. Patient was subject to profuse, watery, painless diarrhœas, and to localized sweats. The scotoma disappeared slowly in the course of six months. The author believes himself justified, from all the symptoms, in classifying the case as an abortive type (form fruste) of exopthalmic goitre. He has been able to find but one similar case in literature, but in this the type of the ocular motion was different.—*Deutsche Medic. Wochenschr.*, No. 8, 1891.　　　　　　　　　　G. J. K.

A CASE OF MULTIPLE SCLEROSIS WITH FACIAL INTENTION TREMOR.—Dr. V. Cohn reports the case of a woman 34 years old, who presented many symptoms of both multiple scleroses and paralysis agitans. Besides intention tremor of the extremities, running walk, and other symptoms of the former, the characteristic contractures, rigidity and rest, tremor of paralysis agitans were present. It was decided to be a case of multiple sclerosis. The most important symptom was a strong intention tremor of the facial muscles, which has been very rarely observed in this disease.—*Deutsche Medic. Wochenschr.*, No. 13, 1891.　　　　　　　　　G. J. K.

A Case of Tetany with Intention Spasm.—A tutor, æt. 22, had undergone severe fatigue in the heat of summer, six months before admission.   He became unconscious and remained so for six days.   Since then, numerous attacks of unconsciousness, one being followed by rigidity of the extremities.   Improvement followed treatment; but a few weeks after another attack of unconsciousness with apisthotonos, rigid flexion of the elbows, and extension of fingers occurred.   He became conscious the next day, slight rigidity of the limbs remaining, together with an intention spasm, Trousseau's, and the facial nerve reflex.   Motor and sensory galvanic excitability increased.   Similar attacks followed for several weeks, and then ceased, leaving the legs rigid.   Proceeding from the fact that Horsley had found mucin in the blood in goitrous tetany, as well as that Wagner and Hammerschlag had produced tetany by the injection of mucin, the author cast about for a remedy to increase its elimination, and injected 3-10 gr. (O. 02 gm.) of pilocarpine.   The result was profuse saliva-tion and perspiration, clonic spasms and trismus.   Patient fell asleep and awoke with the rigidity gone.   A relapse was cured by a similar dose.—Dr. Th. Kasparek, *Wien. Klin. Wochensch.*, No. 44, 1890; *Neurol. Centralbl.*, No. 2, 1891.

G. J. K.

————

Idiopathic Tetany in Childhood.—Prof. Escherich (*Wien. Klin. Wochenschr.*, No. 40, 90) found symptoms corresponding to the tetany of adults in thirty children, admitted to hospital for laryngeal spasm.   Trousseau's phenomenon and mechanical hyperexcitability of the facial nerve were present in all cases.   In one-half the cases, intermittent tonic spasms of upper or lower extremities could be proven, either by the history or observation.   In two cases the laryngeal spasm caused death.   The number of spasms reaches a maximum at the end of the first week, the average duration being three weeks.   All the patients were well nourished children, gastric disturbances being absent throughout.   E. does not

consider the relation of the symptoms as accidental, as laryngeal spasm also occurs in the tetany of adults, and in these
cases it kept step with the other symptoms.   Almost all the
cases were observed during the spring months.—*Neurolog.
Centralbl.*, No. 2, 1891.   G. J. K.

A CASE OF PARTIAL DUPLICATION AND HETEROTOPY OF THE
CORD.—In a cord from a case of ascending myelitis the author
found, beginning near the upper end of the lumbar cord, an
incomplete duplication of the left half of the cord.   Above
this, the gray matter of the cord presented a great variety of
figures, but never the normal one.   It was split, drawn out
into bands of various shape or direction, or contracted into a
circular field.   It would be impracticable to give a closer
description without the aid of illustrations.   The author
believes that we would find more of these abnormalities if all
cords were examined.   The owner of this cord was 55 years
old at death.   Twelve similar cases have been reported in
literature.—DR. L. JACOBSOHN, *Neurolog. Centralbl.*, No. 2,
1891.

ALTERNATING SCOLIOSIS ACCOMPANYING SCIATICA.—A scoliosis
of the lumbar spine, with the convexity toward the painful
side, has frequently been noticed.   Several of Charcot's
pupils have claimed this to be the rule, and to be due to the
effort to remove as much weight as possible from the painful
limb.   Dr. Remak reports a case in which the patient could
produce a scoliosis toward either side, and claimed that the
change afforded him relief from pain.   The sciatica was on
the left side.   When he carried his trunk to the right, the
distance from the ribs to the crest of the ilium was, right
6 cm., left 10 cm.   When he changed to the left leg the corresponding distances were, right 8 cm., left 6.5 cm.   Consequently it cannot be claimed the symptom is pathognomic.—
Deutsche Medic. Wochensch., No. 7, 1891.   G. J. K.

The Theory of Dyslexia.—Prof. Pick presents an ingenious theory of the cause of this symptom. He compares it to the intermittent lameness of horses, which is due to an arterial spasm, with consequent ischæmia, and is generally a premonitory symptom of the obliteration of the vessel involved. As dyslexia usually occurs as a premonitory symptom of severe cerebral affections due to arterio-sclerosis, he suggests that it is due to a spasm and partial obliteration of the vessels supplying the centre for reading, which may later become permanent.—*Neurolog. Centralblatt*, No. 5, 1891.

G. J. K.

A Case of Thomsen's Disease.—Patient is an engraver, 25 years old, with no hereditary history, although a brother is similarly affected. Health is otherwise good. The disturbance of inervation is worse after prolonged rest, exposure to cold, fatigue, exertion, mental excitement, and fright. Muscles, especially of buttocks and legs, much hypertrophied, while the strength is comparatively slight. Sensation and cutaneous reflexes normal. Tendon reflexes vary with the muscular tonus. Mechanical excitability of motor nerves normal; of the muscles, increased; electric excitability of nerves normal; of muscles, increased and changed. With median and high intensities slowness of contraction and tetanus sets in, which increases with the current. On frequent irritation the contractions become more lively, at last normal. Stabile application of very strong faradic currents causes irregular undulating contractions. With very strong galvanic currents, rhythmic contraction waves running from Ka to An, were produced. Examination of a fragment of the biceps muscle shows enormous hypertrophy of the fibre, increase in nuclei, indistinct striation, and slight increase of connective tissue.—Dr. Paul Seifert, *Neurolog. Centralbl.*, No. 2, 1891.

A Case of Aphasia of the Cerebral Hemispheres.—Prof. Grawitz reports the case of a child born at term, which died

on the 13th day of pneumonia.   The head was very small and
the sutures prominent, the chin being retreating.   The hands
were abducted, the ulnar muscles being contracted, the radial
paralyzed.   The frontal and parietal bones of the right side
projected over those of the left.   The anterior fontanelle was
very small, the posterior could not be felt.   The sutures were
ossified, the dura adherent.   The falx was absent, but instead
a membrane passed across the cranium at the level of the
parietal bosses.   This membrane was continuous with the
dura.   The brain was covered by a pia, the left hemisphere
being a large sac with thin walls, which in the frontal
region showed some traces of cerebral substance.   The right
hemisphere was quite small and connected in front with the
left, posteriorly it was not connected.   The hemisphere was
converted into a sac whose walls were from 1-12 to ⅙ inch
thick.   Within this sac a membranous rudiment of a corpus
callosum was seen.   The ventricle was greatly dilated throughout.   The fornix was present.   There was on this side a well
formed sylvian fissure.   The arteries were normal, as were
the nerves, with the exception of the olfactories; the left one
being hollow, the right partly so.   Cerebellum was very small,
as was the pons; pyramids thin.   The substantia nigra in the
peduncles indistinct; third ventricle, fornix, and foramen of
Monro normal; corpora quadrigemina small; aquæduct and
fourth ventricle narrow.   It is not possible to determine at
what time the arrest of development took place.—*Deutsche
Medic. Wochensch.*, No. 4, 1891.   G. J. K.

ECHINOCOCCUS OF THE CEREBELLUM.—A young man of 25,
in August, 1889, suddenly complained of diplopia and crawl-
ing in the limbs.   Both radials were pulseless, and gangrene
of some of the fingers resulted.   Four weeks later an aneurism
of the left axillary artery was found, containing an echinococcus cyst.   In May, 1890, the patient was obliged to give
up work on account of violent headache and vertigo.   Six
days before death sleepiness and loss of co-ordination were

observed. V. was diminished. Speech was difficult. Death suddenly occurred, with disturbance of respiration, on June 15, 1890. Autopsy showed an echinococcus of the left hemisphere of the cerebellum, of the size of an egg. The vermis was not involved. The embolism probably occurred at the time of the first attack, the aneurism being due to an echinococcus of the arterial sheath. No other cysts were found.—DR. SONNENBURG, *Deutsche Med. Woehenschr.*, No. 8, 1891.

G. J. K.

THE BRAIN OF LAURA BRIDGMAN.—H. H. Donaldson publishes an elaborate study of the anatomical observations of the brain of this person. At two years of age, Laura was attacked by scarlet fever and the sight of both eyes and hearing of both ears were lost; taste and smell were impaired. The sense of smell seems to have been subject to some variation. During the first few years after her sickness it was apparently completely wanting, but later it manifested very slight activity. She could not talk. The weight of her brain was 1389 grms., the probable volume was about 1160 c. c. The occipital and temporal lobes were found to be small; the right cuneus also being much smaller than the left. The surface of the *insula* exposed on the left side was nearly three times that exposed on the right, which, of course, indicates incomplete development. In this connection it should be remembered that there was partial sight in the right eye for five years after her illness, and after she had become totally blind in the left eye.—*Amer. Jour. Psychology*, Sept.

A CASE OF ACROMEGALY FOLLOWING SEVERE FRIGHT.—A girl, æt. 24, in blooming health, received a very severe fright in March, 1889. The next day she complained of headache, and pains and uncomfortable feelings in various parts of the body. There was asthenopia, eyes being intact, and great mental depression. No nervous ancestry. Menstruation was absent since the day of the shock. Soon after the fright, her friends noticed that her head was becoming larger, and she soon found that she could not get gloves or stockings large enough. The lower jaw projected considerably, the mento-occipital diameter being 10 in., the corresponding circumference being 28 in. The nose was large and flattened, as well as both lips. The hands and feet, as well as the lower sections of the fore-arms and legs, were enormously, though symmetrically enlarged. Length of hands 8⅛ in.; of middle finger 4⅞ in; of index 4 in. Length from acromion to tip of fingers, 32 in. Both patellæ, the crests of the ilia, both clavicles, and the spine were enlarged, the thorax being normal. This is the first case reported in which a definite etiological factor has been determined.—PROF. P. K. PEL, *Berlin Klin. Wochensch.*, No. 3, 1891.   G. J. K.

THE REAPPEARANCE OF THE PATELLAR REFLEX.—Dr. S. Goldflam reports two cases of tabes in the pre-ataxic stage in which the patellar reflex reappeared as the result of cerebral lesions. Case 1 was that of a man æt. 34, who had acquired syphilis 17 years before. Specific treatment had no influence on the typical symptoms present (absent, pupillary and patellar reflex disturbance of rectum and bladder, lightning pains). He suddenly developed hemiparesis of the right side, with difficulty of speech, probably due to a specific arteritis. This soon improved to some extent, but not wholly. Six weeks after the apoplectic attack, both the knee and ankle reflex were present on the right, absent on the left side. This con-dition remained until his death by suicide. The second case, 57 years old with no syphilitic history, suddenly developed an apoplexy of which he died. Although the attending physi-cians claimed that the patellar reflex had been absent upon repeated examination for several years, the author, who examined him in the terminal coma, found them exaggerated. In explanation, the author advances the theory that the absence of the patellar reflex in tabes is not due to an inter-ruption of the reflex arc, but to an inhibitory action of the degeneration upon some part of the arc.   In cerebral troubles in which the reflex activity of the cord is heightened by removal of the higher inhibitory centers, the inhibition of the degenerated patches is not strong enough to abolish the reflex. —*Berl. Klin. Wochenschr.*, No. 8, 1891.   G. J. K.

DISTURBANCES OF SENSATION IN MULTIPLE SCLEROSIS.—The later German text books (Strümpell-Seligmueller and Hirt) state that the absence of sensory disturbances is characteristic of this disease.   Dr. C. S. Freund has found such disturb- ances in 29 out of 33 consecutive cases (88 per cent.).   In 14 cases, these symptoms were temporary and fugitive, in six permanent, in six others subjective only, and in three cases could not be classified.   All but six had had some subjective symptoms, though only five had had lancinating pains.   The paræsthesias consisted mainly of numbness and creeping sen- sations.   A temporary girdle sensation was observed once, and reported once.   In 13 of the 27 cases the unpleasant sen- sations were confined to the hands and feet, more especially to the terminal phalanges of the fingers and toes.   This dis- tribution was also noticeable in the objective symptoms. Sensation to touch was disturbed in 18 cases, temporarily in 11, permanently in six, two cases showed hemianæsthesia. In four, both, and in one, only the left lower extremity was involved.   The pressure sense was altered in 17 cases, in one of which, sensation to touch was normal.   In five, it extended over both, in two, over only one lower extremity, two cases showed hemihyparæsthesia.   Sensation to pain was abnormal in 17 cases.   In 10, complete analgesia, in 12, hyperalgesia and in three, temporary hyperalgesia was present.   Out of 17 cases which showed alteration of temperature sense, nine showed hyperalgesia, four analgesia.   Five of these showed diminished sensation for heat only—not for cold.   The mus- cular sense was abnormal in eight cases.   Disturbance of stereognostic* sensation was found transiently in two cases,

*Ability to recognize objects by touch.

which also presented disturbance of finer mobility and distinct, though not lasting, ataxia. In the majority of cases all the different forms of sensory disturbance was localized in the hands and feet, and more especially in the terminal phalanges of the fingers and toes, sometimes to a single digit. In five cases the subjective and objective disturbances involved the same parts, the subjective disturbance beginning a little earlier and lasting a little longer. Several of the patients complained of a reduction of taste, smell, or hearing, which was, however, only temporary. The sudden onset and instability of the sensory disturbances agrees with what is known of the motor and ocular symptoms. In seeking for an anatomical explanation of the symptoms and of their temporary character the author adopts the view of Charcot, that certain nerve fibres survive the sclerotic process, and again becoming clothed with a sheath resume their function. During the time the myelin sheath is absent, the function is, of course, absent, if a *restitutio ad integrum* do not take place permanently.—*Archiv F. Psychiatrie*, Bd. XXII., H. 2 and 3.   G. J. K.

CLINICAL INVESTIGATIONS UPON THE INNERVATION OF THE LARYNX.—Exner has claimed to prove that in animals the motor innervation of the larynx is derived from both the superior and recurrent laryngeal nerves. Dr. Neumann reports a case in which the recurrent nerves on both sides were destroyed by cancerous glands. The glottis showed the cadaveric position. Autopsy showed that the superior laryngeal nerve was absolutely normal on both sides. All the laryngeal muscles except the crico-thyroideus anterior were almost wholly degenerated. He relates another case, in which a suicide cut the superior laryngeal nerve on the right side. The only motor disturbance noticed by the laryngoscope was that the right vocal cord was not as tense as the left, resulting in hoarseness. From these two cases Neumann concludes that Erner's theory of a double motor innervation of the larynx cannot be applied to man.—*Berlin Klin. Wochenschr.*, No. 6, 1891.   G. J. K.

HEADACHE IN CHILDHOOD.—Dr. Simon (*Gaz. des Hop*, March 24—26) divides these into: headaches from growth, from mental overstrain, from digestive disorders, of nervous origin, of anæmic and toxic origin, of sensory disease origin, and of meningitic and cerebral lesion origin. Headaches from growth are far from rare. They occur in the morning, occupy the frontal and superior vertical region, are increased by motion and diminished by repose. The child complains also of joint growing pains, but no arthritic lesion is present. The heart is often slightly hypertrophied. Tonics, diet, and repose intellectual and physical are the chief indications in treatment. Headaches from mental over-pressure result in precocious children pushed too far in studies. The headaches are violent in type. The suspension of mental pressure is at once indicated. Headaches from digestive disorder are of varied œtiology; all related to the gastro-intestinal canal. Headaches of nervous origin are due to nervous over-excitation of a child placed in unfavorable surroundings. They occur in the evening, produce acute suffering vividly described by the child, and are followed by evening anorexia insomnia, nocturnal agitation, and night terrors. The child should be removed from its surroundings as the first element of treatment. Headache precedent to oncoming hysteria presents itself in the form of painful points. Radiations of these into the eyes or neck, as well as sensations of weight, or a ball in the throat, are rarely absent. There is usually excessive sensibility to light and noise. The patients are hyperæsthetic. The headaches of epilepsy occur toward the ages of four or five in a sudden manner and last about an hour and a half. Some cerebral torpor results thereafter. Headache of chorea is of a similar type. Headaches of gouty or rheumatic origin in children occur in those with these ancestral taints. The pain is intense and accompanied with nausea, photophobia, hyperacousthesia, and agitation. Headaches from anæmia and toxic causes present nothing special in type or treatment. The headaches

of sense origin vary in type and character according to the
extent and nature of the original lesion.   There is nothing
specially distinctive about the meningitic headaches.

NEUROTIC COMPLICATIONS OF UTERINE DISEASE.—This was
the subject of a lecture by Dr. Playfair before the Nottingham
Medico-Chirurgical Society.   He states that his own professional work has taught him that in treating diseases of women
the great danger is in developing neuroses.   The gynæcologist
ought thoroughly to appreciate the highly strung and nervous
organization of female patients, and the strong impression
produced upon them by disease, or by the supposed disease
of the uterus.   It may be difficult for one to understand the
amazing influence upon a woman of being told that she has
some uterine trouble.   The practitioner may mean very little
when he states the fact, but for the patient it may be a very
serious matter and may lead to a train of morbid thought that
will eventuate in chronic disease.   In almost all women there
is a strong nervous tendency and physicians should be careful
to avoid anything that may increase it.   In some cases the
treatment of actual disease in women aggravates this tendency
to nervous disorder.—*The Lancet*, April 25.

STUDIES OF SYPHILIS OF THE SPINAL CORD (by Dr. M.
Mœller).—In the etiology of nervous diseases syphilis occupies a prominent place.   Its importance with regard to
diseases of the brain is unquestionable, but its etiological
relation to those of the spinal cord is still rather vague.
There are several reasons for this.   One is, that before sufficient knowledge of the anatomy and physiology of the spinal
cord was acquired as a foundation for the study of its diseases,
attention was chiefly directed to the brain, and disorders
which of old were regarded as nervous, were referred to this
organ.   In the old literature on the subject, there is therefore
a distinction made between syphilis of the spinal cord and
that of the brain.

The influence of syphilis on the nervous functions had early attracted the attention of the old writers. A few years after the epidemic outbreak of " morbus gallicus," they observed that it affected different parts of the body and that it arose from unclean sexual connection.

During the sixteenth century this new disease came to be regarded by most medical men as being the cause of all sorts of ailments, particularly diseases of the nervous system, such as paralysis, apoplexy, epilepsy, etc. It remained, however, an open question whether diseases of the nervous organs were due to syphilis itself, or to the results of excessive use of mercury. Later functional derangements of the nervous system were attributed to processes in adjacent bones: caries, necrosis, and gummata. In 1610, Gudrinoni found syphilitic gummata in the brain of a man, who suffered from headache, epileptic fits, deafness, and somnambulism, and who had been treated with guaiacum. In other respects the seventeenth century affords but little new for anatomy and symptomatology. Writers of this period concern themselves principally about the origin and treatment of syphilis. Medicine was for the most part speculation, and this continued till the time of Astruc. He describes a host of nervous disorders from syphilis: various forms of headache, giddiness, convulsions, epilepsy, paralysis, chorea, etc. As a cause of these he mentions exostosis, or caries in the bones of the cranium, nodi, or ganglia in the pericranium, abscesses, or gummata in the brain itself, and disturbances of the circulation. He explains the symptoms of giddiness, paralysis, convulsions, etc., depending on such changes in the blood, lymph, and the tissues, as will bring about movements of what he calls " ésprits animaux," which he presumes exist in the lumina of the nerve trunks.

Mortgagnis' investigations led to the discovery of gummata in the cortex of the brain, thickening of the membranes, and their growing together with cortical adhesions. He also found syphilitic processes in the arteries of the brain, as for instance in a man of 59 years of age, who had several distinct attacks of syphilis, and who had suffered from brain trouble. De Horne demonstrates thickenings in the vessels of the membranes of the brain of three younger syphilitic persons.

Observations like these were, however, very few, and even as late as at the end of the eighteenth century, the theory— from lack of anatomical and pathological knowledge—of syphilis in the nervous system, as well as in the viscera in general, was a very shaky edifice built up of generalizations of more or less obscure histories of diseases. Vigarous, Carrère, Swediaur, and Hufeland referred nearly every chronic disease to the "lues venerea" as a possible cause.

The exaggerations of Astrue's school gave rise to a reaction which was brought about by Hunter. In his work, "A Treatise on the Venereal Disease," 1876, he says: "It seems as if some organs were much less susceptible to "lues venerea" than others, and many are, in my experience, not at all susceptible to it. I have never seen the brain nor the heart, the stomach, the liver, the kidneys or other viscera affected, though cases are described by writers." His authority in this respect, as in that of the identity of the poison of gonorrhœa and syphilis, prevailed for a long time, so that his ignorance of the "lues visceral" served as a proof of its non-existence. This assumption stopped the progress of the syphilitic theory for half a century. A few cases were reported in periodicals in connection with nervous derangements, but they were explained on the old principle that they had their root in the syphilitic products of the surrounding bones. Syphilitic changes in the internal organs were, however, again affirmed by the works of Ricord and his school. He himself demonstrated a case of gummata in the brain to the Académie de Médicine in 1846. On post mortem examination of persons afflicted with tertiary syphilis and nervous disturbances, Lallemand, Rayer and others found abscesses and tumors in the brain. It was, however, above all, in the mode of origin, transmission, and external phenomena of syphilitic diseases, that rapid and genuine investigations were made in the thirties and forties, but the visceral affections remained obscure to a certain extent. In 1858, Virchow laid the foundation of the anatomical character of the syphilitic products, which is still prevalent. He showed that the anatomical process in the internal and external organs was the same, and that the former were not free from the irritating processes which generally corresponded to syphilis in its early stage. But as to the existence of this disease in the nervous system Virchow had no particular illustrating case to record. During the next decade important works on this subject were, however, published by Oedmanson and Lunggren, and in 1874, Heubner introduced a new factor into the discussion on nervous syphilis, which had hitherto been overlooked, viz.: the changes in the the walls of the vessels. All the discoveries made in the seventies, showed that syphilis was a near or remote cause of disorders in the nervous system, and that not only in the brain, but also in the spinal cord. It was in reality not until then that the existence of the disease in the latter was acknowledged, though doubtful cases were reported some time before. Thus in 1846, Ricord says that he was convinced that a great number of cases of paraplegia were originally due to tertiary syphilis. Knorre in 1846, described a case of paraplegia and paralysis of the bladder and rectum which recovered under treatment of iodic mercury. Referring to this in 1864, Zeisel seems inclined to think that the paralysis was caused by syphilitic changes in the corresponding spinal nerves. At any rate, he does not say a word about syphilis in the cord.

Ojör mentions several cases of paraplegia, arguing against the theory once adopted that it almost always is due to pressure from exostosis.

Speaking of a case of syphilitic paraplegia, Steenberg says that there was nothing abnormal in the brain or its membranes, but the spinal cord was softened in the region of the third and fourth vertebrae. In the great works on syphilitic diseases of the nervous system by Lagneau, Zambaco, Le

. 4

Gros and Lanceraux, published at the beginning of the sixties, there are also several cases mentioned of paraplegia from syphilis. Zambaco in particular, refers to no less than eleven, of which five died. On dissection of four of these, one was found to have a gelatinous effusion of a gumlike consistency in the cord, which under the microscope showed the usual structure of gummata.

In the other three the result was negative, and this Zambaco considers to be an unquestionable proof that syphilitic paraplegia of the greatest intensity may exist without appreciable lesions, and he is inclined to think that this is usual. Virchow also says that sometimes negative results are obtained in apparently syphilitic paralysis. A step forward was made by Lanceraux in his great work of 1866. He says that syphilitic lesions in the spinal cord only differ from those in the brain by position and less frequency. They may be situated either in the membranes or the substance of the cord. As a proof of the correctness of this view Potain states that in the case of a twin child, who died three days after birth, suffering from hereditary syphilis, the spinal cord was small, firm, and presented the appearance of a fibrous string of reddish gray color. Under the microscope no ganglion cells nor distinct nerve fibres could be discovered. Other writers corroborate the opinion that syphilitic paraplegia may exist without appreciable changes in the spinal cord, whereas some have found degenerations, tumors, or hyperæmia, with the arteries excessively filled. Mere microscopical examinations are thus insufficient to establish a case of syphilis of the spinal cord.

They do not give an absolutely conclusive result any more than do the symptoms of paraplegia which are principally the same whether due to syphilis or other causes.

The author next describes with considerable minuteness of detail 24 cases gathered from the literature, chiefly giving facts of pathological and anatomical interest, referring only briefly to their history, ignoring altogether numerous cases of syphilitic myelitis reported in connection with them, as the disease does not seem to have any *sui generis*.

In 50 cases on record where syphilis of the cord had been demonstrated he found that the time elapsing between the infection and the symptoms of paraplegia was:

In 4 cases ½ year.
In 16 cases between ½ and 1 year.
In 10 cases between 1 and 2 years.
In 4 cases between 2 and 3 years.
In 3 cases between 3 and 5 years.
In 4 cases between 5 and 8 years.
In 5 cases between 8 and 15 years.
In 3 cases between 15 and 18 years.
In 1 case over 20 years.

We thus see that functional disturbance due to syphilitic lesions of the cord mostly appear during the first year after infection, although the author allows that this frequency of morbid manifestations during the early stage of the disease may be apparent and not real, inasmuch as both physician and patient are apt to overlook the true nature of the disease whenever the symptoms are developed later in the course of the disease.

The author concludes his valuable paper by a careful report of five cases of his own, describing the microscopical as well as macroscopical change found in the diseased portions of the cord. The bulk of the literature relative to the microscopical changes found in syphilitic lesions of the spinal cord, seems to be directed mainly to changes in the nervous tissue, while Dr. Mœller's investigations are directed more prominently to changes found in the vascular structure of the cord.

Investigations thus far made seem to demonstrate that the endeavors to find definite and characteristic pathological and anatomical processes in syphilitic lesions of the spinal cord, have led to divergent results. What one investigator has considered constant and peculiar, others have regarded as being of less importance. There is as yet nothing absolutely certain with pathological and anatomical examinations alone, without further clinical information. We can, as a rule, only arrive at a probable diagnosis and nothing more.—*Nordiskt Med. Arkiv.*, Vol. XXII., No. 22.

GULSTONIAN LECTURES ON LEAD POISONING.—By Dr. Thos. Oliver. It may be considered that the insidious introduction into the human system of lead in minute quantities is followed by symptoms of a more serious and persistent character than when taken in larger quantities and for a shorter period. So insidious are the symptoms developed that physicians sometimes fail to discover it, and even medical men, the authors of pamphlets on lead poisoning, have suffered from the malady without being aware of it. The lead miner never suffers from lead poisoning. In smelting of lead men are frequently poisoned. Cattle grazing in the fields near smelting furnaces are poisoned with lead. Dogs that drink water from the streams running from lead factories, or that have slept upon their master's coats, have been poisoned. When equally exposed, women are more easily affected with lead than men. Some waters act upon lead readily, others not at all. Water containing acids produced by vegetable decomposition acts upon lead readily. One of the first and most important symptoms in lead poisoning is the development of anæmia. In all of the lead workers red blood cells are diminished in number. The bones may contain lead. The blue line on the gums is an important indication of poisoning when present, but it may be absent. An early and common symptom is pain in the abdomen. Dr. Oliver believes that the colic is caused by the action of lead upon the nerve ganglia primarily. He considers that as a rule the seat of the lesion of lead palsy is central, though in some cases it may be a peripheral neuritis. Optic atrophy is a frequent consequence of lead poisoning. Attacks of acute lead encephalopathy are common, and are sometimes preceded by hysteria. People addicted to drink are more easily poisoned by lead when exposed.—*Lancet,* March 7, 14, 21.

———

ANGINA PECTORIS.—Dr. R. Douglas Powell considers that angina pectoris has a broader neuro-pathological foundation than is generally supposed. It may be simply a functional disorder, or it may be a symptom of fatal disease of the cardio-vascular system. The first degree of angina pectoris he considers is vaso-motor, and in this condition from slight causes the systemic arterioles contract, raising the blood pressure and causing the heart to labor painfully. High arterial tension is an essential element in the majority of cases of angina pectoris. At first this high tension may produce only functional disturbance, but if it becomes habitual, organic lesions of heart and vessels result. The mechanism of vaso-motor angina pectoris is paroxysmally increased blood pressure from spasm of systemic vessels. The next form of angina pectoris is that associated with degenerative cardiac disease. Another form the author calls syncopal angina; this is associated with gout and occurs in persons over sixty-five years of age.—*The Practitioner*, April.

———

Total Paraplegia of two Years Standing Cured by Hypnotic Suggestion.—A woman, forty years of age, of previously good health, without history of nervous troubles, other than that of somnambulism during childhood and a few hysterical manifestations during later years, had lost in the space of a few days, complete use of both lower extremities. This condition of paraplegia had lasted two years when she presented herself for treatment to Dr. G. Raymondan of Limoges. A diagnosis was made of hysterical paraplegia, and treatment by hypnosis and suggestion was begun by him. Not until after a great number of sittings did improvement manifest itself, but in the end the result was perfect, or nearly so, the woman being able to work, and be about on her feet nearly the whole day.—*Revista Medica y Cirugia Prácticas*, Dec. 22, 1890.

———

Alcohol and Longevity.—E. McDowel Cosgrove, *Dublin Med. Journal*, July 1890 (abstract in *Schmidts Jahrb.*) discusses the statistics of the "Friendly Societies" in Great Britain on the relations of intemperance to longevity.

Amongst the "Friendly Societies" is the order of the Rechabites founded in 1835, the members of which are pledged to total abstinence. The figures of these cover 37,802 individuals and 127,269 years of life. Against these are placed the fraternities of Foresters and Odd Fellows, who take upon themselves no such obligations. The following table shows the expectation of life:

At the age of years	Odd Fellows. Years.	Foresters. Years.	Rechabites. Years.
20	41.3	40.2	45.1
30	34.0	32.9	37.3
40	26.7	25.8	29.1
50	19.9	19.1	21.2
60	13.6	13.2	14.2
70	8.5	8.3	8.5
80	5.0	4.9	4.9

In the higher extremes of normal longevity, the three groups are very much alike, so that one may perhaps conclude that in old age, total abstinence brings with it no special advantage. The greatest gain to the abstainers, is seen in early maturity, as in that period they have a greater expectation of life by five or six years. In other words, out of 1000 abstainers at twenty years of age 165 may expect to reach extreme old age, of Odd Fellows 134, and of Forresters only 118.

In the year 1840 a quaker by the name of Warner formed a society of teetotalers. In the year 1847 he included a section of "moderate drinkers" under the name of "General Section" who, together with the original abstainers, formed "The Temperance and General Provident Institution" of the United Kingdom. It was found between the period of 1866 and 1889 the number of deaths in the temperance section was 3198 as opposed to a pre-expected number of 4542, and in the "General Section" the deaths were 6645 against an expectation figure of 6894. The moderate drinkers, therefore, fell below the expected figure about four per cent., while among the abstainers, deaths were thirty per cent. less than had been calculated upon. It is not stated whether there was any difference in the former vital conditions of the two groups that might act in favor of one or the other, but it may be fairly presumed that there was none.—*Bannister.*

SCHEME FOR A PATHOLOGICAL INDEX.—Dr. Howden, of the Royal Asylum at Montrose, Scotland, gives a scheme for a pathological index which he has used for thirty years. It comprises a list of all possible pathological conditions of the various organs in a tabular form which is intended for use in institutions. If this scheme were adopted it would facilitate and simplify pathological observations in asylums, and render the records of cases complete.—*Amer. Jour. of Insanity*, April.

CO-OPERATION IN RESEARCH AMONG PATHOLOGISTS IN INSANE HOSPITALS.—Dr. Spratling, First Assistant and Pathologist in the Insane Asylum at Morris Plains, N. J., suggests that the pathologists of the American hospitals for the insane form an association for mutual co-operation and study. The advantages of such a scheme would be: exchange of specimens, liberal exchange of ideas, methods, and discoveries, and uniform method of practice in work, all of which would lead to a more thorough and exhaustive study of insanity and its pathology.—*Amer. Jour. of Insanity*, April.

THERAPEUTICS.

THE TREATMENT OF INFLUENZA-NEURALGIAS BY THE TURKISH BATH.—Dr. A. Frey, of Baden Baden, states the number of cases of neuralgia following influenza, coming to that resort in the last season, has been surprisingly large. As most of the patients try the thermal springs as a last resort, he states that it may be presumed that the therapeutic resources of the home medical adviser have been previously exhausted. The Turkish bath was successful in almost all cases, if continued for from four to six weeks. The frequency of bathing is not stated.—*Deutsche Medic. Wochensch.*, No. 12, 189. G. J. K.

ERGOT.—Dr. Hemmeter in experimental and clinical study of this remedy concludes that:

1. Ergot reduces the number of pulse beats per minute.

2. In the isolated frog's heart it reduces the force of the contractions.

3. It exerts a local poisonous influence on the heart of the batrachian, as well as on that of the mammal, when injected into the jugular vein.

4. Its main action, however, is exercised through the influence of the central nervous system.

5. It raises arterial pressure when injected into the jugular vein of mammals. The rise is preceded by a primary depression due to the local action on the heart.

6. It is impossible at present to decide whether this local action is due to an influence on the heart-muscle, or on the cardiac ganglia.—*Med. News*, Feb.

UNTOWARD EFFECT OF DRUGS.—This is the title of a series of articles now appearing in *The Chicago Medical Standard* by the editor of that Journal, Dr. Jas. G. Kiernan. The first article deals with neurotics, narcotics, antipyretics, and hypnotics. The action of the remedies is considered as affecting the lungs, heart, brain, cord, eyes, ears, throat, skin, liver, kidneys, and bladder. The names and actions of the remedies are arranged in a tabular form, which indicates a vast amount of labor with accurate and painstaking observation. It is impossible to summarize the article, as it is itself a summary. Every physician, however, should obtain copies of this journal, as the articles are valuable.

EPILEPSY CURED BY ANTIPYRIN.—Drs. McCall, Anderson, and W. R. Jack report a case. The patient was a boy nine years of age, who had been subject to fits two and one-half years. The patient had had at times as many as 57 fits in one day. Antipyrin was given at first in five grain doses three times a day. The patient appeared to have entirely recovered. There was no effect produced upon the fits until the antipyrin was increased to twenty-five grain doses, when the fits were stopped. When the antipyrin was reduced to twenty grain doses the fits began again.—*Amer. Jour. Med. Sciences*, May.

ANTIPYRIN.—Among spasmodic affections, asthma is generally relieved by this remedy at first, but it unfortunately soon loses its effect. In whooping cough, antipyrin is undoubtedly beneficial. For a child five years of age the dose should not exceed four grains. It is an excellent remedy in chorea; also in laryngismus stridulus.—*Dr. Jno. Aulde, in New York Med. Rec.*, Feb.

ATROPINE IN LOCALIZED MUSCULAR SPASM.—Dr. Leszynski read a paper on this subject before the New York Neurological Society, Feb. 10. He says he has used this remedy in a number of cases with good results. In a case affecting the right sterno-mastoid muscle an initial dose of $\frac{1}{80}$ of a grain was given, which was gradually increased until $\frac{1}{6}$ of a grain was taken daily. On the fourth day the patient was well. In another case, in which the spasm was on the right side of the neck, atropine was used hypodermically. The patient recovered in two weeks. Another case treated also recovered.

GALVANIC CHANGES DURING IDEATION.—In 1883, Prof. Fleischl deposited with the Vienna Academy of Sciences a sealed manuscript, which was lately opened at his request. Its main points are as follows: If unpolarizable electrodes are applied to symmetrical spots on the cerebral hemispheres, and connected with a delicate galvanometer, the deflection observed is small, or nil. If any organ, over whose center one of the electrodes is applied, be put into action, a deflection is observed. This can be easily demonstrated by putting the electrodes on the spots designated by Munk as the centers for optic perception, and alternately illuminating each eye, although similar spots can be found for other nerves. In chloroform narcosis no deflection results, thus showing that

this drug produces a paralysis of the cortex. It is not necessary to expose the cortex, as the deflection can be obtained
from the surface of the dura, and even from the bare bones of
the skull.—*Wien. Med. Presse*, No. 6, 1891.   G. J. K.

SURGERY AND TRAUMATIC NEUROSES.

DIAGNOSTIC VALUE OF SOME SYMPTOMS OF TRAUMATIC NEU-
ROSIS.—Geo. Güth (Thesis, Berlin, 1890) has investigated the
value of the contractions of the visual field and the irritability
of the pulse. He finds the first symptom to occur in such
order and regularity that its simulation would only succeed
after a long course of special training. In cases of unilateral
appearance of other symptoms, the contraction is greater on
this side. The functional cardiac irritability is not due to the
presence of the physician. G. does not believe a voluntary
acceleration of the pulse possible. After this symptom has
lasted sometime, an increase in the area of cardiac dullness
can be made out in some cases. *Neurolog. Centralbl.* No. 1,1891.
                                                    G. J. K.

SHOCK.—Dr. H. E. Bunts has an article on this subject in
the *Medical Record* for May 2. He says: "Regarded as an
abstract term, the word 'shock' conveys to the mind an
impression varying largely with the experience and personal
observations of the individual. One has seen sudden collapse,
and even death, result from a blow upon the epigastrium, and
recognizes it as a shock. Another looks upon the victim of a
railway accident and sees the crushed muscles and nerves
and comminuted bones, and is not surprised that here, too, is
shock; and though the complete, and sometimes fatal collapse
which may follow unwelcome news, or even joyous tidings, may
impress the same observer as something to be accounted for
on psychical rather than on physical grounds, yet he is forced
to recognize it as one of the undoubted forms of shock."

Here, then, we have an occult agent whose baneful effects are shown in the most varied range of cases, and yet presenting such a characteristic group of symptoms that we are forced to believe it identical in each, and it is to this grouping of symptoms, all indicative of a profound impression upon the nervous or vascular system, or both, that the name of shock has been given.   The author holds that shock is the sudden and violent impression upon some part of the nervous system acting not alone, as some say, upon the heart, but upon the entire blood vascular system.   While all the organs of the body may sympathize in the prostration which ensues, these disturbances are due to the derangement of the nervous and circulatory systems and must be regarded as secondary. Goltz's experiments would indicate that shock results from reflex paralysis of vascular nerves, especially of the splanchnics.   The stasis in the abdominal veins, and the afflux of a large portion of the entire mass of blood into the dilated vessels of the intestinal tract in consequence of the reflex paralysis of the intestinal vaso-motor nerves, causes anæmia of the brain, skin, and muscles, and thus leads to all the grave symptoms of shock.   Age is an important factor in shock: thus in youth it is rapid and extreme, but under favorable circumstances the reaction is equally rapid and pronounced.   In old age shock is apt to be less pronounced, while reaction, owing to the unfavorable condition of the patient, is often slow. Sudden and severe injuries causing contused and lacerated wounds are among the most frequent causes of shock. Speaking in a general way the nearer the wound is to the great centres of intellect, circulation, and digestion, the greater will be the shock.   Among the favorable prognostic symptoms are increased strength of pulse with diminished frequency, greater regularity of respiration, disappearance of cyanosis, return of bodily warmth, voluntary motion, lessened irritability of the stomach.   After reaction appears to be established relapses may occur and the exhausted and over stimulated heart give way and death ensue.   In the treatment of shock, pulse, temperature, and respiration are to be primarily considered. Stimulants are indicated. Hot baths, hot plates applied to the chest, abdomen, and extremities are important. Even when life seems extinct the patient may be revived by heat. Stimulants are indicated. The best are brandy or whisky, digitalis and strychnine. Injections of ether are also used. Morphia and atropin are of value when restlessness or delirium is present. Transfusion is said to be of service.

TREPHINING FOR TRAUMATIC APHASIA; RECOVERY.—At the March meeting of the New York Surgical Society, Dr. McBurney and Dr. M. Allen Starr reported a case of unusual interest. The patient was a physician, aged thirty-three, who, in August, 1889, was thrown heavily from his buggy. When picked up he was partially conscious, becoming entirely so shortly afterward, and being then sufficiently recovered from the shock to administer a hypodermic to one of his patients, and to converse rationally. Toward the evening of the same day, however, he became entirely unconscious and was hemiplegic on the right side. He remained unconscious for three days, during which time he lost control of his sphincters. On consciousness returning, it was found that he was suffering from aphasia, with paralysis of the right leg and the lower portion of the right arm. After three weeks there was some slight restoration of motion in the right foot and the ability to make slight extension of the forearm. Control over the sphincters was normal and the appetite was good, all the other symptoms persisting. At the end of three months there was some ability to utter sounds, and apparent intelligent appreciation of printed matter, but articulation was utterly unintelligible, and the intellectual faculties were undoubtedly greatly disturbed. The patient had come under the care of the speaker three months after the date of the injury. At this time he could make some motion in the right foot, and, when supported, some effort at locomotion. The right pupil was diminished in size, but the right side of the face was otherwise unaffected. There was apparent partial apprecia-tion of what was said to him, but the intellect was very much obscured.

The speaker had called on Dr. Starr to assist him by making out the exact character of the lesion, and without his invalu-ble aid he should have felt unwilling to go on with any sur-gical treatment of the case. The conclusion was that hæmor-rhage had occurred, shortly after the accident, on the left side of the head, affecting that area of the brain which controlled the faculty of speech, and those of motion and sensation in the right arm and leg. It was also believed that this hæmor-rhage existed on the surface of the brain, and not on the internal capsule, and that the general symptoms indicated a cortical, and not a deep lesion.

An operation was performed a year ago last December with a view of exposing the speech center, and, if necessary, of following up the exploration to the arm and leg centres, in the search for the suspected superficial clot. After making a large flap, the trephine was applied seven-eighths of an inch behind the angular process of the frontal bone on the left side. The middle meningeal artery ran directly across the space thus exposed.

There was no clot outside the dura mater, which was then incised and the middle meningeal artery ligated. There then protruded a soft, pulpy, cystic-looking mass which it was decided must be œdematous pia mater. Anterior to this œdematous portion the pia and the brain presented the normal appearance, but behind and above this the pia was dark red and looked as if it covered the thin edge of a distinct clot. The opening in the skull was enlarged in an upward and back-ward direction by means of the rongeur forceps, and the dura laid open by suitable incisions until the entire surface of the clot was exposed. Thus an irregular piece of bone 3½ x 2¾ x 1¾ inches was removed. The long axis of this area was along the fissure of Rolando, and the greatest breadth was backward from this fissure. The area of densest blood clot was in the course of the fissure of Rolando, and along this the pia was incised from the lower extremity of the fissure to the highest point uncovered by bone. An old blood clot was found dipping down into this fissure, and spread over the neighboring convolutions, especially the middle part of the postcentral convolution. This portion consisted of a very delicate layer of coagulum, and was removed by gently rubbing the surface of the brain with a soft wet sponge. In this way the clot which extended down between the convolutions was removed, as well as that on the surface, and with no apparent laceration of the brain tissue. All hæmorrhage was carefully checked by applying ligatures of fine catgut to bleeding points. The dura was sewed up with fine catgut, except at the middle of the incision. The scalp incision was closed, except at the middle, with a continuous silk suture. Loose iodoform packing was introduced, forming a soft mass filling the cavity from the surface of the brain to the level of the cutaneous surface. An absorbent antiseptic dressing was applied, so arranged as to exert but little compression. After the brain had been exposed by the removal of so large an area of bone, the frontal portion had settled down and back, leaving a remarkably large air space between the brain, covered by pia, and the skull lined by dura. On the evening of the same day the patient was able to utter the first words spoken by him since he had become unconscious, some hours after the injury. At the end of a week he was able to move his leg, and in three months to walk with the aid of a stick. His right arm was now improved, but its action was sluggish. The aphasic symptoms had steadily diminished and the patient was constantly acquiring new words, was now able to converse, and had resumed his professional work, driving his horse alone. His intellect was now absolutely clear. That this improvement was to continue to the point of physiological re-establishment of all the normal functions of the parts implicated in the hæmorrhage the speaker was unwilling to assert, but he did look for much improvement still to come, and believed that the patient's chances were extremely good.

Dr. M. Allen Starr said that the case had seemed to him specially interesting because of the possibility of making a pretty positive diagnosis. If an aphasia was at first total, and then partial recovery ensued, the possibility arose of the existence of subcortical lesion, or lesion in the course of the speech tract between the third frontal convolution and the medulla. But in this case the aphasia had remained total for three months, and hence it was thought that the lesion must be cortical and located in the posterior part of the third frontal gyrus. The location of the paralysis pointed to the central convolutions in their middle portion as the seat of the lesion. Another interesting feature in this case, which had served to throw light upon a disputed point, was the fact that there had been throughout a slight degree of anæsthesia in the paralyzed limbs. This pointed to the acceptance of the theory that the sensory and motor areas in the cortex coincided. The lesion found thus threw a light on the localization of the tactile sense.

The success which had so far attended the case was no doubt due to the prompt action taken by Dr. McBurney when the patient came under his care. The reason why the patient was not completely restored was the delay of four months between the time of injury and the operation. It could therefrom be deduced that, where a cortical lesion could be diagnosticated, the case should be dealt with surgically within four weeks after the injury, and that the sooner the operation was performed the better were the chances of recovery. There was no doubt that degeneration had progressed in a number of cells and of fibres in the speech area, and therefore he was not yet quite well. The case was also interesting for observation as to whether there would be complete regeneration in the motor tract; such regeneration was apparently in progress, as the power was constantly improving.

In reply to a question by Dr. Stimson as to whether other lesions of the brain would cause aphasia, Dr. Starr said that, according to the teachings of Broca in 1861, it was necessary, in order to produce disturbances of speech, that the third frontal convolution should be injured. This conclusion had, however, been modified. At present it was necessary to distinguish three separate forms of aphasia and lesions producing them. In the first place, motor aphasia, such as the patient had suffered from, was due to a lesion of Broca's convolution, in which there must be ability to recognize the meaning of words, but inability to articulate. The power of understanding a word when uttered was different from the power of articulation, and had been shown to be a function of the first and second temporal convolutions, about an inch behind and below Broca's centre. Lesions in this locality give rise to sensory aphasia, or word-deafness. This form was often met with by surgeons in patients with disease of the mastoid cells and complicating abscess of the brain; and in it the patient could not speak, because he could not recollect the words with which to express himself. The third form was associated with lesions situated farther back in the occipital and parietal lobes, and was accompanied by that form of aphasia known as word-blindness, in which the individual was unable to recognize printed language. All cases of word-blindness were complicated in part with word-deafness. The connection between the occipital and temporal lobes was so close that one could not be injured without inhibiting the action of the other. Hence a lesion in three different localities gave rise to three different forms of aphasia.

At the same meeting Dr. McCosh reported a case of *Laminectomy for Injury of the Cervical Vertebræ*. Dr. McCosh presented a man, aged 33, who had, while a sailor, been severely injured eighteen months before, when he was supposed to have sustained fracture or dislocation of the fourth, fifth, or sixth cervical vertebra. He had been bedridden for the nine months following the accident. The injury was caused by the fall of a shackle on board ship. This had struck the man upon the head and doubled him up, bearing his head down upon his breast and flexing his thighs upon his abdomen. He had remained unconscious for the first twenty-four hours and during the first week had been more or less delirious. He had also been completely paralyzed over the entire body below the clavicles. In May, 1890, the patient had come under the speaker's charge for surgical treatment. There was then marked atrophy of all his muscles. He was unable to raise himself in bed, and when he was held up his head fell forward upon his chest. When he was placed upon his feet he could stand with assistance for a few minutes, but could not walk. The left forearm was flexed on the arm and the whole left upper limb was in a condition of spastic paralysis and utterly useless. He had some use of his right arm, but could not move the fingers nor hand. He was never free from pain, which was most severe in his upper limbs, especially in the left shoulder and arm. Any movement aggravated the pain. An incision was made starting at the occipital bone and exposing the vertebral column. A distinct curvatute was found of the cervical spine to the right, and the fourth vertebra was found displaced an inch and a quarter to the left of the median line. No sign of fracture could be discovered. The laminæ of the fifth cervical vertebra were divided and the posterior arch was removed. The dura mater in this region was found to be of a dark red color and very much thickened. It bled profusely on the removal of the arch of bone to which it had been adherent. The dura was not opened. The wound was left open with the expectation of doing further operative work if the symptoms did not improve. It would be seen from the man's present condition that a great improvement had taken place. He could now get about quite comfortably, could use his arms, and was altogether in a fair condition. He has walked as far as four miles, and could sign his name with his right hand. In the left upper limb flexion and extension at the elbow were perfect and he could raise his arm as far as a fibrous ankylosis at the shoulder allowed.—*N. Y. Med. Jour.*, May 2.

PSYCHOLOGICAL.

PATHOLOGY AND SYMPTOMATOLOGY.

AUTOMATIC MUSCULAR MOVEMENTS AMONG THE INSANE. Dr. C. P. Bancroft has an article on this subject in the February number of *The American Journal of Psychology.* There is an intimate relationship between automatic functional activity of the central nervous mechanism and the muscles. Muscular movement represents cerebral cell activity. A certain constant and regular transmission of nerve force from central cells to muscles is natural in a state of health. An outlet for the constantly accumulating energy is found in muscular activity in growing animals and in the young. In adult life the potential energy of cerebral cells is manifested in more practical and useful ways; then it is under conscious control. As the individual develops, useless and superfluous movements are one by one checked, eliminated or co-ordinated. The close relationship between automatic muscular movements and the inhibitory power renders a study of the latter essential to the understanding of the subject of automaticity in health and disease.

Inhibitory power varies with the individual, and varies at different periods in the life of the same person. It is a power that is incorporated in the nervous mechanism of the individual, and is largely a matter of individual growth. It is associated with the highest faculties, and disturbance of these involves disorder of inhibition. In health the inhibitory power is exercised in checking irrelevant trains of thought, and in restraining undesirable muscular movements. In

diseased conditions of the brain, morbid irritation of delicate
nerve centres may overcome inhibitory power, and thus liberate
an excess of nerve force that may seek an outlet in the usual
way by appearing as muscular movement. Chorea and epilepsy are illustrations in which brain cells are in an unstable
condition. In the functional and organic brain disturbances
that accompany insanity, morbid muscular activity and inactivity are constant symptoms. The muscular disturbances of
insanity are three fold:

1. *States of excessive muscular activity due to central irritation.*
The constant muscular activity of acute mania, *melancholia
agitata*, the active stages of paretic excitement, and of many
cases of recurrent mania, illustrate this form of motor activity.
The central irritation is so great as to overcome all inhibitory
resistance, and the patient, therefore, paces to and fro, runs,
jumps, and engages in purposeless acts and spasmodic movements.

2. *States of deficient nerve-muscular activity due to central
degeneration.* This is illustrated in conditions of permanent
mental enfeeblement, as terminal dementia.

3. *States of automatic cell activity in the cerebrum, occurring
not infrequently in acute insanity, and quite constantly in the
chronic forms of the disease.*

All nerve structures manifest a tendency to muscular automatic activity. This law is especially noticeable in the lowest
forms of life. In man, a high volitional power distinguishes
him from the lower animals. In man the voluntary activity
of the highest centres may be disarranged by insanity and
their normal exercise suspended; at the same time the basal
ganglia being uncontrolled, act in their normal automatic
manner, and, as a result, there is purposeless thought and
action. Every insane hospital has illustrations of this automatic action of the nerve centres. One patient walks backward and forward mechanically for hours together; another
picks at his clothing for an indefinite period; another gives
utterance to strange and meaningless noises; another repeats

a peculiar word or phrase; another makes strange motions with arms and hands, or maintains singular attitudes. Prolonged automatic activity in either speech or muscular movement among the insane suggests a serious lesion in the higher centres. The more mechanical and purposeless the acts and words of the patient, the graver is the prognosis. We see evidences of this tendency to automatic action of the nerve centres in advanced stages of fever, where purposeless repetition of words and muscular movements are noticeable; and this condition is taken as an indication of serious impairment of the higher brain functions. Two laws underlie every form of nerve activity:

1. The discharge of the nerve centres occurs along those tracts that offer the least resistance.

2. The more frequently does the discharge occur the easier does its repetition become. Among the insane, a delusion, or hallucination of sight or hearing, may conform to the patient's surroundings, and, being unarrested, lead to the establishment of the habit. Probably in some such simple and fortuitous way are developed the pulling out of the hair, tearing the clothing, walking in a beaten path, making singular movements, or uttering meaningless words. It matters little whether the performance of these habits be painful or disagreeable, the morbid route having been once selected is pursued in their lack of self control. The importance of early breaking up bad habits among the insane will be readily understood. Apparently the same tendencies to automaticity of action may be utilized in a good, as well as in a bad direction. By careful supervision and employment we may break up useless and vicious habits and thus retard mental deterioration.

———

Dr. Cowles closes his last article on *The Mechanism of Insanity* with the following summary:

1. The first obvious fact presented is, that every new mode of action of the mechanism gives it a functional disposition to repeat the organic process. This under the law of use and

1891.]   PATHOLOGY AND SYMPTOMATOLOGY.   55

practice constitutes the law of habit, which prevails in all the activities of the mechanism, bodily and mental, and is shown to have a physiological basis. It is seen that this law works with equal force in fixing dispositions to repeat disorderd actions.

2. The extension of the law of habit shows that it is fundamental to the law of association which is apparent in all the activities, and is of special importance in the mental sphere. The study of its processes develops the fact of there being a regulating and controlling influence which may be broadly designated as inhibition, and tnat the higher inhibitory control, in the ideational processes, manifests itself through the attention, which is closely related to the motor apparatus; in its natural form it is spontaneous or reflex, and in the highest development of inhibitory power, is under the control of the will and acts as voluntary attention. It is seen that among the earliest indications of fatigue and exhaustion of mental activities is weakening of the inhibitory and directing power of voluntary attention, and that this has an important signification in mental disorder.

3. The mechanism being thus constituted and subject to certain laws that regulate its activity under due control, must have an indwelling motive force. The energy of muscle and nerve comes into play under the physiological law of storage and expenditure according to the fundamental law of all cell activity. Rest, sleep, and nutrition contribute to the building up of complex cell compounds, subject to physical and chemical laws; and the discharge of energy is accompanied by destructive chemical changes that yield toxic waste products. These processes probably occur in both peripheral and central mechanisms.

4. The mechanism when put into continuous use under the foregoing conditions, does not go on indefinitely upon being stimulated into activity, but manifests the phenomena of fatigue when subjected to prolonged exercise. The results of such use within normal limits may be regarded as normal

fatigue; this is wholesome, and power is gained in accordance with the law of physiological use. But there are harmful results from disuse and overuse constituting conditions of pathological fatigue.

5.   Normal use with normal fatigue of both peripheral and central mechanisms are inseparably accompanied by chemical changes in the parts exercised, and toxic products are formed in the tissues. The effects of fatigue can only be studied as including the direct results of the discharge of energy plus the toxic influence of the products of cell activity, which emphasize and produce in part, the fatigue and exhaustion; these elements exist in various proportions and under varying conditions. It is seen that, as between peripheral and central mechanisms, fatigue in one will produce its phenomena in the other; and that the blood must be studied as the carrier not only of nutritive material but of toxic elements from one part to another.

6.   In the normal mechanism, under normal use, there is maintained the balance of waste and repair; and a healthy activity is sustained by the removal of the autogenous waste products and the supply of nutritive material under due conditions of rest and sleep.

7.   In studying the processes that produce the phenomena of fatigue and the graver degree of exhaustion, four factors must always be kept in mind as possibly in operation. Two of them are positive, viz: the direct results of discharge of specific cell energy, and the reinforcing effect of the toxic products of tissue activity. The other two are negative, viz: the withholding of nutritive material in the circulating medium, and the presence of toxic material in it that lessens the power of assimilation in the tissues.

8.   The mental mechanism presents a definite series of elements for ordinary clinical observation as to divergencies from the normal in the functions of its activities. These are subject to examination in due order, as in their sequence as elements of reaction time; and they may be tested to some

extent, experimentally, by psycho-physic methods. In an ordinary clinical examination they may be observed to show marked differences from their normal reactions, presenting different symptom-groups under different bodily conditions. The phenomena of habit, association, and control, appear; and there are different psychical manifestations of fatigue, independent of, or correlative to, conditions of fatigue in the peripheral mechanisms. It may be possible also to discriminate between the relative influence of pure fatigue from discharge of energy, and the effect of toxic substances, in the causation of conditions of fatigue and exhaustion and concomitant disorder of psychical processes.

Next in order is the consideration of the effects of disuse and overuse in the mechanism of pathological fatigue or nervous exhaustion, with special reference to the significance of the accompanying mental symptoms.—*Amer. Jour. of Insanity*, April.

SELECTION OF CASES OF INSANITY FOR PRIVATE CARE (by H. R. Stedman, M. D., Boston). This is an excellent article and an abstract does it scant justice. In secluding the insane we should keep in mind the necessity for the fullest freedom that is compatible with the welfare of the patient. Considering the prejudices against insane hospitals, and the odium that attaches to those who have been committed to such institutions, it is better for insane patients who can afford private care to be treated outside of the public hospitals. When the insanity is pronounced, patients should not be advised to travel, but in carefully selected cases of this disorder in its incipient stage a certain amount of travel may be advisable. The author's experience coincides with that of other specialists, that in cases of melancholia in particular, travel is injurious. In considering the home care, the relation of the patient to his family and to his business should be taken into account. If the patient keeps the household in turmoil, and wears out his relatives, home is not the place for him; or, if he has an

increasing antipathy towards any of them, or dangerous delusions, or is suicidal, then he should not be treated at home.
In treating these cases outside of an institution the physician
should see the patient as regularly as if he were treating him
for some acute disease, and should give the most minute directions for his care and treatment.   Insanity connected with the
puerperal state can often be treated at home.   It is important
that the child should be removed from the mother.   In those
cases where visits from friends are inadvisable it is better for
the patient to be taken some distance from home, as the proximity to home and family may cause the patient to worry to
such an extent as to interfere with recovery.—*Amer. Jour.
Med. Sciences*, April.

DIAGNOSIS OF INCIPIENT MELANCHOLIA.—Dr. S. Grover
Burnette delivered a lecture at the Kansas City Medical
College upon this subject.   His classification of melancholia
is that of Dr. Gray of New York which is: simple melancholia,
melancholia agitata, melancholia attonita, and melancholia
with stupor.   Simple melancholia and melancholia agitata are
the forms that so frequently escape an accurate diagnosis in
the early stage of the disease.

Too frequently melancholia is allowed to go on in its incipi-
ency unrecognized as to the true nature of the affection
present, but instead is diagnosticated as "neurasthenia," ·
"nervous exhaustion," etc., a condition entirely foreign to
the one that should not be overlooked, and with such a diagnosis the patient is treated in a like indefinite manner until
the realization of the true nature of the disease is forced upon
us through the medium of some tragic affair, either homicidal,
suicidal, or both.   This peculiar tendency of the disease
renders its early recognition of the utmost importance.

In simple melancholia the hallucinations and delusions are
rarely definite enough to be of diagnostic value.   In the early
part of this form the intellectual faculties are seemingly
unimpaired.   In this condition there are usually three well

defined symptoms to which Dr. Gray has called attention, viz: insomnia, "post cervical pain," and depression.- Dr. Gray has found the characteristics of this form to be: indifference, slow mental reflexes, occasional history of terrifying delusions and hallucinations.   In melancholia attonita and melancholia with stupor, the three symptoms mentioned are not so easily demonstrated, owing, perhaps, to the more profound mental affection.—*N. Y. Med. Jour.*, May 2.

PERIPHERAL NEURITIS IN PARETIC DEMENTIA.—Pick, *Berlin Kl. Wochensch.*, 47, 1890, (abstract in *Schmidt's Jahrb.*) discusses a case of general paralysis with autopsy in which there occurred a typical peroneal paralysis in the course of the disease, due to peripheral neuritis, in which peculiarity the disorder in this case was comparable to certain cases of tabes dorsalis.   The same complication was also observed in a second case, occurring in a woman thirty-eight years of age.

                                                          G. J. K.

ANOMOLIES OF THE SCALP IN THE INSANE.—Dr. C. Paggi has studied the peculiarities observed by himself and others in two or three of the Italian asylums, of a wrinkled, or folded condition of the scalp in the posterior region of the head, where it corresponds with a platycephalic condition of the superior occipital and sometimes of the parietal.   It is not exactly of the hypertrophic nature, as he had formerly held in 1884, but was rather due to an arrest of development of the posterior portion of the skull, while the scalp itself developed normally; being not put upon the stretch by the bones beneath, folds develop in the skin.   All the subjects of this peculiarity being affected by congenital mental weakness, he considers this as an important point to be noticed.   From a study of 150 cases of his own observation, and 50 collected by Dr. Pianetta, he concludes:

1.   That in the insane the vortex of the brain at the obelion is situated most frequently on the right side, next on the median line, and least frequently on the left side.

2. The distance between the vortex and the median line is greatest when on the right side, and in that case when it is also at a level much lower than the obelion, is a sign of degeneration.

3. A double vortex specially and regularly disposed (one low and the other high, or one central and the other lateral) is a sign of physical degeneracy.

———

CONFUSIONAL ACUTE PSYCHOSES AND PARANOIA.—Dr. Rosenbach (*Ann. Medico-Psych.*, Jan.–Feb., 1881), raises the question of the relation of the acute confusional insanities to paranoia. The acute confusional insanities develop on a temporary neuropathic basis, which arises from essential fevers, operations, anæsthetics, and other drugs and conditions of exhaustion. They are characterized by an hallucinatory confusional state, with agitation which closely resembles in some particulars, the period of transformation of paranoia. The present article discusses chiefly the literature antecedent to further analysis of the subject.

———

SYSTEMATIZED DELUSIONAL INSANITY IN THE DEGENERATE.— Dr. Magnan, (*Prog. Med.*, March 28, 1891,) claims that paranoia in persons from ancestral or congenital taint differs from that occurring in the degenerate, from the fact that systematized delusions appear in adult age, while in the degenerate their appearance has long been preceded by character anomalies, emotional phenomena, intellectual and moral defects, imperative conceptions and impulses, which from an early age reveal the mental instability and produce very striking mental phenomena. The systematized delusions sometimes appear early. The ambitious delusions may appear even in childhood. They become impressed on the mind of the patient and remain unchanged for a prolonged period of the patient's existence, which separates them from the progressively evolved systematized delusions. These are the fixed ideas analogous to imperative conceptions by which indeed they are some-

times produced. If sometimes they appear to have been produced by more or less plausible reasoning, more often they
occur without any preparation, without hallucinations and
without delusional interpretations of any kind. Magnan cites
several cases, but these certainly show that there is a progressive evolution; indeed he admits that one of them forms a
connecting link between the two species of paranoia.

SYSTEMATIZED DELUSIONS IN THE DEGENERATE.—Dr. Magnan (*Prog. Med.*, March 28, 1891), claims that long before the
full evolution of systematized delusions in the degenerate,
these are preceded by anomalies of character, emotionalism,
moral and intellectual lapses, sometimes imperative conceptions and impulses which, early in life, give striking evidence
of disturbance of mental equilibrium. The delusions may
appear very early, even at the age of ten or twelve. These
fixed ideas long remain the same, and are not always the
product of gradual evolution.

EVOLUTION OF ROMANTIC LOVE.—It is discussed by Richet
in the current "*Revue des Deux Mondes*" and by Kiernan in
the April "*Alienist and Neurologist*" who both agree that it
was a check on explosive manifestations of the low animal
type, and like all checks on egotism, a moral factor. Kiernan
claims that love originated in protoplasmic hunger, which,
through admiration of personal beauty, desire to please the
loved object, and love of offspring developed into romantic
love; the mere material enjoyment of sexual society became
subordinated to intellectual pleasure. Fink, Geddes, Buchner
and others, have demonstrated that romantic love exists in
birds and animals. The Wanderoo monkeys pair for life,
and the paired die if separated from one another. Kiernan
points out that the value of woman as laborer among the
lower savages sank the human female into a mere beast of
burden, and temporarily checked the evolution of romantic
love. Sex equality in labor sank woman below the animal

and degraded man in that respect. The race has since had to regain the romantic love evident in animals. Dr. Kiernan expresses the opinion that the abhorrence of sexual crimes, now evident in the highest races, shows a decided evidence of advance, and justifies hope for the future, since these crimes are returns to primeval conditions and hence demonstrate the pathway along which the race has advanced.

PARETIC DEMENTIA OF LUETIC ORIGIN. — Dr. Camuset (*Annal-Medico-Psych.* Jan.–Feb. 1891) asks for solution of following problem: Is lues an ætiological factor in paretic dementia; if it be a factor, is it a potent factor; how does it act, directly or indirectly? In other words, is paretic dementia a tertiary, or hereditary accident? His own cases do not solve this, but he is of opinion that there is a type produced by syphilis characterized by slow progress, prolonged duration, and very often by frequency and duration of the remission as well as by a tendency to dementia, or depression, rather than expansion. He cites one case, that of a female, which was twelve years under observation. Another case, that of a male, is still living, the paretic dementia having become demonstrable in 1876 immediately after a luetic eye disease. He cites other cases varying from five to ten years in duration in which several seemingly complete remissions occurred from time to time.

THROAT IMPERATIVE CONCEPTIONS.—Under a title of which this is a translation. Dr. V. Galippe describes (*Arch. de Neur.*, Jan., 1891), a condition which frequently crops up in laryngological literature, under the title of "Imaginary Foreign Bodies in the Throat." Dr. Galippe points out that these phenomena have a certain illusional basis. These conditions are frequent in hysterical women, and some precede the onset of neuroses like paretic dementia. There is often coincident lingual and buccal disorder. Treatment of a moral nature has been found of most value when conjoined with that indicated by local condition.

FALSE SECONDARY SENSATIONS.—Under this title, Dr. F. S. de Mendoza, (*Arch. Internat. de Laryn., de Rhen et d' Otologie,* March, April, 1891,) places mental perceptions false, though physiological, of the special senses which in themselves have no real existence, but arise from an actual perception of another sense. These secondary sensations occur with all the senses, but chiefly with that of light. These phenomena, Dr. Mendoza entitles physiological pseudo æsthesia, and divides them into five classes: Pseudo-photo-æsthesia, false visual secondary sensations; Pseudo-acou-æsthesia, false auditory secondary sensations; Pseudo-phre-æsthesia, false olfactory secondary sensations; Pseudo-gou-æsthesia, false gustatory secondary sensations; Pseudo-apsi-æsthesia, false tactile secondary sensations. The most frequently observed are false sensations of color, then follow in order of frequency, olfactory, gustatory, auditory and tactile. Each of these five classes may be subdivided into six, according to whether the primary actual association be (in order of frequency) visual, auditory, olfactory, gustatory, tactile, or occur merely in the intellect. Dr. Mendoza calls attention to the fact that noises may in certain railroad men create false perceptions of actual colors.

THERAPEUTICS.

HYPNOTISM IN THE NEUROSES.—Luys (*Jour. de Med.*, March 15, 1891) reports the results he has obtained from hypnotism in the neuroses. The great majority are hysterics, or hystero-epileptics, many of whom had been cured of the disorders for which they had specially plsced themselves under treatment, such as frequent crises, contractures, neuralgias, nervous coughs (which in Luy's opinion are due to diaphragm spasm). The hysterics, who readily yield to transfer phenomena, have found in hypnotism a means of throwing off their nervous exaltation and checking their crises. Three cases of this type have been utilized in his clinic, yet the patients, far from becoming exhausted, find in this a sedative which lessens the intensity and decreases the number of their crises. Nine cases of paralysis agitans have been cured and so remained for a year. Other cases, while improving in many respects, retain the hand motions which, however, are somewhat lessened. Epileptics have been benefitted chiefly by the transfer method. The cases accompanied with vertigo have been most rapidly improved. Neuralgias, cephaliac, sciatic, and brachial, have been gradually cured on transferrence. In the treatment of "writers cramp" by this method there has been produced not only a cramp in the opposite arm, but also a loss of motor power in the leg of the same side. This, in Luy's opinion, indicates that the cramp is not limited to the muscles of the arm and forearm, but is a complex disorder affecting the nerve centre themselves, shown by the similar reaction in the arm and leg of the subject of transferrence. Tremblings, contractures of various origins, hemieplegias or paraplegias, vertigoes with titubation, and agoraphobia have been cured, or much improved. Cerebral anæmia, and cerebral torpors, due in young subjects to school overpressure, followed by hypnoleptic states have been cured. This transferrence has also been found of value in melancholiac, depression, and emotional disturbance.

CHLORALAMID IN INSANITY.—Dr. Morandan de Monteyel . (*Jour. de Med.*, March 15, 1891) reports that he has submitted sixty patients to chloralamid. Twenty-five were cases of pure insanity, no taint. In half of these the results were valueless, while in the remainder the drug acted well. In the cases with nocturnal agitation the drug was useless. The same was true of the vesanic cases. Where nocturnal agitation was not present the drug could be discontinued and the sleep habit formed by it would continue. In five per cent. only of the agitated cases did chloralamid give good results. Vomiting and diarrhœa were the chief untoward effects observed and to these collateral conditions contributed.

ANTIPYRIN IN MENTAL DISEASES.—Berarducci and Agostini report the results of a trial of antipyrin in eighty-five patients, subjects of various forms of mental disorder; in the *Archivio Italia no per la Malatie Nervase* XXVIII., I., II., Jan. and Mar. 1891. They conclude as follows: Making deductions from numerous and prolonged trials, covering a period of eighteen months, with eighty-five patients, we are able to say that antipyrin has a decided effect in irritative hyperæmic meninga-encephalic states and while modifying little, or not at all, the progress of the mental disorders, it diminishes the hallucinations, quiets the agitation occasioned by the painful stimuli, manifesting altogether a purely analgesic, and not a hypnotic action.   It may have beneficial effect in:

1. Epileptic *petit mal*, and notably in the accessory psychic disturbances.

2. In hallucinatory delirium.

3. In chronic mania.

4. In agitated dementia following mania.

5. In melancholia agitata.

It is without action in

1. Simple and hypochrondriacal mania and melancholia.

2. In hysteria.

3. In epileptic *grand mal*.

We may say in conclusion that while antipyrin was not an essential medicament in the cases recorded, it may be used with advantage as a substitute for the ordinary sedatives when their employment is for any reason contra-indicated.

THE TESTICULAR JUICE OF MAMMALS IN THE THERAPY OF MENTAL DISEASE.—Drs. Venturi and Frondo *Il Manicomio*, 1 and 2, 1890 (Abstr. in *Archivio Italiano*, I and II., 1891). The following are the author's conclusions:

1. The subcutaneous injection of the testicular juice of the lower mammals in the treatment of insanity of neurasthenic origin, has given only negative results, or at best only transitory and unimportant betterment.

2. The majority of the patients experimented upon showed no change whatever, or a very slight and transient one. In a few there was observed a sudden increase of excitability that disappeared with the suspension of the injections, and gave place to the former depression.

3. As regards the reinforcements of the spinal functions (in vesical paralysis) there was only proven a slight amelioration during treatment, or at most for a short time after.

4. It was constantly observed that the peripheral temperature was lowered from one to five degrees, that the radial pulse diminished in frequency and increased in force, that the breathing became slower and more superficial; all these facts were observed very soon after the injection, and gradually disappeared in from six to eight hours.

5. The temporary character of all the resulting psychic and physical symptoms, even after the treatment has been continued for weeks, lead the authors to admit the greater probability that the injection of the testicular juice excercises on the nervous centres, especially the spinal ones, only a stimulant and exciting action, rather than increases their power of action as was believed by Brown-Sèquard.

NUTRITION IN HYPNOTISM.—Dr. Gilles de la Tourette (*Progrés Medical* Dec. 1890) is of opinion, from urological researches, that, so far as these furnish any evidence, hypnotism is certainly a pathological state.

STOMACH-WASHING IN INSANITY.—Dr. Voisin (*Bulle-Gén de Therap*, Jan. 30, 1891) cites in detail several sitiophobiac cases of depressed emotional states in which he used washing of the stomach in the treatment of sitiophobia. He concludes that washing of the stomach gives the best results in the depressed states accompanied with gastro-intestinal torpor, whether this be the cause or consequence. In non-hereditary cases rapid results of benefit are obtained. Hereditary cases are improved. In emotional depression of paretic dementia with sitiophobia good results are also obtained. Voisin has found that ideas of negation, suicidal tendencies, delusions of poisoning, and hallucinations, are accompanied with gastro-intestinal torpidity, hence he concludes that this is the patho-anatomical substratum on which refusal of food depends. Treatment of this sitiophobia by stomach-washing will fulfill all the indications and cause its disappearance.

Dr. Seppilli concludes an article on *The Therapeusis of Mental Diseases by Means of Hypnotic Suggestion* as follows:

1.   Therapeutic hypnotic suggestion cannot be instituted as a general means of cure in the treatment of mental diseases, owing to the difficulty of hypnotizing the insane.

2.   Hypnosis succeeds most readily in the hysterical and epileptic.

3.   The most certain results of hypnotic therapeutic suggestion have, up to the present time, been obtained in the psychoses depending on hysteria and dipsomania.

4.   Hypnotic suggestion may be employed when the insane submit to it of their own accord, and derive benefit from it. The physician should use it with great caution, and take account of the hurtful effects which, in certain cases, may be produced.

5.   Therapeutic suggestion made in the waking state is the most reliable and effective means of cure in mental diseases, and to it almost solely are due the beneficial effects of the asylum, which represents a real suggestive surrounding.

6.   In cases of melancholia without delirium, cases of fixed ideas, cases of alcoholism, and in slight forms of stupor, suggestion methodically repeated in the waking state, in order to combat the morbid phenomena, may prove effectual.

7.   In the chronic forms of paranoia suggestion has never given favorable results.—*Amer. Jour. of Insanity*, April.

6

MEDICO-LEGAL.

Dr. Clark concludes an article upon *Crime and Responsibility* in *The American Jour. of Insanity* for April, as follows:

1.   The natural history of crime shows that the bráins of chronic criminals deviate from the normal type and approach those of the lower creation.

2.   That many such are as impotent to restrain themselves from crime as the insane.

3.   That immoral sense may be hidden from expediency by the cunning seen even in brutes, until evoked by circumstances.

4.   No man can shake himself free from the physical surroundings in which he is encased.

5.   Crime is an ethical subject of study outside of its penal relations.

6.   Insanity and responsibility may co-exist.

7.   Some insane can make competent wills, because rational.

8.   The mono-maniac may be responsible should he do acts outside the line of his delusion, and which are not influenced thereby.

9.   Many insane are influenced in their conduct by hopes of reward, or fear of punishment, in the same way as the sane; the rudiments of free will remain.

10.   Many insane have correct ideas in respect to right and wrong, both in the abstract and concrete.

11.   Many insane have power to withstand being influenced even by their delusions. Therefore, irresponsibility and insanity do not always cover the same ground.

MODIFIED RESPONSIBILITY.—Wille, *Ztschr. f. Schweizer Strafrecht*, III. 1. 1890, (abstr. in *Schmits Jahrb.*) criticises the use of the term dubious as applied to mental conditions affecting responsibility as employed by Flemming and other writers on the medico-legal questions of insanity. On the other hand he says there are well marked conditions in which responsibility is modified, viz:

Richet—Experimental Studies in Thought-Transference and Clairvoyance. Translated by A. v. Schrenck-Notzing (adv.).

Lippman—Medical Care of the Insane outside of Institutions (adv.).

Franke-Hochwurt—Tetany (adv.).

Flechsig—Handbook of Balneotherapy (adv.).

Forel—Hypnotism, its Psycho Physiological, Medical, and Formsic Importance (adv.).

Demme—On the Influence of Alcohol on the Organism of the Child (adv.).

Laehr—Asylums and Hospitals for the Insane in German-Speaking Countries (adv.).

Ballet—Internal Speech and the various Forms of Aphasia. German by Bongers (adv.).

Albert—Compression of the Brain. (Review.)

Schringer—Epilepsy and Female Fertility in their mutual Relations. (Review.)

Schmaus—Compression Myelitis in Caries of the Spine (adv.).

Ufer—Nervousness and the Education of Girls (adv.).

Wilbrand—The Hemianopic Visual Field and the Optic Perception Centre (adv.).

———

MASSAGE PRIMER BY SARAH E. POST, M. D.; LECTURES BEFORE THE TRAINING SCHOOLS FOR NURSES CONNECTED WITH BELLE-VUE HOSPITAL, MT. SINAI HOSPITAL, ST. LUKE'S HOSPITAL, AND CHARITY HOSPITAL, NEW YORK.

———

This is the best book on the subject that we have seen. It is not as voluminous as some, and that is one of its chief merits, another of which is that it is illustrated. To give instructions for massage without demonstration or illustration, is like attempting to teach lawn tennis in didactic lectures. This little book is of value to any one who wishes to learn the methods, and we are glad to know that it is meeting with a large sale.

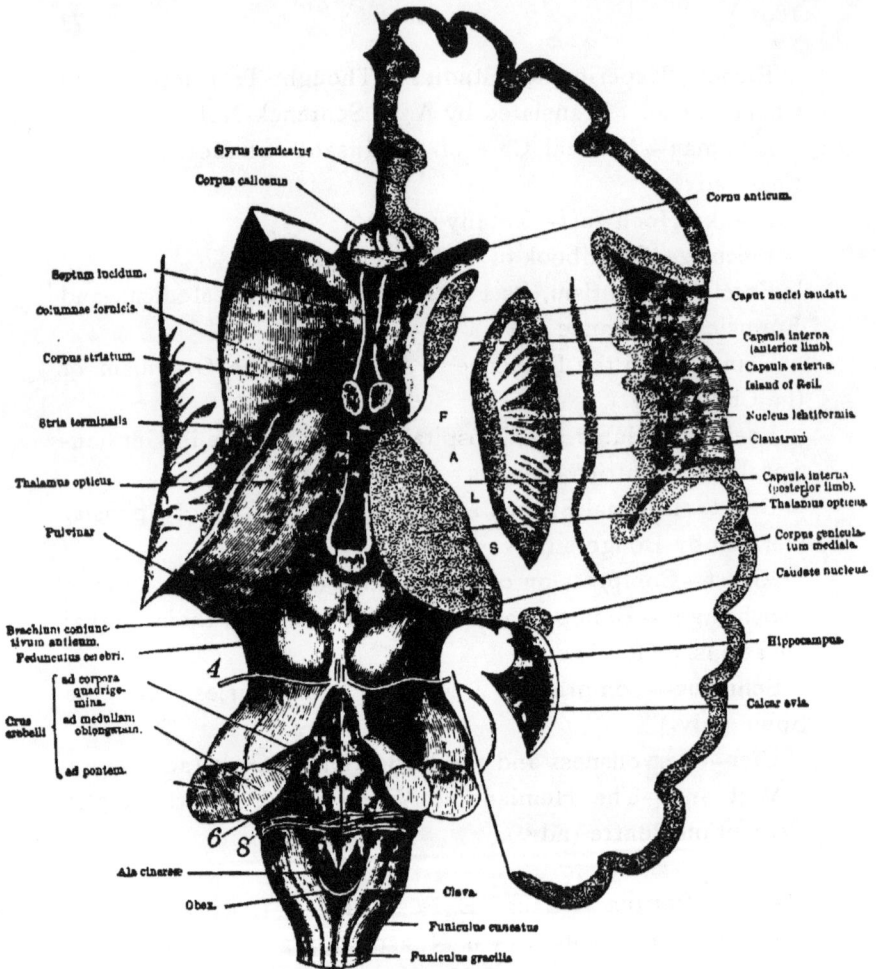

Fig. 5.

Human brain, with the hemispheres removed by a horizontal incision on the right side. 4, trochlear; 8, acoustic nerve; 6, origin of the abducens; F, A, L, position of the pyramidal (motor) fibres for the face, arm, and leg; S, sensory fibres.

GENERAL INDEX.

1. In certain periods of life, such as the age of uncertain accountability and old age.

2. In certain physiological conditions in females, such as the menstrual period, in pregnancy, in child-birth, and during the change of life.

3. In conditions of weak mindedness and in the deaf and dumb.

4. The influence of certain nervous diseases and conditions, such as hysteria, hypochondria, epilepsy, somnambulism, and hypnotism.

5. The condition of alcoholism and morphinism fever and traumatism.

6. Under the influence of heredity, and after a former attack of insanity.

All these conditions, permanently or temporarily, either alone or in connection with other external excitations, may influence, or rather diminish, the freedom of action of the individual. In all such cases the jurist should take into account the question of responsibility.   G. J. K.

CRIMINAL DEGENERACY.—Dr. Penta read a paper before the Italian Congress of Alienists at Novara, and reported in *Il Pisani* XI, 1890. He had examined a large number of the criminal class and found very frequently anomalies of development, as had already been found to be the case by other observers. To a somewhat less extent he had found similar anomalies in the insane, the more frequently as there was a history of heredity. Beside numerous other analogies the author, with Prof. Virgilio, found that in the insane and in criminals, febrile diseases such as phthisis, erysipelas, and rheumatism, took on an apyretic character. He also found in criminals, as Virgilio and Marro had in the insane, numerous anomalies of the internal organs. He had also undertaken a systematic examination of a large number of neuropathic individuals and found in them also, though less marked, anthropological abnormalities.

The diseases which most frequently cause death in criminals are, according to extensive statistics which he reported, of the phthisical, diathetic, and nervous classes. According to his study also of the fecundity and vitality of the delinquent classes, it would appear that they tended to extinction more than other degenerate classes. He considered that all the delinquent and insane belonged to one general type. The following is his summing up:

1. It is not correct that anthropological anomalies are peculiar to the delinquent classes, since they may appear in others, but this condition is due to a serious morbid heredity which may develop insanity, phthisis, or neuropathic conditions and may even develop in criminality, all these being only circumstances, or phenomena of degeneration, not the primary condition itself. This, however, depends upon a defect in the progenitors, while being transmitted to the offspring during embryonic life, causes arrest of organic development.

2. This being the case, crime is really a disease as has been already stated by Vergilio.

TRANSLATORS.

ITALIAN.

H. M. BANNISTER, M. D., Asst. Physician Eastern Illinois Hospital for Insane, Kankakee.

GERMAN.

G. J. KAUMHEIMER, M. D., Milwaukee.
H. M. BANNISTER, M. D., Kankakee.

RUSSIAN.

T. KACZOROUSKI–PORAY, Chicago.

FRENCH.

J. G. KIERNAN, M. D., Chicago.

SPANISH.

HORACE M. BROWN, M. D., Milwaukee.

SCANDINAVIAN.

M. NELSON VOLDING, M. D., Asst. Physician State Hospital for Insane, Independence, Iowa.

NEWS.

In the wild enthusiasm over Koch's discovery it has been apparently forgotten that Dr. Sam'l G. Dixon, of Philadelphia, practically anticipated Koch by a year in his method of treating tuberculosis. Dr. Dixon's discovery was announced in the *Medical News* in 1889, and his remedy and that of Koch is essentially the same. Though experience has not sustained the expectations held concerning the remedy, we yet believe it to be a great discovery. It at least indicates the probability of a remedy for bacterial diseases and points the way for future investigations. We believe a remedy for these diseases will some day be found, and when that is done, Dixon and Koch will be regarded as having been the first to demonstrate the possibility of such a thing. We call attention to this matter now, chiefly for the purpose of emphasizing Dr. Dixon's prior discovery. While American physicians are screaming themselves hoarse over Dr. Koch, we trust they also uncover to this modest fellow countryman, who antedated the great German's discovery by at least a year.

———

The *Cleveland Med. Gaz.* says that Mr. Edwin Cowles, editor of the *Cleveland Leader*, who died last March, had a peculiar form of deafness. He never heard the sound of a bird's note, and until he grew to manhood he always thought the music of the bird was a poetical fiction. "You may fill the room with canary birds," he once said, "and they may all sing at once, and I would never hear a note, but I would hear the fluttering of their wings. I never heard the hissing sound in the human voice; consequently, not knowing of the existence of that sound, I grew up to manhood without ever making it in my speech. A portion of the consonants I never hear, yet I can hear all the vowels. About a quarter of the sounds in the human voice I never hear, and I have to watch the motion of the lips and be governed by the sense of the remarks in order to understand what is said to me. I have walked by the side of a policeman going home at night and seen him blow his whistle and I never could hear it although it could be heard by others half a mile away.   I never heard the upper notes of the piano, violin, or other musical instruments, although I would hear all the lower notes.''

Prof. Kræpelin of Dorpat has accepted a call to fill the chair of Prof. Fürstner at Heidelberg, who succeeds Prof. Jolly at Strassburg as professor of psychiatry.

Dr. D. R. Brower has been appointed Prof. of Materia Medica and Therapeutics in Rush Medical College.   Dr. Sanger Brown has been appointed Prof. of Hygiene and Medical Jurisprudence in the same college.

Dr. C. H. Hughes, editor of the *Alienist and Neurologist*, is president of the Mississippi Valley Medical Association.   The annual meeting occurs in St. Louis in October.

We are sorry to learn of the death of Dr. Gundery of the Maryland Insane Hospital.   Dr. Gundery had been connected with institutions for many years, and had shown himself to be a man of superior ability.

NEW BOOKS.

Wichman—Chronic Articular Rheumatism and its Relations to the Nervous System (adv.).

Peyer—Spinal Irritation and its Relations to Diseases of the Male Genitals (adv.).

Adam Kiewicz—On the Pachymeningitic Process in the Spinal Cord.

Wengo—Effects of Irritation of the Nerves by Intermittent Galvanic Currents (adv.).

Oppenheim—Further Studies on Traumatic Neuroses, with Especial Regard to Simulation (adv.).